Craft Treatises and Handbooks: the Dissemination of Technical Knowledge in the Middle Ages

DE DIVERSIS ARTIBUS

COLLECTION DE TRAVAUX
DE L'ACADÉMIE INTERNATIONALE
D'HISTOIRE DES SCIENCES

COLLECTION OF STUDIES
FROM THE INTERNATIONAL ACADEMY
OF THE HISTORY OF SCIENCE

DIRECTION
EDITORS

EMMANUEL
POULLE (†)

ROBERT
HALLEUX

TOME 91 (N.S. 54)

BREPOLS

CRAFT TREATISES AND HANDBOOKS: THE DISSEMINATION OF TECHNICAL KNOWLEDGE IN THE MIDDLE AGES

Edited by

RICARDO CÓRDOBA

BREPOLS

COMITE SLUSE

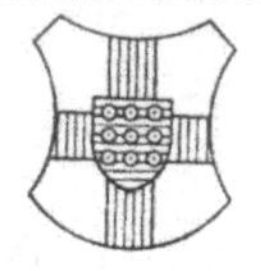

Publié avec le soutien de la Région Wallonne.

D/2013/0095/264
ISBN 978-2-503-54439-7

Printed on acid-free paper

TABLE OF CONTENTS

FOREWORD

The book that the reader has in his hands is the result of the papers presented at International Symposium about Craft Treatises and Handbooks held at Cordoba, Spain, in 2005. The congress was organized by the Department of Sciences of Antiquity and Middle Ages of the Cordoba University and the Spanish Society for Medieval Studies, under the Research Project "Dissemination of scientific and technical knowledge in the Middle Ages: technical literature in medieval Spain" (Ref. BHA2002-00739), funded by the Ministry of Science and Technology of Spain, partly with FEDER funds. And counted for funding with the help of the Spanish Ministry of Education and Science (Complementary Action HUM2004-21319-E) and the University of Cordoba (10th Program Aid to Research).

The conference aimed to bring together a group of specialists interested in the study of medieval manuscripts of a technical nature, those texts that provide information about how they took out certain manual activities related to the textile industry, hide works, metallurgical techniques and other crafts. The pressing technical exigency and professional qualification of occupations such as coinage, and gold and silver assayer, the cloth and silk dyer, leather tanner or illuminator determined the need to rely on handbooks and treatises, which permitted the knowledge and propagation of the set of techniques used in each sector to be made easier. For this, a technical literature emerged, which has been conserved in the form of manuscripts, more or less complete or in a loose form, and of treatises which appeared in alchemy, medicine, arithmetic and trading texts and other technical and scientific activities of those times. And which form an ensemble of inestimable value for the knowledge of the written culture of medieval artisans and work techniques employed in the diverse trades of the Age.

Technical manuscripts have special interest themselves for knowledge of production and work techniques used in medieval times. But also knowing how to transmit and disseminate technical and scientific knowledge and how it carried out the teaching of crafts. Throughout history, all human societies have provided themselves with means and instruments to conserve and transmit their knowledge; and in each era, Learning, in general, and each of its "skills" and "techniques" in particular, have possessed specific means for their dissemination. Although it is thought that during the Middle Ages most craftsmen were illiterate and the knowledge of their trade techniques, acquired empirically, was transmitted only by word of mouth, actually medieval workers had written texts at their disposal, which helped them to disseminate the new techniques and convey them to other artisans.

Researchers who contributed to this work from both the University and independent research field, in several European countries (Spain, United Kingdom, France, Germany, Netherlands, or Portugal) and in the United States of America. The work includes studies devoted to the analysis of manuscripts containing technical descriptions associated with various crafts, but the greater volume is occupied by the works devoted to the consideration of recipes and manuals of artistic, detailing the techniques used by writers, illuminators, painters and sculptors of Medieval Times and Renaissance. Mark Clarke has done a study divided into two parts, on manuscripts of artistic techniques known to the European Middle Ages. Then there are papers devoted books recipes for illuminators used in the German fifteenth-century (D. Oltrogge), the recipes included in the manuscript 19 of the BNE de Madrid (S. Kroustallis), the colours used for the illumination of manuscripts in the English fifteenth-century (C. Porter), the recipes in the *Book of how to make the colours* (L. U. Afonso, A. J. Cruz, D. Matos), the information contained in a medieval manuscript from Siena to light golden letters (A. Wallert), recipes the preparation of *verdegris* pigment (N. Sancho, M. San Andrés), the gold and its manipulation in medieval treatises (E. López, C. Dalmau, S. Bucklow), pigments and inks used in the Hebrew manuscript *Prato Haggadah* (S. Centeno, N. Stavisky) and Castilian recipes for the manufacture of writing inks (T. Criado).

A number of studies are devoted to metallurgical issues, such as latemedieval metallurgical recipes (A. Williams, D. Edge), etching iron (A. Stijnman), the manufacture of firearms and gunpowder in Europe in the 15th and 16th centuries (S. Walton), techniques of bronze casting in medieval Italy (G. Helms) or elaboration of *luto sapiente* used to coat the vessels in which metals are melted (N. Thomas). The works devoted to the techniques of tanning leather in Italian medieval manuscripts (R. Córdoba), the lute making techniques in the medieval Islamic world (A. Margerum) and the linguistic analysis of the famous *Liber Illuministarum* (M. Lautenschlager) complete this book.

In short, the proceedings of this conference meet varied contributions written by leading experts in the matters addressed, and united by the common bond of written texts in the Middle Ages. These instruments provide remarkable developments in the field of art history, and also for the study of medieval written culture. And meet for the first time in a work that allows contact, on the whole, the world of medieval manuscripts devoted to teaching the techniques of craftsmanship and its transmission.

Ricardo Córdoba de la Llave
Córdoba, Mayo de 2012

The earliest technical recipes: Assyrian recipes, Greek chemical treatises and the *Mappae Clavicula* text family

Mark Clarke

Introduction

Written sources that inform us about ancient technology are as old as writing itself. In the following few pages I shall attempt to give a brief outline of the earliest technical craft recipe texts. Not only does an examination of these earliest 'recipe books' set the scene for the study of the mediaeval craft treatises, but we shall see that in many respects these ancient texts were the direct ancestors of the mediaeval ones: in style, in format, in methods of composition and transmission, and even on occasion in specific technical content.

The oldest written recipes

Writing, when it began in its earliest form in Mesopotamia, (some time before the earliest finds at Uruk *c.* 3400 BC), was concerned primarily with lists of materials, e.g. in accounts and inventories. This is convenient for the historian of technology, as these lists tell us what materials were being produced, used, stored and traded, often in great detail and with a great subtlety of detail. There were, for example, some 300 words for types of bread.[1] In addition to inventory-type lists, from *c.* 2300 BC there survive bilingual wordlists, in which scribes tabulated equivalents between the archaic Sumerian and contemporary Akkadian words.

Knowing what materials were used, however, is only of limited use in working out *how* they were used. The earliest surviving technical recipes or procedural texts also come from Assyria. These are Mesopotamian recipes written on cuneiform tablets. They are written in the Akkadian language, but in a lit-

1. J. Bottéro, "The Oldest Cuisine in the World," in J. Bottéro (ed.) *Everyday Life in Ancient Mesopotamia*, (tr. A. Nevill) Edinburgh University Press, 2001, pp. 45 ff. (Originally published as *Initiation à l'Orient ancien*, Paris: Editions du Seuil, 1992).

erary form that often uses archaic words from the older Sumerian language. The earliest of all, a middle-Babylonian period tablet, contains recipes for coloured vitreous materials; it has long been thought to be from the seventeenth century BC, although it has been argued that it may be considerably later.[2] This is followed – after a substantial lacuna – by a number of recipe tablets from the seventh century BC for cookery,[3] for dyes (*c.* 600 BC),[4] and further recipes for coloured vitreous materials.[5] These last-mentioned 'glass' texts are spread across a number of fragmentary tablets, and in survive in multiple versions.[6] All these Mesopotamian recipe texts consist of simple, brief lists of ingredients with brief descriptions of processes, but with no discursive material. These texts are recognisable to us as true recipes: they have a title, they list ingredients and quantities, they note processes and manipulations such as mixing or oven preparation, and they describe what signs indicate that the final product is ready. They contain little that appears magical or mystical.[7]

"An Assyrian Chemist's Vade-mecum": dye recipes

By way of example, we shall now look briefly at Assyrian dye recipes. They are of particular relevance here as their content parallels that found in the Greek chemical texts of the third century AD, and subsequently in mediaeval examples; this will be outlined below. Certain clues to dyeing are contained

2. For photographs, editions, English translations and commentaries for all these "glass" texts, see: A. L. Oppenheim, R. H. Brill, D. Barag, A. von Saldern, *Glass and Glassmaking in Ancient Mesopotamia: An Edition of the Cuneiform Texts Which Contain Instructions for Glassmakers With a Catalogue of Surviving Objects,* The Corning Museum of Glass Monographs Vol. III, Corning New York: The Corning Museum of Glass, 1970. This volume supersedes previous editons and translations of these texts published between 1925 and 1938 by Thompson, Gadd and Zimmern, and Meissner. See notes 6 and 7 below for details.

3. Bottéro 2001 *Oldest Cuisine.*

4. British Museum Department of Western Asiatic Antiquities tablet WA 62 788 + 82 978. It contains recipes that are apparently for indigo, madder, and their combination to make purple. It was first noticed in 1993, and its publication is in progress, by the archaeological dye expert Hero Granger-Taylor and the assyriologist Irving Finkel of the British Museum, and should eventually appear in the journal *Dyes in History and Archaeology* as "Neo-Babylonian recipes for dying wool." These tablets are illustrated and discussed in J. Balfour-Paul, 1998, *Indigo*, London: British Museum Press, p. 17, and in D. Cardon, 2007, *Natural Dyes*, London: Archetype Publications Ltd., p. 4.

5. The exact nature of the materials whose manufacture they describe is unclear. They may be frits and glazes, or opaque glass, or even transparent glass. Some recipes are for making vitreous substitutes to imitate precious stones. For convenience they are referred to here as 'glass'.

6. All known glass texts are included in Oppenheim *et. al*, *op. cit.* It has often incorrectly been stated that these recipes for vitreous materials are recipes for pigments. It may as well be stated here that a further often-repeated statement, that the recipes for red glass use gold as an ingredient, is also based on an incorrect reading and interpretation of a single damaged word in a single damaged recipe. See: M. Clarke, "Is gold an ingredient in Assyrian-Mesopotamian written recipes for red glass?" in: M. Clarke, A. Stijnman, J.H. Townsend (eds.) *Art of the Past – Sources and Reconstructions.* London: Archetype Publications, 2005, pp. 24-32, in which paper are given details of the earlier publications of these texts.

7. Much commentary has been written about the use of 'embryos' or 'images' in these 'glass' recipes, but it seems that this may be based on incorrect readings of the tablets. Compare, for example, 'embryos' in R. Campbell Thompson *On the Chemistry of the Ancient Assyrians*, 1925, London: Luzac & Co. (p. 57), with '*Kubu*-images' in Oppenheim *et al.*, *op cit.*, p. 32.

within a seventh century BC Mesopotamian word-list. This is mainly in the form of an index or key, containing only brief instructions for use, with only occasional prescriptions or recipes *per se*.[8] The text is arranged in two columns, with the keyword on the left, and the meaning or translation on the right, linked by the imprecisely understood word *ina* which therefore seems to mean something like 'use for' or 'that is'. Approximately 130-140 'drugs' and synonyms are described. The text covers the use of these materials for medicine, food, dyeing, tanning, and other miscellaneous uses, and includes a key to certain synonyms and what their translator terms "alchemists' jargon." Three of these short entries are quoted below, in their entirety.

– "Lichen of tamarisk in alum." This suggests lichen dyes, alum being a common dye mordant. Thompson speculates as to the exact species of lichen. Those he favours, *Roccella montagnei*, Del. [*sic.*] and *Lecanora tartarea*, Ach., [*sic.*] are now, subsequent to standardisation in lichen nomenclature, both known as *Rocella* species.

– "*Kamme gurgurri* on grey hair." Here *kamme gurgurri* has been shown by Thompson to be ferrous sulphate.[9] This item surely refers to ferrous sulphate solution making a black iron-tannin dye analogous to iron-gall ink. Recipes for blackening grey hair with iron-gall ink are also found in mediaeval collections.

– "*^{u}Ha-za(l)-lu-nu* in iron and pomegranate." Thompson initially thought this referred to some plant, possibly alkanet, later revising this to the purple-producing shellfish. Whether *^{u}Ha-za(l)-lu-nu* refers to a *Murex*-type shellfish or to some plant substitute, this text surely refers to dyeing, where a red-purple colorant is combined with a blue-black iron-tannate dye. Iron (whether in the form of metallic iron, e.g. as iron filings or rust, or as ferrous sulphate) produces a black dye when combined with the tannins present in pomegranate skins. This would darken the purple from the shellfish or plant, to achieve the appearance of the fabulously expensive and prestigious deep-dyed genuine shellfish purple.

Passing references

The Assyrian examples of true recipes are followed chronologically by a number of references to craft processes. Passing references to craft materials often survive in otherwise non-craft texts, e.g. the references to woad as a dyestuff in the Egyptian medical *Ebers Papyrus* of 1552 BC. Many such references appear in works by Classical authors who apparently practiced the crafts (Vit-

8. For these extracts, with English translation and commentary: R. Campbell Thompson, "An Assyrian Chemist's Vade-mecum" *Journal of the Royal Asiatic Society of Great Britain and Ireland*, 1934, pp. 771-785.

9. Also known as 'vitriol', 'green vitriol', '*misy*', '*kalkanthon*', '*atramentum sutorium*'.

ruvius), by those practicing other professions but who refer to craft practices *en passant* (Dioscorides), and by those who were probably non-practicing observers and encyclopaedists (Theophrastus, Pliny). While these are not on the whole technical recipe texts *per se*, nevertheless they are useful sources of technical information. These are sufficiently well-known not to elaborate on them here.[10] One tantalising reference seems to be lost to us: Pliny noted that there existed several treatises on making imitation precious stones, although he preferred not to name their authors,[11] but a clue survives in Seneca, who explicitly names one 'Demokritos' (on whom see below) as the discoverer of this process (in *On Philosophy* and in *Moral Epistle 90*).

Early Greek technical treatises

The first late- or post- classical recipe texts were written in Greek. The *Corpus chemicum grec* may be divided into chemical recipes, alchemical authors, and commentators; these survive as collections of extracts, and we do not have any one complete treatise.[12] The technical recipes are contained principally in two papyri: the *Stockholm Papyrus* and the *Leiden Papyrus X*, to which may be added parts of the *Physika kai Mystika* and a few other fragments.

Physika kai Mystika[13]

Recipes for purple dyes that use iron are found in the *Physika kai Mystika (Physical and Mystical Matters)* attributed to Demokritos the alchemist.[14] Dating from *c.* 100-300 AD it is perhaps the earliest true alchemic text.[15] Parts of this interesting treatise are clearly an ancestor of parts of the Stockholm and

10. Lists of the classical sources and the materials discussed in them are given in M. Clarke, *The Art of All Colours: Mediaeval recipe books for painters and illuminators*, London: Archetype Publications, 2001, pp. 40-49.

11. *Quin immo etiam exstant commentarii – quos non equidem demonstrabo –, quibus modis ex crystallo smaragdum tinguant aliasque tralucentes, sardonychem e sarda, item ceteras ex aliis; neque enim est ulla fraus vitae lucrosior. Nat. Hist.* XXXVII.lxxv.196.

12. R. Halleux, *Les Alchimistes Grecs*, see below, note 18, pp. xi-xii.

13. Edited with Italian translation and commentary M. Martelli *Pseudo-Democrito: Scritti Alchemici, con il commentario di Sinesio* (Textes et Travaux de Chrysopœia 12). Paris: S.É.H.A and Milan: Archè, 2011. This supersedes the edition of M. Berthelot, *Collection des Anciens Alchimistes Grecs*, 3 vols. Paris: Georges Steinheil, 1887-8. French translation: vol. II, pp. 43 ff.; Greek edition: vol. III, pp. 41 ff.

14. Or '*pseudo*-Demokritos', sometimes identified with Bolus of Mendes, (*fl.* 145-116 BC), i.e. not the philosopher Demokritos of Abdera (*c.* 460 BC-*c.* 370 BC). For the identity of the *ps.*-Demokritus/Bolus, the attribution of a possible medical-technical recipe collection by him entitled Χειρόκμητα (see n. 66 below), together with a plausible case for an interest in dyes, colour and metallurgical operations on the part of Demokritos of Abdera himself, see J.P. Hershbell "Democritus and the beginnings of Greek alchemy," 1987, *Ambix* 34.1, pp. 5-20. The *P.Holm* attributes one of its recipes to Demokritos (#2).

15. The dating is difficult and controversial. See Martelli, *op. cit.*

Leiden papyri. It begins with recipes for purple dye reminiscent of the Assyrian texts. One recipe apparently uses lichen (orseille, i.e. *Rocella* species), and one apparently uses *Anchusa tinctoria* (anchusa, alkanet). It is unclear, but certainly iron slag or dross is also an ingredient. It continues with recipes for using and imitating gold and silver.

Stockholm Papyrus[16] *and Leiden Papyrus X*[17]

Probably our two most important sources, also written in Greek, come from Egypt, and date from *c.* 300 AD. They are the well known *Stockholm Papyrus* and the *Leiden Papyrus X*, both written in codex form and both dating from *c.* 300 AD.[18] (These will be abbreviated below as *P.Holm.* and *P.Leid. X* respectively). Although these papyri are often mentioned in secondary literature, and their importance acknowledged, I have long suspected that they are actually very infrequently read, as for a long time their texts have only been available in hard-to-obtain and out-of-print publications, and even Halleux's 1981 edition and translation is not easily obtained.[19] The texts of these papyri are of immense importance and use: they are among the earliest chemical and technical texts to survive from anywhere, they are clearly and comprehensibly written, and they cover a wide range of techniques. Their influence continued long after they were written and, as we shall see, quotations from them occur in a number of later mediaeval compilations. (Since these papyri were presumably buried and lost, we can thus deduce the existence of other copies, now lost.) To these important two papyri may be added several contemporary fragments and extracts, some of which are discussed below.

16. Stockholm, Kunglige Biblioteket, Handskriftsavdelningen, Dep. 45, also known as the *Papyrus Holmiensis* (Formerly in the Victoria Museum Uppsala).

17. Leiden, Rijksmuseum van Oudheden (Museum of Antiquities), Museum catalogue no. I 397, inventory number A.MS 66. *P.Leid. X* is also known commonly as *P.Leid. X* and as *P.Lugd.Bat.* J397 (X) (after Leemans 1843-85, see below, note 18).

18. Note that the Leiden papyrus *sigla* 'V' and 'X' are roman letters, not roman numerals.

19. R. Halleux, *Les Alchimistes Grecs* Vol. 1: *Papyrus de Leyde, Papyrus de Stockholm, Fragments de Recettes*, Paris: Société d'Édition "Les Belles Lettres," 1981. (Reprinted 2002.) pp. xi-xii. The *editio princeps* of *P.Holm.* is the almost-unobtainable O. Lagercrantz, *Papyrus Graecus Holmiensis* Uppsala: Almquist and Wiksells, 1913, which includes a German translation. The *editio princeps* of *P.Leid. X* is the equally difficult-to-find C. Leemans, *Papyri Graeci Musei Antiquarii Publici Lugduni-Batavi*, 2 vols. Leyden, 1843-85, which includes a Latin translation. Furthermore, before Halleux's 1981 publication (and since!), the most often cited publications of these papyri have been the somewhat inadequate editions and translations of Berthelot (French) and the English translations of Caley (which were made not directly from the Greek, but from the French of Berthelot and from the German of Lagercrantz): E. R. Caley, "The Leyden Papyrus X: An English Translation with brief notes," *Journal of Chemical Education*, III (10), 1926, pp. 1149-66; and "The Stockholm Papyrus: An English Translation with brief notes," *Journal of Chemical Education*, IV(8), 1927, pp. 979-1002. The chemist Berthelot published several versions of his translation in collaboration with the Greek scholar M. Ch.-Em. Ruelle: e.g. M. Berthelot, *Archaeologie et Histoire des Sciences, avec publication nouvelle du papyrus grec chimique de Leyde ...* Paris, 1906 (Reprinted Amsterdam 1968).

While their provenance is unknown it seems clear that *P.Leid. X* and *P.Holm* were written in Egypt, very possibly near Alexandria or Thebes (now Luxor), around 300 AD.[20] It has been argued *pro* and *contra* whether the papyri originally formed part of a single codex written by a single scribe.[21] My own examinations of the two papyri lead me to conclude that, while clearly they are closely related and perhaps even from the same workshop or individual, they did not originally form two halves of a single codex.[22] Certainly there are many similarities and few small differences. The most obvious similarity is their unusual subject matter. Small differences in terminology usage between the two papyri have been suggested.[23] The hands and formats are similar but not identical.[24] It is interesting that they should be in codex form; the rise of the codex has been closely linked with the rise of Christianity, and it is not impossible that the Papyri could have belonged to a craftsman who was, or who worked for, or at least lived among, Christians. The textile dye recipes (especially for mixed purples) correspond extremely closely with the dye analysis of Coptic textiles contemporary with the papyri.

Content of the papyri

As with the *Physika kai Mystika* the papyri are concerned with purple, gold and silver. *P.Holm.* contains recipes for purification of metals and pearls, precious and semiprecious stones, changing the colours of stones to resemble other stones (including *en passant* recipes for verdigris), removing writing from papyrus, and seventy recipes for dying wool and linen. *P.Leid. X* contains recipes for purification of metals, artificial gold, chrysography, and the same for silver, and a further ten dye recipes, ending with brief extracts from the *Materia Medica* of Dioscorides describing the qualities of ten useful ingredients (see below). A wide range of materials appear in the recipes, many of them with their geographical origins specified (e.g. salt of Cappadocia, copper of Cyprus, copper of Galacia, earth of Chios, cadmia of Thracia, Macedonian verdigris, stone of Paros). These materials come from all over the Mediterranean and near east, as can be seen from the map in figure 1. The recipes in the

20. There has also been a certain amount of disinformation transmitted as to their provenance. Most common, and relatively harmless, is the assertion (based on a speculation by Berthelot) that they were found in a grave or in a mummy; this speculation is often cited as fact, for example, by the usually authoritative E.J. Holmyard, in his popular history *Alchemy*, Harmondsworth: Pelican, 1957, p. 27. The papyri *P.Leid. X* and *P.Holm.* were acquired by the Dutch and Swedish governments from Johann d'Anastasi in 1828 and *c.*1828-1832 respectively. He acquired them from an unknown source, perhaps some secret excavation, while consul of Sweden and Norway at Alexandria.

21. Halleux concluded that *La structure des deux codices exclut donc qu'ils soient les deux moitiés du même*. Halleux 1981, *op. cit.*, p. 8.

22. It is possible that they were bound together after writing (and then separated again after finding); although this is more typical of *mediaeval* manuscripts.

23. Hammer-Jensen, cited in Halleux 1981, *op. cit.*, p. 38 and note 3.

papyri are similar in style to the Assyrian recipes, being short and concise, with little explanatory material, and devoid of magic.

As with the Assyrian examples, it is unclear for whom they were written, or why. It has long been debated whether the papyri are a fair copy of an artisan's workshop notes, or whether they were compiled by or for a reader with a more general or theoretical interest such as an alchemist. Indeed, an objection to their belonging to a craftsman would be precisely the wide range of techniques that is described, as opposed to a concentration on a single craft. It has been argued that they are not really practical recipes, but pure alchemy, and the exact opposite has also been argued.[25] Holmyard has stated that "Closer examination of the recipes, however, has shown that they could not yield any practical results, and it may therefore be that they were snippets collected by an alchemist."[26] He does not state who carried out this work, or what their

24. *P.Leid. X* and *P.Holm.* were originally codices, not rolls. No trace of either binding survives, although I have observed sewing holes on *P.Leid. X*. Both written in a good, clear hand, in uncial script (almost completely without accents and diacriticals) typical of that from the late third and early fourth century ad, using dark brown-black ink throughout, without other colours used, on plain papyrus. A small amount of self-correction is evident, in each case executed in the same ink and hand as the main text.*P.Leid. X* survives as 20 sheets (20 folded bifolia, i.e. 40 sides or pages) 170 x 295 mm. It is written with between 40 and 47 lines per page, in one column. The first 2 and last 10 sheets were blank. (The current glass mounts have been numbered such that 'Blad 1' is the first *written* page.) The incomplete filling could mean that it was either an original text or compilation, rather than an exact copy of a single text (if the scribe did not know how many pages would be needed, and was simply filling up a blank codex or quire), although it could also mean that it was an unfinished copy or indeed a personal notebook. All the text and corrections appear to have been written in a single hand in the same black ink and same pen. It is important to note that the section of extracts from Dioscorides is continuous and homogeneously written with the rest of the papyrus. The hand and margin go a little adrift for the last six lines of page 9 but recover for the last part of the same recipe at the head of page 10.

The *P.Holm.* also was a codex originally, and survives as 14 separate sheets (i.e. 28 sides or pages) with abraded edges, a little smaller than 170 x 297 mm. It is written with between 40 and 48 lines per page, in one column. Page 28 is blank. The section on dyes in *P.Holm.* starts after a decorative page division on p. 14, and is in the same hand but apparently a separate writing session, *c.f.* in *P.Leid. X* after the ruled line that precedes the Disocorides extracts. The paragraph headings and separation methods are similar in *P.Holm.* and *P.Leid. X*, although in *P.Leid. X* a paragraph heading always is centred and on a new line, whereas in *P.Holm.* it is sometimes centred and on a new line, and sometimes placed to the right of the last line of the previous paragraph. The most conspicuous difference is the page numbering. In *P.Holm.* each page (side) is numbered in Greek letters (alpha=1, beta=2 etc.) centred at the head of each page; on the first couple of pages the pen of the page numbering is somewhat fatter than that of the main text, suggesting the numbering was added a little later, but further on in the papyrus the hand and nib and ink seem identical to that of the main text, suggesting they were added at or soon after the time of writing. Halleux says that numbering is exceptional for the period (1981, *op. cit.*, p. 8) and suggests that perhaps the pages had become disorganised at some point soon after writing and that someone had subsequently added the numbering. However, the numbering could well be original. Similar codex form, *mise en page*, and laminated papyrus waste bindings are found in the contemporary Egyptian papyrus *Nag Hammadi* Codices, of which Codex I also has page numbering, also centred at the head of each page, and apparently original, although in the Coptic alphabet. See M. Clarke (2009) "Codicological indicators of practical mediaeval artists' recipes," in *Sources and Serendipity: Testimonies of Artists' Practice* Edited by Erma Hermens and Joyce H Townsend. London: Archetype Publications, pp. 8-17 and plates 3-6.

25. Halleux 1981, *op. cit.*, pp. 24 ff.

26. Holmyard 1957, *op. cit.*, p. 27.

Fig. 1.

conclusions were based on, but unfortunately this dismissive and demonstrably incorrect statement has been widely circulated. The debate continues: *R. Pfister en a tiré que ces recettes n'avaient aucune valeur pratique et qu'elles relevaient de la recherche pure.*[27] Pfister, a chemist and Egyptologist working in the 1930's, based his conclusions on analyses of Egyptian textiles, in which he did not find the dye combinations described in the papyri. However: *K. Reinking a montré qu'on les retrouve encore dans des textes* [*sic*] *coptes.*[28] It is certainly the case that the mixtures of dyes described in the papyri for imitating shellfish purple *have* been found (since Pfister) by chemical analysis of Coptic textiles. Reinking, and other since, also concluded that the recipes in *P. Holm.* for dyeing blue with woad conform well with the fermentation woad vat.[29] The combinations of dyes for producing purple have now been identified, frequently and with confidence, in numerous Coptic textiles, i.e. in textiles from Egypt approximately contemporary with the papyri. The papyri (if the translations are correct) describe the use of alkanna, lichen, insect red and a combination of blue indigo/woad + red madder. Analyses of the textiles found the same dyestuffs: alkanna, lichen, insect red, and indigotin + madder. The chemical analyses of these finds are summarised conveniently (including discus-

27. Halleux 1981, *op. cit.*, p. 46.

28. Halleux 1981, *op. cit.*, p. 46.

29. *P.Holm.*: Halleux §109 = Caley §104. K. Reinking, *Die in den griechischen Handschriften aus den Altertum enthaltenen Vorschriften für Wollfärberei* IG-Farbenindustrie, Frankfurt, 1938, cited in J. Hofenk de Graaff, *The Colourful Past*, London: Archetype Publications and Riggisberg: Abegg-Stiftung, 2004.

sions of where they supersede those of Pfister) by Hofenk de Graaff.[30] These findings are confirmed in an authoritative (and very readable) book by Dominique Cardon.[31, 32]

The physical form of the papyri is a possible source of clues as to their intended purpose: the fine quality of the script has been used as an argument against their craft use, while their convenient codex format has been used as an argument for it. They are devoid of workshop stains, but this is not necessarily a contraindication: they would probably have been handled only by a master or supervisor.[33] There is other evidence that can be examined to support the practical/theoretical arguments. For example, there is a clear interest in materials *per se*. For example, *P.Leid. X* ends with ten extracts correctly introduced as "Dioscorides, from the [book of] *Materia* [*Medica*]."[34] Editors and translators of *P.Leid. X* with the exception of Leemans did not include these passages, presumably as Dioscorides was considered a 'standard text'. However, if the compiler of the papyrus considered it worthwhile to include them, it seems worthwhile to ask why. The extracts discuss several materials common to art and medicine (e.g. orpiment, vitriol, alum, and cinnabar). Each extract tells how to identify the best quality material, and how to refine it. Dioscorides then continues by describing the medical uses of each material, but the compiler of *Papyrus X* leaves off each extract at this point. Clearly he was not concerned with these medical uses, but with establishing how to identify and prepare good quality materials for craft use; i.e. not *materia medica* but just *materia*. The papyrus is therefore *not* a composite of craft and medical texts.

That the content of the papyri resembles alchemic texts is surely no argument for dismissing their technical relevance – practical/physical uses and spiritual uses/interpretations were not mutually exclusive but were two aspects of one world; it is indisputable that the theoretical/spiritual/mystical form of alchemy had its origins in the practical forms of craft/alchemy/chemistry. Bucklow has written recently on the role of alchemic theories of matter in recipes for artists' materials.[35]

Even if we exclude the philosophical and spiritual, the intention is unclear for the recipes 'making', improving or doubling gold, silver, gems and purple

30. Hofenk de Graaff 2004, *op. cit.*, pp. 44-5, 266, 275.

31. D. Cardon, *Le monde des teintures naturelles*, Paris: Belin, 2003, (e.g. pp. 15, 480-1, and *passim*).

32. English translation with updates: *Natural Dyes*, 2007, London: Archetype Publications Ltd.

33. One might compare the nineteenth century recipe books of the English colourmen Winsor & Newton, which were without doubt genuinely used by manufacturers of paints, but the majority are spotless.

34. *P.Leid. X*: page 14 line 15 to the end of the papyrus on p. 16.

35. For details of Bucklow's publications see Clarke, "Late medieval artists' recipes books," in this volume. See also Bucklow's paper in this volume.

dye. Were they intended to deceive (forgery), or were they explicitly false (for imitation or paste jewellery)? One *P.Leid. X* recipe for diluting *asem* (silver alloy) with tin and copper concludes: “It will be asem of the first quality, which will/can deceive even the artisans”[36] and it seems that there was indeed a market for *ersatz* ceramics, metalwork, jewellery and dyed textiles.[37] Certainly a couple of centuries earlier Pliny had stated firmly his opinion that gem imitation was fraudulent and for profit.[38]

Alternatively, if it was not fraudulent, did the user believe that he was truly making new gold (transmutation)? As historians of ideas we need to consider this, but as historians of technology it is irrelevant: either way, whatever was in the mind of the craftsman, the recipes can nevertheless reflect true craft practice. Their concentration on imitation gold or superficial colouring of metals has been cited as a possible reason for the destruction or hiding of such chemical papyri.[39]

Halleux (and others) have justly concluded that *il est extrêmement difficile de distinguer une recette technique d'une recette alchimique.*[40] They are essentially at the junction. Whatever the *intention* behind writing them, the recipes are on the whole very clearly stated, and (despite the sometimes considerable problems of interpreting the technical terminology and the names of ingredients) it has proved possible in recent years to actually follow many of these recipes in the workshop and reconstruct these ancient techniques.

Magic

It is instructive to consider the place of magic in the Greek chemical texts. A further papyrus from Leiden, *P.Leid. V*, provides an interesting illustration.[41] *P.Leid. V* was originally a roll, with one side (the oldest) written completely in Egyptian Demotic, and one side (more recent) written mainly in Greek with a few Demotic sections. Like *P.Holm.* and *P.Leid. X*, it is also from Egypt, and the Greek text is of a similar but slightly later date, probably *c.* 300-350 AD,

36. *P.Leid. X*: Halleux §8 = Caley §8.

37. Halleux 1981, *op. cit.*, p. 26.

38. Quoted in note 11 above.

39. In *c.* 296 AD the Roman emperor Diocletian ordered chemical texts in Egypt to be burned, apparently to prevent the manufacture of gold and silver by alchemic means. Clarke 2001, *op. cit.*, note 59.

40. Halleux 1981, *op. cit.*, p. 29.

41. Leiden, Rijksmuseum van Oudheden (Museum of Antiquities), Museum catalogue no. I 384, inventory number AMS 75. Edited with Latin translation in Leemans 1843-85, *op. cit.*, Vol. II, pp. 1-76, and edited with German translation in K. Preisendanz, *Papyri Graecae Magicae*, 3 vols. Berlin: Teubner, 1931. 2nd edition 1973. Also known as *P.Leid. V* and as *P.Lugd.Bat.* J384 (V) (after Leemans) and as *PGM* XII (after Preisendanz). There is a new edition and photographic facsimile of this papyrus: Daniel, Robert W. (ed.), *Two Greek Magical Papyri in the National Museum of Antiquities in Leiden: A Photographic Edition of J 384 and J 395 (= PGM XII and XIII)*, series *Papyrologica Coloniensia* (Opladen: Westdeutscher Verlag) vol. XIX (1991).

however, it does not contain technical texts but magical ones. Among the magical spells are instructions for making amulets consisting of magical texts written on various supports such as papyrus or bark. Some of these amulet-making instructions are precise as to the ink that should be used, and the ink recipes are given. One spell is particularly interesting as it contains the earliest complete recipe for an iron-tannin ink, i.e. the first to list *all* the three main components, namely galls, vitriol, and gum.[42]

> "To gain favour and friendship forever: Take a pasithea or wormwood root and write this name on it [...] The formula: 1 dram of myrrh, 4 drams of *misos*, 2 drams of *kalkanthou* , 2 drams of oak gall, 3 drams of gum arabic."[43]

In fact the recipe is apparently hyper-complete, including *kalkanthon* and *misy*[44]. Another ink recipe from *P.Leid. V* seems to be primarily a *ritual* ink:

> "Drawing made with Typhonian ink: A fiery red poppy, juice from an artichoke, seed of the Egyptian acacia, red Typhon's ocher, asbestos, quicklime, wormwood with a single stem, gum, rainwater."[45]

42. The earliest text that describes the metallo-tannate reaction, using an aqueous infusion of galls and 'kalkanthos', is Philo of Byzantium (*c.* 250-c.200 BC), *Compendium of mechanics* (*Mekanike syntaxis*), where the reaction is used to make an invisible ink for secret communications. Edited with German translation: Diels, H. and E. Schramm 1919. 'Exzerte aus Philons Mechanik', *Abhandlungen der Preussischen [Deutsche] Akademie der Wissenschaften zu Berlin*, Philosophisch-historisch Klasse No. 12, p. 79. See also Monique Zerdoun Bat-Yehouda, 1983, *Les Encres Noires au Moyen Âge (jusqu'a 1600)*. Documents, Études et Répertoires Publiés par l'Institut de Recherche et d'Histoire des Textes. Paris: Éditions du Centre National de la Recherche Scientifique (CNRS), pp. 91-2, citing J.R. Partington *A History of Chemistry*, London, 1970, vol. I, part I, p. 205.

43. Edited in Preisendanz col. XII line 399-400 (= Leemans col. XXIIA lines 16-17); my translation, based on H.D. Betz (ed.), *The Greek Magical Papyri in Translation*, Chicago & London: Chicago University Press, 1986, 1st ed. p. 167 (2nd edition 1992, reprinted in paperback 1996.) Betz has: "1 dram of myrrh, 4 drams of truffle, 2 drams of blue vitriol, 2 drams of oak gall, 3 drams of gum arabic" (p. 167). Although *misos* is indeed also used to mean "truffle" by Theophrastus, it seems far more likely that in this case its alternative meaning of 'vitriol' is correct: Zerdoun Bat-Yehouda more plausibly, translates these ingredients as (i) myrrh, (ii) *misy* (not "truffle"), (iii) vitriol, (iv) gall-nuts, and (v) gum respectively (*op. cit.*, p. 94). Preisendanz, more specifically, translates *misos* as "Vitriolerz" and *kalkanthou* as "Kupfervitriolwasser" (Preisendanz vol. II, p. 83).

44. Presumably there was at that time some distinction between the two vitriols *kalkanthon* (presumably, due to the *kalka-* prefix, this is copper vitriol) and *misy* (here presumably iron vitriol). The names used for copper and iron sulphates in classical and mediaeval written sources are not used consistently. Various attempts have been made to definitively assign terms that clearly describe vitriols to either copper or iron salts, but there are always texts that go contrary to any given definition. A commonly adopted solution is to use the neutral term 'vitriol' to translate these terms into English. Even where a salt can be identified with confidence, it is unlikely that early vitriols would be pure; blue vitriol (copper sulphate) would often have contained green vitriol (iron sulphate) as an impurity, and *vice versa*. Indeed, if these recipes are followed using modern, chemically pure, blue vitriol, which lacks these iron sulphate impurities, the black ink forming reaction does not take place. The presence of some iron sulphate in the 'other' vitriol thus explains the success as inks of some otherwise inexplicable recipes.

45. Leemans col. IIIA line 23 = Preisendanz col. III lines 96-106; tr. Betz, p. 156.

Other inks specified in spells are less clear, but seem also to be ritual inks, e.g.:

> "Zminis of Tentyra's spell for sending dreams: Take a clean linen cloth, and (according to Ostanes) with myrrh ink draw a figure on it which is humanlike in appearance."[46]

P.Leid. V also contains an unclear recipe apparently describing the refinement or imitation of gold, which may be connected with the items immediately following it which treat the fabrication of a magical ring.[47]

It is perhaps important briefly to consider the place of magic in *P.Holm*, in which the 14 sheets of recipes are followed by a separate sheet containing a prayer, spell, or incantation. Based only on this one sheet Caley builds quite a picture of the personality of the writer or owner of the papyrus. Calling it the "last recipe in the collection," he concludes "If it [i.e. the magical sheet] belonged to the owner of the remainder of the collection, and it is probable that it did, then it tends to show that the chemical arts in ancient Egypt were largely in the hands of the priestly caste, a fact that has been deduced from other sources and of which this is the earliest direct evidence yet shown."[48] However it can be shown easily that sheet 15 is clearly separate and does not belong.[49] Consequently this deduction does not hold.[50] It seems important to emphasise this: if we can distance sheet 15 from the rest of the contents of the papyrus, then we can distance the magical from the technical, at least for this papyrus. Leicester' comments are surely nearer to the truth: "Magical practices are not suggested [in *P.Holm.*], except on a separate leaf that contains an invocation ... This may have been part of another papyrus, but its existence with that from Stockholm implies that magic practices were found even among the practical artisans ..."[51] While not proving anything, this is not an unreasonable conjecture. The case of the ink recipe in *Leiden Papyrus V* may be compared with this: in that case the technical information is incidentally embedded in an oth-

46. Leemans col. IVA line 15 = Preisendanz col. IV lines 121-143; tr. Betz, pp. 157-8.

47. Preisendanz col. VI lines 193-201; tr. Betz, pp. 160-1; ed. and tr. Halleux, pp. 163-6.

48. Caley 1927, *op. cit.*, p. 1001.

49. The papyrus of this sheet is of a coarser quality; the handwriting is different, the text is written using a broader pen, and it lacks the page numbering at the top.

50. I have suggested elsewhere that this intrusive sheet is a detached sheet of a laminated papyrus waste binding cartonnage: Clarke 2009 *Codicological*, p. 14. Similar laminated papyrus waste bindings are found in the contemporary Egyptian papyrus *Nag Hammadi* Codices. Having re-examind the papyrus recently, I now consider that the presence of dirt and the lack of paste or glue residue perhaps argues against this. Furthermore the side of sheet 15 on which is written the spell is very dirty and worn. A more likely alternative is, therefore, that the *P.Holm*,booklet originally had one or more outer bifolia of scrap or blank papyrus added to act as a wrapper. The rough and dirty nature of sheet 15 tends to support the hypothesis that it originally formed the outer wrapper of such a booklet.

51. H.M. Leicester, *The Historical Background of Chemistry*, New York: John Wiley, 1956. Reprinted Dover Publications, New York, 1971, pp. 38-39.

erwise clearly magical text, not *vice versa*. Consequently this does not suggest any magical component of craft practices, but rather it is a clue to the practical details of actually performing magic.

Another example of intermingled technical and magical recipes may be seen in the *Demotic Magical Papyrus of London and Leiden (P.Mag.LL)*.[52] This is a papyrus with similar content to the magical *P.Leid. V.* It has the same date and provenance as *P.Leid. V* and *P.Leid. X*, however it is written in Demotic with some Greek and Coptic glosses and some Hieratic words.[53] Unlike the rituals in Books of the Dead, the spells in *P.Mag.LL* are intended for use of the *living*, i.e., it contains spells that would be performed using actual earthly materials. This papyrus contains one magic spell of particular interest for us. It is for "Divination by vessel to see the bark of Ra." The following procedure is prescribed in column X,[54] and is repeated later in the papyrus at column XXVII where it survives in a more complete form:[55]

> "The Preparation: blood of a Nile goose, blood of a hoopoe, blood of a nightjar, 'live-on-them' plant, mustard,[56] 'Great-of-Amoun' plant,[57] *qs-'nh* stone,[58] genuine lapis lazuli, myrrh, 'footprint-of-Isis' plant.[59] Pound, make into a ball, and paint your eyes with it ..." (column XXVII lines 9-11).[60]

It is possible that here the various 'blood' ingredients are not what they seem. Much has been written speculating on 'secret' or 'coded' names for

52. *P.Mag.LL* is now split in two, divided vertically through recto column X. Part 1 = recto cols. I-X is London, British Museum Papyrus 10070 (*was*: Anastasi 1072); Part 2 = continuation of recto col. X, to the end, is Leiden, Rijksmuseum van Oudheden (Museum of Antiquities) museum catalogue no. I 383, inventory number A. MS 65. It is transcribed into the roman alphabet, with an English translation, in: F.L. Griffith and H. Thompson, *The Demotic Magical Papyrus of London and Leiden*. London: H. Grevel & Co, 1904. (Reprinted as *Leyden Papyrus: An Egyptian Magical Book* New York: Dover Publications, 1974), which see for details of earlier publications and facsimiles of this papyrus. Betz, *op. cit.*, also provides an English translation of both Greek and Demotic portions of this papyrus.

53. It is in fact possibly the latest Egyptian manuscript to use Demotic script.

54. Recto column X, lines 31-2.

55. Recto column XXVII. The complete spell starts on line 1.

56. Literally "bread of heaven plant," identified as mustard in Betz, p. 237, n. 512.

57. Griffith and Thompson suggest that "Great-of-Amoun" plant is flax (p. 80, note to col. X line 32, cited in Betz, p. 213, n. 249.)

58. *qs-'nh* stone is probably haematite. Betz, p. 200, n. 65, citing J.R. Harris *Lexicographical Studies in Ancient Egyptian Minerals*, Deutsche Akademie der Wissenschaften zu Berlin, Institut für Orientforschung, 54 (Berlin: Akademie-Verlag, 1961), pp. 233-34.

59. Griffith and Thompson suggest that "footprint-of-Isis" plant is *pittaxis* (p. 80 note to col. X line 32, cited in Betz, p. 213, n. 252.), which Liddell and Scott translate as the fruit of the cornelian cherry.

60. Translation of col. XXVII from Betz, pp. 236-7. (The translation of the almost identical col. X recipe is given in Betz, p. 213.) The translations of this recipe and its duplicate, both published in Betz, are by Janet H. Johnson, following the edition and translation of Griffith and Thompson. The Griffith and Thompson translation of the col. X recipe (essentially the same as their translation of the col. XXVII duplicate) is given here for reference and comparison: "Formula: blood of a *smun*-goose, blood of a hoopoe, blood of a n[ightjar], *ankh-amu* plant, [*senepe* plant], 'Great-of-Amen' plant, *qes-ankh* stone, genuine lapis lazuli, myrrh, 'footprint-of-Isis' plant, pound, make into a ball, [you paint] your [eyes] with it ..." (Griffith and Thompson, *op. cit.*, Vol. I, p. 81).

ingredients in alchemical and other recipes. There is, however, a convenient almost-contemporary text that explicitly lists such deliberately misleading terms: it appears in *P.Leid. V* lines 401-444:

> "Interpretations which the temple scribes employed, from the holy writings, in translation. Because of the curiosity of the masses they [i.e. the scribes] inscribed the names of herbs and other things which they employed on the statues of the gods, so that they [i.e. the masses] – since they do not take precaution – might not practice magic, [being prevented] by the consequence of their misunderstanding. But we have collected the explanations [of these names] ... "

Many of the code-words are bloods: e.g. "Blood of a snake: hematite," "Blood of an eye: tamarisk gall," and directly relevant here: "Blood of a goose: a mulberry tree's 'milk'."

While the interpretation of technical terminology and ingredients is often difficult and uncertain (compare note 43 above, on 'truffle'), it seems that there is no controversy surrounding the translation of such a common material name as 'lapis lazuli', and so this spell becomes interesting to us for two reasons. Firstly, the making of a ball with a resin (myrrh) and pounded lapis lazuli is reminiscent of the mediaeval methods of refining ultramarine. Secondly, although much earlier artefacts survive that use lapis lazuli as a semi-precious stone, including very early Egyptian artefacts, this magical formula text seems to be the earliest evidence for the use of lapis lazuli in a *paint*.

Although *P.Leid. V* is a magical papyrus, and contains recipes for peculiar inks that probably had no application outside magical practice, nevertheless the iron-gall ink recipe it contains does reflect contemporary practice; it is therefore just possible – although by no means inevitable – that the use of ground lapis lazuli as a component of paint in *P.Mag.LL* also reflected contemporary practice. (If this is so, when ultramarine is detected on late Egyptian artefacts, we should be careful not to continue to automatically assume that it is a later retouching).

Parallels between Assyrian and Greek recipes: purple dye

To illustrate the continuity of recipe texts, let us consider some more recipes for purple dyes. The same dye materials as were described in the Assyrian example above (iron, tannin and purple from alkanet, lichen or shellfish) also appear in some of the recipes for purple dyes in the *Leiden papyrus X* and *Stockholm Papyrus*. In *P.Leid. X* and *P.Holm.* the iron is present as ferrous sulphate (*kalkanthon* or *misy*). Ferrous sulphate is also described as a general mordant.[61] One recipe states that *kalkanthon* and bark of pomegranate are used

61. *kalkanthon* in *P.Holm.*: Halleux §92 = Caley §87.

in dyeing.[62] In one recipe for purple dye the tannin comes from pomegranate bark or rind, and the redness from alkanet.[63] In yet another the tannin source is gall-nuts and the purple source is alkanet.[64] For "Genuine Purple" pomegranate blossoms and chalcanthum are used.[65] Ferrous sulphate is used in recipes for purple dyeing with alkanet: "Alkanet in company with *kalkanthon* dyes linen as well as cotton. With *kalkanthon* alkanet red changes into purple."[66] All three ingredients – an iron source, a tannin source, and a purple dye source – are, as in the Assyrian example, all present together in one recipe: "Take the juice of the upper part of the alkanet liquid. Roast a gall-nut in the oven. Having ground it with the addition of a little iron sulphite, mix with the juice, boil, and make the purple dye."[67] There are several further examples of the use of these materials in the papyri.

The earliest mediaeval Latin recipes: the *Mappae clavicula* text-family[68]

From the third to the fifth century AD texts of all kinds were preserved in Greek. During the fifth to tenth centuries many texts were translated from Alexandrian Greek first into Syriac and later into Arabic, in which languages they were preserved and transmitted through to the early twelfth century, at which time they began to be re-imported from the Arab and Persian world back to Europe *via* translations out of Arabic and into Latin that were made in Italy and Spain. Nevertheless, there seems to be a long gap in the text record for practical texts following the *c.* 300 AD *P.LeidX* and *P.Holm.*

The gap is eventually broken by what may conveniently be termed the *Mappae clavicula* family of recipe compilations. The *Mappae clavicula* contains about 300 more or less accurate recipes and descriptions of miscellaneous chemical and alchemical operations but especially many detailed technical recipes for crafts, including painting, metalwork, dyeing, pigment preparation (minerals and lakes), staining of bone and leather, varnishes, glues for many purposes, niello, chrysography and gilding, as well as other decorative tech-

62. *P.Leid. X*: Halleux §92 = Caley §94.

63. *P.Leid. X*: Halleux §95 = Caley §97.

64. *P.Leid. X*: Halleux §100 = Caley §98.

65. *P.Holm.*: Halleux §100 = Caley §95.

66. *P.Holm.*: Halleux §94 = Caley §89. Other examples include *P.Leid. X*: Halleux §94 = Caley §96 (*kalkanthon*) and *P.Holm.*: Halleux §108 = Caley §103 (*misy*).

67. *P.Leid. X*: Halleux §98 = Caley §100.

68. For an overview of mediaeval recipe books for artists, together with a catalogue of over 400 mediaeval European manuscripts that contain such recipes, with a bibliography of their editions, translations, and commentaries, see Clarke 2001 *Art of All Colours*. A new study of the *Mappae clavicula* by R. Halleux is forthcoming.

niques including mosaic and decorative metal-working, together with remarks on some minerals, metals and some unrelated recipes. Again these are more-or-less concise, and clearly originally served as recipes for a practitioner; for example, weights of ingredients are generally specified. Most of the recipes are plausible, but with a few 'magical' recipes (or at least, ones that will not work, such as for an arrow poisoned with the sweat of a horse which 'has been properly proved').

From internal evidence it seems that the core of this compilation of concise practical artists' and craftsmen's texts was probably assembled around 600 AD, very possibly in Alexandria. Although the early history of this compilation is unclear, it contains items traceable to earlier classical texts, and indeed it shares some recipes with the papyri.[69] A Greek-language origin (or at least a Greek-language stage or component) is evident from certain of the terminology (and indeed one MS contains a recipe which is written in Greek but using Roman letters) and perhaps from the title.[70] Subsequent to this, it is likely that it was translated into Latin in Italy, around 750-800 AD. This makes it the earliest Latin alchemical or technological manuscript.

The earliest MS of this family to survive is the *c.* 800 AD Latin *Lucca manuscript*,[71] but the name of the group is derived from one lost MS from *ante* 821-822 AD,[72] from one surviving 9th-10th century Latin manuscript,[73] and

69. e.g. compare the proportions and exotic ingredients in the artificial chrysography recipe in the *Lucca MS* that is practically a word-for-word translation of a recipe in *P.LeidX*. [*Lucca MS*: Hedfors, p. 30 'N 13-19' = Caffaro edition, p. 112, item 80; *P.Leid X*: recipe Halleux §72 = Caley §74]. See below for details of editions.

70. *Lucca MS*: 'De crisorantisia' Hedfors, p. 61 '∂ 18-22' = Caffaro, p. 160, item 151. It has been convincingly suggested that the curious name of the *Mappae clavicula* ('the little key to the little cloth') might be explained as an imperfect transmission or imperfect translation of a Greek title, e.g. κλειδίον χειρόκμητα (approximately 'the key to the trick-of-the-trade' or 'the key to artifice'), resulting in *Mappae clavicula* when χειρόκμητον ('thing made by hand' or 'knack') became corrupted to χειρόμακτρον ('hand-towel'). Halleux, Robert and Meyvaert, Paul (1987) 'Les origines de la *Mappae Clavicula*' *Archives d'histoire doctrinale et littéraire du Moyen Age*, LXII, pp. 7-58. See pp. 11-13. See also n. 14 above on the possible Χειρόκμητα text of the *ps.*-Demokritos.

71. Lucca, Biblioteca Capitolare Feliniàna, Codex 490. Also known as the *Compositiones ad tingenda* and as the *Compositiones variae*: 'Compositions on colouring mosaics, skins and other things, on gilding iron, on minerals, on chrysography, on making certain adhesives, and other documents relating to the arts'. There are several editions of this MS (listed in Clarke 2001 *Art of All Colours.*, p. 90). It was one of the first MSS to attract attention as a technological source, being published by Muratori as early as 1739. The best study of the Lucca manuscript remains Johnson, R. P. "The 'Compositiones variae' from Codex 490, Biblioteca Capitolare, Lucca, Italy," *Illinois Studies in Language and Literature*, XXIII: 3 (1939), which exhaustively details parallels and occurrences of sections of the Lucca manuscript texts in other manuscripts. Commonly cited is the edition of H. Hedfors, 1932, *Compositiones ad tingenda musiva*, Inaugural-Dissertation. Uppsala: Almqvist & Wiksells Boktryckeri-A.B. The most recent edition of the recipe parts of the MS, together with a colour facsimile and an Italian translation, is A. Caffaro, *Scrivere in oro. Ricettari medievali dárte e artigianato (secoli IX-XI) codici di Lucca e Ivrea*, Napoli: Liguori, 2003.

72. Listed in the library catalogue of Reichenau for 821-2 AD, but now lost. For details see Clarke, 2001, *Art of All Colours* item 3020.

73. Sélestat, Bibliothèque Humaniste (was: Bibliothèque municipale) MS 17 (*was*: MS Latin 360, *was*: MS 1153 bis). Clarke, 2001, *Art of All Colours* item 3100.

from one surviving 12th century manuscript,[74] all sharing the name *Mappae clavicula.*[75] The *Lucca MS* was written in Italy, while the others were from north of the Alps. A further important witness is the 12th century *Codex Matritensis.*[76,77] In addition to these principal manuscripts, extracts from this text family appear in a great number of other manuscripts.[78] Certain of the *Mappae clavicula* MSS are prefaced by a distinct text on suitable and unsuitable combinations of pigments, the *De coloribus et mixtionibus,* which has been identified as a separate text with its own history.[79] In addition to the Alexandrian or Graeco-Byzantine elements of the compilation, influences have been suggested as diverse as Assyrian, Arabic and Hindu. In addition to the parallels with the papyri, certain parallels may be seen with Theophrastus, Dioscorides, Pliny, and Vitruvius. The compiler did not, however, just rely on earlier written material, but also on the oral traditions of Italy and elsewhere, as several additions testify.

Mediaeval technical treatises are most commonly compilations. Until the Herculean task of producing a stemma of these recipes is complete it may be useful to think in terms of a super-corpus consisting of all the individual recipes, sub-sets of which occur in individual manuscripts. Certain manuscripts do contain sets of texts that are similar, or contain texts that are largely confined to a small family (e.g. the Theophilus MSS, MSS of Heraclius books I and II, the Strasbourg MS family, or *De coloribus et mixtionibus*). If the *Mappae Cla-*

74. Corning (New York State), Corning Museum of Glass, MS 5 (*was:* Phillips MS 3715). Perhaps the fullest MS, it also preserves evidence that the text passed at some point through England: it contains a table of Anglian runes, and some words that seem to be corruptions of English plant names. Clarke, 2001, *Art of All Colours* item 2260.

75. Edition: "Letter from Sir Thomas Phillipps, Bart., F.R.S., F.S.A., addressed to Albert Way, Esq., Director, communicating a transcript of a MS. Treatise on the preparation of Pigments, and on various processes of the Decorative Arts practiced during the Middle Ages, written in the twelfth century, and entitled Mappæ Clavicula," *Archaeologia*, xxxiii, 1847, pp. 183-244. English translation with extensive commentary and a photographic facsimile of the Sélestat and Corning MSS: C. S. Smith and J. G. Hawthorne, "Mappae Clavicula: A Little Key to the World of Medieval Techniques," *Transactions of the American Philosophical Society* New Series Vol. 64: 4 (1974), whole issue.

76. Madrid, Biblioteca Nacional, MS A.16 (*Was: Ms. A.19*). Clarke, 2001, *Art of All Colours* item 2070. Edition and English translation: J.M. Burnam, *Recipes from Codex Matritensis A.16 (ahora 19),* University of Cincinnati Studies Series II, Vol. VIII, part 1, 1912. A better edition of the *Codex Matritensis,* with a Spanish translation, is S. Kroustallis *Edicion critica y estudio de un tratado medieval de tecnologia artistica: Codex Matritensis 19*, unpublished PhD dissertation, Universidad Complutense de Madrid, 2005.

77. A concordance of correspondences between the MSS appears in Smith and Hawthorne 1974.

78. See for example R. P. Johnson, "Notes on some manuscripts of the *Mappae Clavicula,*" *Speculum*, X, 1935, pp. 72-81 and "Some continental manuscripts of the *Mappae Clavicula,*" *Speculum*, XII, 1937, pp. 84-103.

79. A. Petzold, "De coloribus et mixtionibus: The earliest Manuscripts of a Romanesque Illuminator's Handbook," *Making The Medieval Book: Techniques of Production Proceedings of the 4th Conference of The Seminar in the History of the Book to 1500 Oxford, July 1992*, L. L. Brownrigg (ed.), Los Altos Hills and London: Anderson-Lovelace, The Red Gull Press, 1995, pp. 59-65. See also Smith and Hawthorne, *op. cit.*

vicula is taken to be the union of the Phillipps-Corning MS, Sélestat MS, Lucca MS and Madrid MS, then around 75 mediaeval manuscripts have been identified that contain *Mappae Clavicula* recipes.

For whom were they written?

Rather than wonder why recipes *per se* did not appear until the second millennium BC, 1500 years after writing was available for writing them, it is perhaps more useful to wonder why – at a time when literacy was limited to those who made it their professional speciality – recipes came to be written down *at all.* Few if any craftspeople would have been literate, and the scribes abhorred all manual labour, and indeed composed satirical texts denigrating all professions other than that of scribe, generally on the grounds that they were physically demanding and dirty.[80] The apparent mutual exclusion of the literate person and the craftsperson clearly continued in Europe until the middle ages, and elsewhere until far later.[81] Today where there is widespread literacy it is understandable that recipes are written down for the benefit of the specialist cook and painter, or for the amateur cook and painter, but why were they written before?

Preservation of recipe texts as literature

In the case of the earliest cuneiform recipes (which survive in two main batches, from *c.* 1700 BC and *c.* 700 BC) there are clues from the language. The language used is often deliberately archaic, scholarly or mannered, using, for example, the Sumerian language, which had not been spoken for centuries, instead of the current language in its Akkadian, Babylonian and Assyrian dialects. The effect was as if one wrote in Latin today. This deliberate archaism was in part a reflection of their traditionalism. It has been argued that these 'recipes' are not true craft recipes, but are part of a literary tradition.

It was noted above that certain craft texts were re-worked (e.g. by the addition of new material) to keep them relevant to workshop practices. However, a contrary form of re-working also took place, making them *less* relevant to the workshop. It is important to recognise the phenomenon of the transmission of recipes as texts *per se*, as opposed to them being copied so as to transmit practical knowledge. This practice continued and continues. Many mediaeval European recipe texts were in fact not original treatises by practicing craftsmen but

80. Although the Egyptian *Instructions of Dua-Khety: The Satire on the Trades*, a popular text surviving in several versions, is far earlier (probably composed 1950-1900 BC) there is no reason to suppose that the condescension it shows had in the least disappeared by *c.* 300 AD, nor indeed has it yet!

81. Although this lack of literacy was not as total as often assumed. See Clarke, "Late medieval artists' recipes books," in this volume.

were compilations made from previous books, in many cases assembled by people who had no contact with the workshop. Some were copied more or less without reflection, as though they were literature or history. Examination of recipe texts clearly demonstrates this.[82] Even the *Mappae clavicula* MSS occasionally include the kind of mistaken and incomprehensible material which suggests that it was copied uncomprehendingly, i.e. by someone ignorant of workshop practice (Again, not an extinct practice: the most famous of all English recipe books, Mrs. Beeton's *Book of Household Management* (1861) was not written by an expert housewife, but was compiled from other books by the young wife of a publisher).

Throughout history there has been a desire to assemble and preserve old texts, in particular by refreshing them by re-copying everything considered worth preserving.[83] It has been argued that this was a vital part of the stream of tradition, and examination of copied cuneiform texts shows that this deliberate maintenance of a corpus of texts was *already* practiced in the second millennium BC.[84] Assurbanipal (668-627 BC), king of Assyria, had the literature of his country collected and re-copied, catalogued and stored in a great library of tablets in his palace at Nineveh (around 5000 tablets, plus duplicate texts: now in 30,000 fragments).[85] In the temple libraries of ancient Egypt, the 'houses of life', the main activities were preparation of texts, recopying old manuscripts, correcting errors and completing lacunae. The libraries of Alexandria from their foundation in the third century BC onwards had the right to seize all books from ships that docked there, to make copies, to return the copies to the ships and to retain the originals; the collection grew thus to hundreds of thousands of volumes. These were not simply stored, but edited and re-edited. Charlemagne (742-814 AD), like Assurbanipal, had all the treasures of antiquity collected and recopied at his palace school in Aachen (Aix-la-Chapelle). (For any given classical text, the earliest surviving manuscript is usually Carolingian.) The result of all this diligent preservation by copying is that the date of a manuscript is not necessarily the date of the text it contains, and therefore any craft technique described in that manuscript is not necessarily that which was current when the manuscript copy itself was written.

82. See Oltrogge, in this volume. For more on the two processes of re-working, see M. Clarke, *Mediaeval Painters' Materials and Techniques: The Montpellier 'Liber diversarum arcium'*, 2011, London: Archetype Publications, especially pp. 13-14 and pp. 44 ff.

83. "Like other antique writings housed in monastic libraries, the Reichenau *Mappae clavicula* was probably a fountain into which the brethren dipped to purify their Latin. ... While it could not train a beginning painter in the practical aspects of his craft, an artist would, nonetheless, value such a work, especially in Carolingian times. For it provides an atmosphere of the classical past ...." Smith and Hawthorne, *op. cit.*, p. 20.

84. R. MacLeod (ed.), *The Library of Alexandria*. London & New York: I.B. Taurus, 2000, p. 23, citing Leo Oppenheim.

85. Simple archiving of existing texts was earlier: e.g. the Mesopotamian tablet houses of *c.*2600-2000 BC or the Egyptian temple libraries. Some of Assurbanipal's texts were originals, not copies, e.g. acquired through donation or as booty. They may be distinguished by different clay.

Assyrian and Egyptian craftsmen probably transmitted practical information orally rather than in the form of written treatises. However, their position within temples, and their association with priests (like their successors in mediaeval European in monasteries), seems to have resulted in the acquisition by the priests of a considerable theoretical knowledge, and it has been argued that there was also a body of ritual texts based on such knowledge. For example, the *Mappae clavicula* may have originated as a key to the practical techniques that were deliberately obscured in such religious texts.[86] When did this body of knowledge start to accumulate? Although there is little information about Egyptian temple libraries (and therefore about any potential technical texts) before the establishment of the Library of Alexandria in the third century BC, it has been inferred that, certainly until the first century BC, they almost certainly contained only religious texts.[87] However, by the time the alchemist Zosimas visited Alexandria *c.* 300 AD – close to the probably date of the two papyri – he was able to report that books on chemical subjects were present in Alexandrian temples[88] and "particularly in the Serapaion of Alexandria."[89] The *Leiden Papyrus X* and the *Stockholm Papyrus* are presumably two precious surviving examples of what these texts might have been like.

Why were the earliest recipes ever written down?

Returning to the earliest recipes, the Mesopotamian ones from *c.* 1700 BC, it seems that they may well have formed part of a literary tradition. Nevertheless, they were once new. Why were they written down *originally?* While we may realise that it would be anachronistic to judge them as we do our own recipe books, we can nevertheless see that they are clearly technical records, not simply literary descriptions. As Bottéro observes of the cooking recipes "Their style is very concise, recalling that of our own works which are the preserve of professionals ... and which keep to the essentials of 'specialist' and ultra-laconic manner."[90] Indeed, in common with most mediaeval recipes, there are many details left out that we would dearly like to know. If they cannot be considered didactic treatises, then what was their purpose?

The Mesopotamian recipes, the Egyptian recipes from the papyri, and many mediaeval recipes all emerged in religious environments. The Egyptian temple libraries and the monastic scriptoria preserved texts that aided their own practices, just as did the workshops attached to both Egyptian temple and European

86. See, for discussion of other examples, Eamon, W. (1994) *Science and the Secrets of Nature*, Princeton: Princeton University Press, pp. 34-5

87. See Partington, J. R. (1935) *Origins and Development of Applied Chemistry*, London: Longmans, Green and Co., p. 18.

88. Partington *loc. cit.*, citing Berthelot *Collection* ... vol. ii., p. 230; vol. iii., p. 223.

89. Partington *loc. cit.*

90. Bottéro, *op. cit*, p. 52.

monastery. Descriptions of rituals (procedures, incantations, speeches) were normative: writing them down ensured they would be performed correctly. I would suggest that the food, or the painted amulet, or the jewel, or the precious metal vessel, was part of the ritual, and thus its preparation was to be carried out according to prescribed formulae. Bottéro argues that (at least for the cooking recipes) recipe texts were *administrative* and normative:

> "They would have been 'written' by order from above, to record and lay down 'what was done in the kitchen', as other tablets ... recorded and laid down what was done at court, according to etiquette; in the temple, according to the liturgy; in the offices of the doctors and apothecaries, in keeping with practice; among certain technicians, according to traditional formulas and procedures. Rather than a culinary *manual*, our texts represented [perhaps a culinary] *ritual:* they codified contemporary practices, the outcome of age-old habits, enriched (but still capable of requiring further enrichment) by constant improvements and inventions."[91]

This argument is very plausible. If these recipes were intended for a ritualised cuisine (palace or temple) then their written form becomes explicable. Alongside what we would term 'practical craft recipes' in the Egyptian papyri and the monastic parchments are found spiritual texts, including prayers, incantations, rituals, amulet texts and so forth. It is probably a mistakenly arrogant and superior anachronism to esteem the 'practical' recipes over the 'impractical'. Surely at the time they were written, each was considered as effective a solution to a present problem or requirement.

In this model, there is no need for a literate craftsman. As was the case in late classical, mediaeval and renaissance medicine and alchemy, there was a master reading from a book, and a handyman doing the mixing, cooking, and surgery. Indeed, if the handyman was working correctly there was no need to read from the book – it was retained as a timeless authority, and as a guide for supervisors. The nineteenth century recipe for the proprietary painting medium of the London colourman Roberson provides a modern example. In nineteenth century paint manufactories it was common to have formulation books that would never be seen on a daily basis. Certain workmen made certain sub-components of the final paint; these were in certain instances identified by a code letter, and were combined by further workmen, who worked with these ready-prepared materials identified to them only by the code letters. The whole process (as recorded in the book) was therefore only understood by the master. The recipe for the special medium, known only to Roberson himself, was sealed 'to be opened after I am dead.'[92]

91. Bottéro, *op. cit*, pp. 61-2.

92. J.H. Townsend, J. Ridge, and L. Carlyle, "Cobalt blue, emerald green, and rose madder in copal-based mediums as used by the Pre-Raphaelites," in Clarke *et al.* 2005, *Art of the Past*, p. 62.

The point is: the recipe is a shorthand for which operations and materials are needed. The overseer uses the recipe to brief the craftsman: 'of all the possible things you know how to do, today we will do this one' – it is left to the skill and experience of the craftsman to fill in the gaps.

What was not written down in recipes

Many details are missing from written recipes of all periods, enough to make following them problematic or even impossible. Reconstructions require interpretation and educated guesses as to what the author found unnecessary to write down. Bottéro again:

> "... given that we have never seen those ancient cooks at work, the difficulty we experience in trying to 'realise' and understand the successive stages in their methods, plus the number of words and technical terms whose precise meaning eludes us, together with the annoying interruptions and breaks in the text are all obstacles to both decipherer and translator. ... it is virtually impossible for us to execute these [Mesopotamian] recipes ... The exact and concrete meaning of certain obviously technical terms and the 'tricks of the trade' they suggested to those using them elude us and risk doing so forever, since in order fully to understand and be able to imitate an action – even if it is not very complicated – one must have seen it performed. Every ... [technique] is made up of 'knack and know-how'."[93]

This issue of knack is crucial.

Conclusions

It is unclear when, for whom, or why the earliest technical recipes were originally written down. When using them as clues to historical craft practices we need to be aware that for at least *part* of their history the texts were *sometimes* recorded, transmitted, and used for purposes other than the simple transmission of techniques: they were enjoyed as literary texts, or as alchemical or philosophical illustrations. Furthermore, the texts are often frustratingly incomplete, and so the severest handicap in interpreting and reconstructing ancient recipes is that we do not have the years of experience that a specialist craftsman would have.[94] Often they contain untranslatable terminology.

This is not to say that comprehension or reconstruction is impossible or futile. If ancient texts are studied alongside technical analyses of artefacts, such as textiles or coloured glass, then each type of resource – text and analysis – provides extremely useful clues to the understanding of the other.

93. Bottéro, *op. cit*, p. 56, p. 63

94. This point is well made in Townsend *et al.* 2005, *op. cit.*, p. 65.

For example a Sumerian beer was recently successfully reconstructed based on textual evidence that is only slender and allusive, such as the *c.* 1800 BC *Hymn to Ninkasi* (the Sumerian goddess of brewing), in conjunction with chemical analysis of the reside of drinking utensils found in King Midas' tomb.[95] While we must be careful not to base too many conclusions on too few incompletely-understood texts, we can nevertheless be very grateful indeed that they have survived.

95. The reconstruction was carried out by archaeologist Patrick McGovern. See: S.H. Katz, and F. Maytag, "Brewing an Ancient Beer," *Archaeology*, 44: 4 (July/August 1991), pp. 24-33. See also the texts in L.F. Hartman and A.L. Oppenheim, "On beer and brewing techniques in Ancient Mesopotamia: According to the XXIIIrd Tablet of the Series HAR.r.a = *hubullu*," Supplement no. 10 to the *Journal of the American Oriental Society*, December 1950, 'issued with vol. 70, no 4'.

LATE MEDIEVAL ARTISTS' RECIPES BOOKS (14th-15th CENTURIES)

Mark Clarke

Introduction

This paper will introduce the late-mediaeval texts that contain recipes for the materials and technical techniques of scribes, manuscript illuminators, and painters. It will not (with two exceptions) discuss iconographic manuals or model books of images, nor will it discuss non-textual sources of information. It will also consider some issues with the interpretation and use of these recipes today, including several problems which are common to all mediaeval technical texts.

These artists' recipe texts consist mainly of instructions for the manufacture of materials: pigments, inks and painting media. Some also describe the preparation of supports, such as panel assembly and priming or parchment manufacture. Recipes particularly concentrate on preparation of paints and inks from the raw materials, including selection and testing of materials, refining, grinding, and mixing of pure materials, and making synthetic materials and compounds such as ink, verdigris or white lead. Some describe suitable mixtures, in particular which medium is best suited to which pigment and which pigment may be mixed with which other. They also warn of unsuitable mixtures of pigments, i.e. combinations that are chemically unstable, resulting in discolouration. The use of gold and its imitations was also clearly a widespread preoccupation. There are also manuals on technique.

It is possible to make a few generalisations about artists' recipe books. Coherent extended treatises are the minority, compilations of short recipes the norm. While there are many individual texts, there is also an immense influence of a few particular texts that survive in multiple copies, especially in extracts, notably Theophilus and the *Mappae clavicula*. They are almost entirely free of 'magic' or ritual. They contain very few implausible or impossible recipes, (synthetic blues being a notable exception, and the magical gold recipes in the otherwise apparently reliable Theophilus[1]). These few impossi-

1. See the papers in this volume by Bucklow and Oltrogge.

ble recipes that survive are worrying, as they cast doubt on the reliability of the corpus. There are occasional pieces of alchemy: a disturbingly large number of the later art technological texts are found in manuscripts that are primarily alchemical. Of course, artists, like alchemists, were preoccupied with gold and its manufacture or imitation, so this pairing should perhaps not surprise us. Many art-technological texts are also found alongside medical texts. It should be remembered that St. Luke was both painter and doctor, and that in certain periods and regions painters would belong to the same guild as apothecaries. This also would suggest literacy of artists was relatively common.

Historical overview

A short historical overview of how artists' recipes developed through time may make it easier to appreciate what distinguishes fourteenth and fifteenth century recipes from those that came before and after.[2]

Mesopotamian cuneiform

The earliest technical recipes that survive are Mesopotamian Assyrian cuneiform texts for making coloured vitreous material (perhaps opaque glass, but perhaps ceramic glaze).[3] These originate from the seventeenth century BC or arguably three or four centuries later. There is then a long gap, until the *seventh* century BC, from when survive further recipes for coloured glassy materials, as well as some recipes for dyeing wool blue and purple, and recipes for cooking food. These Mesopotamian recipe texts are very brief, being little more than lists of ingredients, with very little explanatory material; in consequence they require careful interpretation and are hard to use, for example, as the basis for a reconstruction. Puzzling questions about these earliest texts apply equally to later artists' recipes:

– what does the technical terminology mean exactly?

– why exactly were they written down and for whom?

– does the fact that they were intended for a specific type of reader affect their reliability and their relationship with actual contemporary workshop practice?

We shall return to these questions below.

2. For a fuller overview of mediaeval recipe books for artists, together with a catalogue of over 400 mediaeval European manuscripts that contain such treatises, with a bibliography of their editions, translations, and commentaries, see M. Clarke, *The Art of All Colours: Mediaeval recipe books for painters and illuminators,* London, Archetype Publications, 2001.

3. These are often cited as if they were pigment recipes, but this is clearly incorrect.

Classical references

The earliest references to artists' materials and practices are Classical; they are not recipes *per se*, but descriptions. Theophrastus (370-288 BC) describes various pigments and their preparation, as do Vitruvius, Pliny and Dioscorides in the first centuries BC and AD. With the possible exception of Vitruvius, these authors were not practicing craftsmen. It is noteworthy that the gap in 'recipe books' corresponds approximately with the Classical period, *c.*700 BC-*c.*300 AD, although the reasons for this gap are unclear.

Early Greek papyri[4]

The first post-classical recipes *per se* are found in the *Leiden Papyrus X* and *Stockholm Papyrus*, written in Greek, coming from Egypt, and dating from *c.*300 AD.[5] These mainly deal with metallurgy (especially gold and silver), purple dye, and the imitation of gold, silver, purple and gems. *Leiden Papyrus X* is especially of interest to historians of manuscript art as it contains recipes for writing in gold and in silver. The recipes in the papyri are similar in form to the Assyrian recipes, being short and concise, with little explanatory material and (almost) no theoretical justifications. As with the Assyrian examples, it is unclear for whom they were written; whether they were a fair copy of an artisan's workshop notes, or whether they were compiled by or for a reader with a more general interest, or with a theoretical interest such as an alchemist. It has been argued that they are not really practical recipes, but pure alchemy, but the exact opposite has also been argued. Whatever the intention behind writing them, they do appear to be clear, and in recent years it has been possible to use these recipes to make practical reconstructions in the laboratory.

Early mediaeval compilations

Next chronologically is a family of compilation manuscripts, the most important and complete examples being (i) the *c.* 750-800 AD *Lucca manuscript*[6] (also known as the *Compositiones ad tingenda* ... and as the *Compositiones variae*), (ii) the 9th-12th century *Mappae clavicula*,[7] (iii) the *c.*1130 AD

4. For further detail on the papyri see the other paper by Clarke in this volume. The most recent edition, with French translation and commentary, is: R. Halleux, *Les Alchimistes Grecs* Vol. 1: *Papyrus de Leyde, Papyrus de Stockholm, Fragments de Recettes*, Paris, Société d'Édition "Les Belles Lettres," 1981.

5. Arguably the *Physika kai mystika* is somewhat earlier; see the other paper by Clarke in this volume.

6. Lucca, Biblioteca Capitolare Feliniàna, Codex 490, ff. 211v, 217r-231r. [Clarke *Art of All Colours* no. 2020].

7. The most complete MSS are Corning (New York state), Corning Museum of Glass, MS 5 [Clarke *Art of All Colours* no. 2260], 12th century, and Sélestat, Bibliothèque Humaniste MS 17, 9th-10th century [no. 2260]. For further detail on the *Mappae clavicula* see the other paper by Clarke in this volume.

Codex Matritensis,[8] and (iv) the *De coloribus et mixtionibus*.[9] These contain recipes for a number of crafts including metalwork, dyeing, and mosaic, and including several recipes relevant to painting and illuminating, for example recipes for artificial copper greens. Many mediaeval manuscripts containing collections of recipes include extracts from this family of recipes. The core of this substantial collection of recipes – some 300 – was probably originally compiled around 600 AD, perhaps in Alexandria, and contains items traceable to earlier classical texts and to the aforementioned papyri. Again they are more-or-less concise in form, and again it is not always clear what materials and processes are being described. For example, there is a recipe for varnish that includes twelve ingredients, one of which is apparently 'varnish'! Most of the recipes are plausible, but with a few 'magical' recipes (or at least, ones that will not work, such as for an arrow poisoned with the sweat of a horse which 'has been properly proved').

The 'Schedula' of Theophilus

From *c.*1100 survives an apparently non-derivative text, the *Schedula* of Theophilus.[10] This systematic treatise covers painting and illuminating, glass work, metalwork and some miscellaneous other crafts. It has been argued convincingly that this is (at least in large parts) truly the work of a practitioner. I have argued elsewhere that the *Schedula* is in fact a compilation, partly practical partly otherwise.[11] The *Schedula* differs from the majority of early mediaeval recipe texts in that it also describes pictorial technique, e.g. the depiction of beards for old or young men. Extracts from Theophilus then enter the text stream and appear in the compilations.

In addition to these compilations, there are a number of useful early mediaeval recipe texts following Theophilus, and all are more or less the same: short practical recipes, mainly concerned with materials, written in Latin. Two particularly useful manuscripts on illuminating, pigments, and book manufacture are the highly informative *De clarea* ('On glair', i.e. egg white medium)

8. Madrid, Biblioteca Nacional, MS A.16. [Clarke no. 2070].

9. Transmitted alongside the *Mappae clavicula*.

10. There are four relatively complete manuscripts (London, British Library, MS Egerton 840A [Clarke no. 1510] MS Harley 3915 [Clarke no. 1600], Vienna, Österreichische Nationalbibliothek, MS 2527 [Clarke no. 3400], Wolfenbüttel, Herzog-August Bibliothek, Cod. Gud. Guelph. 69 2° [Clarke no. 3560]), and a great number of fragments.

11. This explanation also resolves the apparent internal anachronisms. See: M. Clarke, "Reworking Theophilus: adaptation and use in workshop texts," 2011, in: A. Speer *et al.* (eds.) *Die 'Schedula diversarum artium' – ein Handbuch mittelalterlicher Kunst?* (in series *Miscellanea Mediaevalia*). Berlin-New York: W. de Gruyter. See also: M. Clarke, *Mediaeval Painters' Materials and Techniques: The Montpellier 'Liber diversarum arcium'*, 2011, London: Archetype Publications, pp. 56-7.

from the 11-12th century,[12] and Audemar's *De coloribus faciendis* (13-14th century).[13]

The Fourteenth and Fifteenth Centuries

While perhaps a more significant change in recipe content is pre- / post-*c.*1100, with the twelfth century re-importation of Arab learning (reflected in the *Mappae clavicula*), and Byzantine practices (arguably as recorded by Theophilus and in the *Codex Matritensis*[14]), nevertheless there is a distinct change in style and content of recipe books after *c.*1300.

In the fourteenth and fifteenth centuries there are many more texts and more copies of texts to choose from. Better known texts include:

Liber de coloribus illuminatorum siue pictorum [illuminating][15]
De arte illuminandi [illuminating][16]
Likneskjusmith [polychrome sculpture in Icelandic][17]
Pseudo-Savonarola [mainly pigments][18]
The *Strasbourg Manuscript* [illumination and miniatures][19]
Archerius/Alcherius [3 texts][20]
Tractatus qualiter quilibet artificalis color fieri possit[21]
The compilation of Le Begue[22]
The *Montpellier Manuscript*[23]
Cennino Cennini *Il libro dell'arte* [a systematic treatise][24]
The *Segreti per colori* or *Bolognese manuscript* [a systematic collection][25]

12. Bern, Cantonal Library (Burgerbibliothek), MS A. 91.17 [Clarke no. 140].

13. Paris, Bibliothèque Nationale, MS latin 6741, ff. 52-64 [Clarke no. 2790].

14. Madrid, Biblioteca Nacional, MS A.16, ff. 199-203, Latin, *c.*1130 [Clarke no. 2070]. See Kroustallis, this volume.

15. London, British Library, MS Sloane 1754 [Clarke no. 1900].

16. Naples, Biblioteca Nazionale Vittorio Emanuele II, MS XII.E.27 [Clarke no. 2250].

17. Copenhagen, Arnamagnean Institute, MS AM 194.8 [Clarke no. 3025].

18. Ferrara, Biblioteca Comunale Ariostea, MS Cl. II. 147 [Clarke no. 582].

19. The original manuscript was destroyed by fire; a 19th century transcription survives at the National Gallery London.

20. Paris, Bibliothèque Nationale, MS latin 6741 [Clarke no. 2790].

21. Paris, Bibliothèque Nationale, MS latin 6749B [Clarke no. 2800].

22. Paris, Bibliothèque Nationale, MS latin 6741 [Clarke no. 2790].

23. Montpellier, Bibliothèque de la Faculté de Médicin, MS 277 [Clarke no. 2090]. Newly edited with English translation: M. Clarke, *Mediaeval Painters' Materials and Techniques: The Montpellier 'Liber diversarum arcium'*, 2011, London: Archetype Publications.

24. The pre-*c.*1500 AD manuscripts are Firenze, Biblioteca Medicea-Laurenziana, MS Plut. 78.23 [Clarke no. 590], and Firenze, Biblioteca Riccardiana, MS 2190 [Clarke no. 1020].

25. Bologna, Biblioteca Universitaria, MS 1536 [Clarke no. 160].

Ambrigio's *Ricepte d'affare più colore*[26]
The *Göttingen Modelbook*[27]
The Hastings manuscripts [Housebook][28]
Livro de Como se Fazem as Cores[29]

and various texts titled *de coloribus*.

What, if anything, changes? How does the content alter in post *c.*1300 recipe texts? In general, the later manuscripts are less concerned with manufacture of materials (e.g. making synthetic pigments); these were by now readily available for purchase. More organic pigments are described than before. More notes on technique appear. Often recipes are longer and more discursive.

We can make a few approximate statistical generalisations. These can only be approximate, as subject categorisation is very subjective, and there is plenty of overlap (an isolated recipe may well be a quotation from a larger treatise, some compilations are multi-lingual, pigments may be used for many purposes etc.) Nevertheless it is possible to draw some general outlines.

As one might expect, the most obvious feature is survival: the great majority of surviving mediaeval recipe manuscripts date from this period. (Less than 40 manuscripts survive from the 13th century, over 100 from 14th and over 250 from the 15th.) This is not surprising, and would be the same for most types of texts, and we need not dwell on it here. As one might also expect, there is a significant increase in the use of the vernacular. While vernacular artists' recipe books are not unknown prior to the fourteenth century, only a couple survive. During the 14th-15th centuries Latin continues to be used, but the vernacular accounts for half the number of manuscripts.

There does seem to be a difference between the content of vernacular and Latin texts.

Content of Latin texts

The majority of the surviving 14th-15th century Latin texts consist largely of copies or extracts from earlier mediaeval treatises such as Theophilus, Heraclius, and the *Mappae clavicula*, or they are of parts of other technical works

26. Siena, Biblioteca Comunale degli Intronati, MS I.II.19 [Clarke no. 3110].

27. Göttingen, Niedersächsische Staats- und Universitätsbibliothek, Codex 8° Uff. 51 Cim. [Clarke no. 1120].

28. San Marino California, Henry E. Huntington Library, MS HU 1051 [Clarke no. 3090].

29. Parma, Biblioteca Palatina MS De Rossi 945. The *Livro de Como se Fazem as Cores*, [Clarke no. 2950] which for a long time was believed to have been dated to AD 1262 and attributed to Rabbi Abraham ben Judah ibn Hayvin (or Hayyin), is in fact more likely an anonymous 15th century text. António João Cruz and Luís Urbano Afonso (2008) "On the Date and Contents of a Portuguese Medieval Technical Book on Illumination: O livro de como se fazem as cores," *The Medieval History Journal* 11: 1-28. See also the paper in this volume by Afonso, Cruz and Matos.

by other named authors that contain material on paints etc. (e.g. by Albertus Magnus or Vitruvius). [See Table 1.] It is possible that these manuscripts, which preserve and circulate older knowledge, may be considered less relevant to contemporary practice, but this is not the moment to consider this. The remaining Latin texts are short isolated recipes, typically for pigments or inks; often these are written in spare blank spaces in books.

	'Independent' treatises		Copies of early mediaeval treatises (Latin)	Isolated recipes (all languages)
	Book arts (all languages)	*Other painting (all languages)*		
before *c.*1300	13 (15 %)	19 (23 %)	30 (36 %)	21 (25 %)
14th-15th centuries	107 (28 %)	126 (34 %)	57 (15 %)	81 (22 %)

Table 1: Relative frequency of types of text.

Content of vernacular texts

Vernacular manuscripts, at first inspection, mostly appear to contain new material. However, translation may include re-ordering or paraphrasing, so much further work is required to establish the extent to which they are in fact simply translations of known earlier Latin treatises. It seems likely (but not inevitable) that, if they are new texts, and in the vernacular, then they were composed from original observation – that is, they more accurately reflect the practices of the time in which they were written. Very few vernacular texts occur in multiple copies. This may indeed imply they were more for personal use.

In the 14th-15th centuries Latin (185 manuscripts) is the most common language for texts, followed by German (67), English (46), Italian (39), and Middle Dutch (16), a few in French (5) (plus one in Occitan), Greek (5), and Hebrew (2), and one each in Provençal, Castilian, Catalan, Portuguese, Polish, and Icelandic.

As well as this difference in content between Latin and vernacular manuscripts, there seem to be some differences in content between texts in the different vernacular languages:

– English recipes, with only a few exceptions, are primarily medium length treatises concerned with book manufacture: ink, illumination, parchment

and chrysography recipes (This is the case from as far back as the 12th century Anglo-Norman recipes).
– Middle Dutch (Dutch, Flemish) recipes are also primarily scribal, with many short isolated ink recipes, with the magnificent exception of the polyglot multi-subject compilation of De Ketham.
– Italian texts tend to be longish treatises, rather than isolated recipes: Cennini, Michele Savonarola, the *Segreti per colori*, and a couple of long untitled treatises of over 150 folios. The recipes are sometimes quite long and discursive. They are about evenly divided between treatises on book arts and other forms of painting.

It may be, however, that this apparent difference in emphasis between languages may in fact only reflect the different preoccupations and nationalities of the individual researchers who have studied mediaeval technical treatises. I myself, an Englishman, working more often in English libraries than abroad, and preoccupied with manuscript illumination, found a great number of English texts on manuscript production; Ploss, a German, was interested in all "Alte farben" including pigments and dyes; Oltrogge, again German, is primarily concerned with works of art on paper; Jansen-Sieben was compiling a systematic *repertorium* of Middle Dutch texts on the seven liberal arts and during this noted many isolated recipes that might otherwise have gone unnoticed; Zerdoun Bat-Yehouda, writing in French, Hebrew and Greek, and concentrating on inks, therefore added to our knowledge of Hebrew and Greek recipes, but only for ink. I am sure that in time many more texts, in a wider variety of languages, will surface.

Content of recipes

Of all surviving texts from before *c.*1300, a third of the texts (but forming the most bulk in terms of pages) consists of long treatises in Latin on several subjects, often by named authors [Table 1]. There are an equal number of shorter or 'independent' treatises, more or less equally divided between book arts and other painting, and a substantial number of isolated recipes (e.g. for just one or two pigments, or for an ink), often infilled into odd blank spaces.

It is interesting to note that this proportion alters somewhat after *c.*1300. The proportion of texts consisting of extracts from named Latin authors halves, and the proportion of 'independent' treatises or collections increases correspondingly. These 'independent' texts remain divided in approximately the same proportion as before between book arts (approximately two fifths) and 'other painting' (three fifths). There is actually in all periods a slight preponderance of texts on 'other' painting; but when one considers the range of 'other' painting – murals, polychrome sculpture, panels and altarpieces, painted textile hangings – clearly the scribes are disproportionately represented.

This does perhaps confirm the suspicion that literate people will compose texts on their own specialities. If this is so, then because scribes must always be literate and (at least early on) painters need not necessarily be literate, one might expect that the proportion of treatises on scribal subjects might to be higher earlier on, i.e. before more widespread literacy. However, this is not what is observed.

The proportion of 'substantial' texts to 'isolated recipes' remains similar. These isolated jottings are surely done for the benefit of individual users. One might expect the proportion to increase, therefore, as manuscripts become increasingly individualised, but this is not what is observed either. This clearly requires further investigation.

Workshop practice *versus* text

Are these late mediaeval recipe manuscripts *more*, or *less*, reliable than their predecessors?

– In *all* periods they are more-or-less magic-free
– In *all* periods the transmission of recipes as literary texts continues. Many of them are found transmitted alongside recipes and texts on alchemy and medicine – rather than, as one might hope, alongside texts on pictorial arts. Let us consider some aspects of these issues.

Preservation of old texts versus composition of new texts

In the mediaeval period scholarship was to a great extent concerned with preserving ancient texts, and in the renaissances of *c.*800 AD, the twelfth century, and the fourteenth century, was concerned with recovering ancient texts (complemented throughout by preparing compilations, florilegia, summaries, or commentaries). Is it correct to say that after *c.*1100 and certainly by the 14th -15th centuries, that the emphasis had changed to writing down personal workshop experience? Concerning the copying and use of older texts, it must be emphasised that the mediaeval reader had *no way of knowing* when a given text was composed – even in the case of named classical or contemporary authors – there would be no sense that a later text superseded an older one, and any 'progress' discernable in recipes can only be discerned by us here and now. A late copy of a classical or early mediaeval text therefore had, for the mediaeval reader, was as 'alive' and had as much usefulness, validity and contemporary current relevance as had a newly compiled text. 'Uncritical copying' was simply the precursor of printing – it is unreasonable to expect every scribe to update or correct every manuscript he worked on. Also, when considering the critical faculties of the mediaeval reader, it should also be remembered that a mediaeval reader probably saw extremely few recipe treatises, if more than one, and comparison was probably impossible.

Compilations

Mediaeval compilations or florilegia abound in all subjects and for all types of texts, and artists' recipes are no exception. By their nature recipes would surely have been attractive to compilers: they are sets of very short texts; and the opportunity to extract, select or interpolate was surely greater than it would have been for a sustained text. Quite often a few short recipes are used to infill a blank space in another manuscript.

One of the greatest problems in studying mediaeval artists' recipes is trying to determine when a text was transmitted independently from workshop practice. Compilations can be used as evidence for and against this: *contra*, that they are simply collections of earlier texts, *pro*, that a practitioner collected items of interest to him.

Contra: evidence for literary use or transmission as a literary text

Examples exist of where recipes have been carried along with other texts. It is often seen that a recipe or a few recipes have been added in a blank space in a manuscript of a more-or-less related text, or indeed in an unrelated text. This is generally in a rough, not a fair, hand.[30] If the manuscript is copied at some point, the recipe is generally copied too; but this time it will appear in a fair hand, uniform with the rest of the text. This is seen, for example, in a certain tenth century manuscript.[31] In this case the addition makes sense: the colour recipes follow on from Vitruvius. In a certain twelfth century manuscript we find the same thing, but here it makes no sense; the extracts from Heraclius and Theophilus follow a portion of Isidore, but not apparently a relevant one.[32] We may deduce that in the *exemplum* the extracts were simply infilling a blank space. This is not confined to *early* mediaeval manuscripts. A fourteenth century example in has a recipe for hair dye added to medical text (although this might be intentional).[33] In another fourteenth century example, where colour recipes are appended to an unrelated text, the rubric of the recipes is in an antique, rustic hand, possibly an echo of the original style of the manuscript.[34]

Pro: evidence for practical use

Oltrogge has drawn attention to the fifteenth century Netherlandish *Cologne modelbook*[35] which consists of three separate fascicules, now bound together,

30. For example: Hattem (Netherlands), Voerman Museum, MS C 5, page 116; Auch (Gers, France) Archives départementales, MS Série I, pièce 4066 [Clarke no. 50], e.g. f. 24v.

31. Leiden, University Library, MS Voss. Lat. folio 88, on f. 106r [Clarke no. 1310].

32. Edinburgh, University Library, MS 123, f.155r-v [Clarke no. 500].

33. Cambridge, Peterhouse, MS 222 iii f.32r [Clarke no. 270].

34. Edinburgh, University Library, MS 131 f.54v [Clarke no. 510].

35. Cologne, Stadtarchiv MS W 8° 293 [Clarke no. 435].

and she argued that this showed a compilation being made and used by a practicing illuminator. The *Göttingen Modelbook*[36] is another exceptional manuscript, in that a clear purpose and audience can be determined. It has been suggested that this step-by-step how-to book on decoration was aimed at semi-skilled illuminators who were needed by early printers. Decoration very close in style has been found in the Göttingen and Mainz library copies of Gutenberg's 42-line bible. This argument is partly based on the assertion that trained scribal illuminators had no need of a manual. Here, therefore, we have a genuinely new text, not simply re-cycled older material.

Brussels Royal Library MS 10147-58 (thirteenth century) contains a fascinating little quire which clearly shows that we should not automatically dismiss as 'purely literary' all occurrences of texts that contain extracts from known popular Latin authors. Six pages form a *Compendium artis picturae* in Latin, containing recipes for painting, tempering, and adhesives. This manuscript includes excerpts of Theophilus and others, but with an amount of original material, added, apparently written by a practicing illuminator. In addition to the textual content, the physical make-up of this manuscript also makes a convincing case for this being a compilation made by an illuminator for himself: it is on small, poor quality, holed parchment, with almost no blank margins, and the scribe has crammed 60 lines into each 15 cm high sheet. A further ink recipe has been fitted into a small space in one lower edge. Surely this is someone writing for himself, on available scraps, rather than someone on commission, who would not be so concerned to save parchment.

It further seems that household compilations or 'housebooks' contain material intended to be used everyday.[37] Housebooks were individual compilations of useful or generally interesting stuff – a known personal 'audience.'[38] I would also suggest that, for example, isolated ink recipes, which are often found jotted in blank spaces or on blank pages of otherwise unrelated manuscripts, were almost certainly written down by people who intended to use them themselves.[39]

36. Göttingen, Niedersächsische Staats- und Universitätsbibliothek, Codex 8° Uff. 51 Cim. [Clarke no. 1120].

37. For example: Maastricht, Gemeentearchief, MS 85 ff.158-9, fifteenth century Netherlandish [Clarke no. 2065]; San Marino (California), Henry E. Huntington Library, MS HU 1051, 14th-15th century English [Clarke no. 3090]; Aberystwyth, National Library of Wales, MS Brogyntyn 2.1 ff. 33r-52v, fifteenth century English [Clarke no. 10]. Images of the Aberystwyth MS are available at http://www.llgc.org.uk/drych/drych_s078.htm

38. Arguably some of these are prescriptive not descriptive – how life *should* be led! See: E.L. Eisenstein, *The printing press as an agent of change*, Cambridge, University Press, 1979, p. 429 [Citation is from paperback edition 1980].

39. For example: Den Haag, Koninklijke Bibliotheek, MS 133 M 107, f.9v, ink recipe, late fifteenth century, east Netherlandish.

Vernacular

Does the use of the vernacular really indicate that a text is a new, personal observation of contemporary practice, rather than simply a 'blind' transmission of a Latin text? I used to think this was so, but now I am not so sure. Let us consider a manuscript compilation from *c.* 1444, from Auch in Gascony, almost on the Spanish border.[40] It is written in an extremely localised dialect, that has been traced to a very specific area (the valleys of the Aure and the Louron, in the Haut-Pyrénées). It is a little like a housebook, but is primarily veterinary. It interests us as it contains recipes for the preparation of vermilion and (as interpreted by Dr. Córdoba) for whitening copper with arsenic. Surely the use of such a localised dialect must indicate personal compilation? And yet, perhaps not. There are odd Latinisms present, more than one might expect in an Occitan-type dialect, and this seems to point to the text being a translation. I have not yet identified the original text, if this is so. This is, I fear and suspect, the case for many vernacular texts. More work is needed to identify any Latin originals for vernacular texts. The process of translation was also often accompanied by interpretation, paraphrasing, commentary, adding or substituting personal observations or readings from elsewhere, all making the identification of the base text extremely problematic. Another problem with treating the vernacular as an indicator of reliability is that we are still extremely unclear about how widespread literacy was in the middle ages, and in the later middle ages we are even less sure. This point will be returned to below. As with literacy, use of the vernacular is not status linked in an obvious way. It appealed to apprentices and lowly folk, but also to 'middle class' merchants, and indeed gentry. Nobles and courtiers were often those who commissioned vernacular translations in the first place. Latin was for scholars and professionals (law, medicine, religion) but many of them were of relatively low status.

Description of new material and processes

What other, less ambiguous, evidence might we find to show that 14th-15th century manuscripts reflected workshop practice? We could look for new materials and processes described in later manuscripts. If they are present, then new knowledge has been recorded. It is known that oil painting (*strictu sensu)* developed throughout the period under consideration; can we find records of it? Furthermore, more and more pigments became used as centuries progressed, particularly organic pigments based on plants. The 14th and 15th centuries in particular showed an increase in the number of pigments being used, especially organic pigments for scribes and illuminators. Are these new introductions reflected in the recipe texts? The answer is: yes and no. Yes, more and

40. Auch (Gers, France), Archives départementales, MS Série I, pièce 4066.

new materials do become described. But there is a time lag, of around a century. For example, ultramarine began to be used (with a couple of possible earlier exceptions) in Northern Europe *c.* 1000, yet we find no description of it in recipe texts for another 100-150 years. Oil painting was only cursorily described in a few lines by Cennino Cennini (*c.*1390) and in a couple of contemporary manuscripts, and we find no full description until after 1500 – well after its use was established and perfected.[41] As Oltrogge pointed out, lead-tin yellow is present in late mediaeval artworks, but absent from the contemporary recipes. (She also pointed out an interesting exception: the case of *marcasita*, which appears in recipes and is also found by analysis in the same period.) This lag is found in later texts, and a similar lag occurs with the coming of print. It is often stated that print brought about a great revolution in the content of books, and it is often implied that old knowledge, that had been preserved solely by the uncritical copying of old manuscripts, was swept away to be replaced by new, contemporary, relevant, correct knowledge. But this is demonstrably untrue; the earliest printed books were simply perpetuating the old knowledge by other, non-manual means of reproduction. So we find in the earliest printed recipe books – which for pictorial art do not appear until mid-sixteenth century – which repeat the usual material.

Of course, as we saw with the Brussels manuscript[42] this repeating of old material does not mean that recipe texts do not reflect contemporary workshop practice. Firstly (and problematically) material may be *coincidentally* correct, because, while it copies centuries-old descriptions and recipes, the processes have not changed in those same centuries. The production of lead white or verdigris did not change significantly for almost two millennia following the descriptions of Theophrastus or Pliny.[43] Secondly (as I have demonstrated elsewhere, especially in the Brussels MS and the Montpellier MS 277) pre-existing texts could be re-worked to make them more up to date, and more relevant to workshop practices.[44]

41. Kieckhefer draws attention to the disparity between gems mentioned in lapidary texts and those found on jewellry of the period [R. Kieckhefer, *Magic in the Middle Ages*, Cambridge, University Press, 1989, p. 105].

42. Koninklijke Bibliotheek, MS 10147-58, item 10152 [*sic*] [Clarke no. 180+185].

43. A striking example of such a cross-temporal coincidence may be the *Livro de Como se Fazem as Côres*. It has been demonstrated that the technique of the Prato Haggadah from *c.*1300 corresponds with that in the *Livro*, which was long believed to date from 1262 (Stavisky and Centeno, this volume). However, it has now been argued that the *Livro* is probably fifteenth century (Afonso, Cruz and Matos, this volume) Nevertheless the correspondance between recipe text and observed illuminator's practice holds good.

44. For more on the two processes of re-working, see: M. Clarke "Reworking and re-use: adaptation and use in workshop texts," 2012, in: S. Eyb-Green *et al.* (eds) *The Artist's Process: Technology and Interpretation*. London: Archetype Publications: 27-31. See also: M. Clarke, *Mediaeval Painters' Materials and Techniques: The Montpellier 'Liber diversarum arcium'*, 2011, London: Archetype Publications, especially pp. 13-14 and pp. 44 ff.

In conclusion: I suggest that two processes of re-working co-existed: the first made texts more literary, the second made them more practical.

Other changes in the 14th-15th centuries

Format changes: the use of paper

In the pages to follow I shall refer repeatedly to the work of Eisenstein on the effects of printing on learning. The importance of text format on content and use is so great that many useful lessons can be drawn, often parallel to the changes observed in post-*c.*1300 recipe texts. Eisenstein suggested that "When ideas are detached from the media used to transmit them, they are also cut off from the historical circumstances that shape them, and it becomes difficult to perceive the changing context within which they must be viewed."[45] When considering a recipe text's likely audience and use, codicological considerations are clearly important. Is it a presentation, fair or rough copy? Is the recipe inserted among unrelated material?[46]

One technological change that may well have altered the nature of recipe manuscripts in the 14th-15th centuries was the increasing availability of paper in Europe. While paper was available in Europe before *c.*1300, it was rare.[47] Paper, and especially the all-paper manuscript, is not much found in Europe before *c.*1400.[48] But once mills became well established in Europe the price of paper fell to around a quarter of the price of parchment. This comparative cheapness of paper surely encouraged the making of books by people who were not professional scribes.[49] The literate man was more likely to be his own scribe.[50] Paper allowed a greater freedom to produce your own compendium,

45. Eisenstein *Agent*, p. 24.

46. For more on this, see: M. Clarke, "Codicological indicators of practical mediaeval artists' recipes," in *Sources and Serendipity: Testimonies of Artists' Practice* Edited by Erma Hermens and Joyce H Townsend. London: Archetype Publications, pp. 8-17 and plates 3-6.

47. For example, the first surviving dated paper in England is from 1307; this was imported, as the first paper mill in England was 1494.

48. For comparison: in tenth century Byzantium parchment was 23-38 % of the cost of a manuscript. Initially paper was half price of parchment, but in the fourteenth century it was much cheaper. See: N. Oikonomides "Writing Materials, Documents, and Books," in: A. E. Laiou (ed.) *The Economic History of Byzantium: From the Seventh through the Fifteenth Century.* [in three volumes, published as number 39 in the series *Dumbarton Oaks Studies*]. Washington, D.C.: Dumbarton Oaks Research Library and Collection, 2002 (This chapter translated by J. Solman), pp. 589-592.

49. Some manuscripts in the 14th-15th century include copies of texts taken from printed books [E.L. Eisenstein, *The Printing Revolution in Early Modern Europe*, Cambridge, University Press, 1983, p. 21]. e.g. Berlin SB, Ms germ. quart. 417, copies Boltz' 1549 *Illuminirbuch.*

50. Survival of a text may well often have depended on an interested reader making his own copy. A problem with technical literature is that copying could not really be left to hired hands who were not subject specialists - for errors were much more likely to multiply in this type of text (especially where numerals or unusual terminology were concerned – imagine for example a technical term in a text that originated in Latin then passed through Arabic, Latin again, ending up in English). Eisenstein argues similarly [*Revolution*, p. 18, *Agent*, p. 465, note 34].

indexed in your own way, of material that interested you.[51] Abundant, relatively inexpensive paper also increases the likelihood of personal note taking and copying, and personal original compositions and memoranda, e.g. – at least potentially – recording ones own observations of workshop practice. Paper, rather than print, may be the key to the shift from reproducing old texts to producing new ones. In addition to making fair copies of manuscripts on parchment for preservation of a text for posterity, paper 'rough copy' manuscripts are written for one's own use, especially those in one's own personal vernacular consisting of translations of extracts and new personal observations (e.g. the Auch manuscript).

Format changes: the effect of print on technical treatises

What changed with the coming of print? There is, of course, a relationship between the coming of print and increased literacy, but again it is not as obvious as it has often been made to seem. After print, the availability of texts allowed the student or apprentice to overtake the knowledge of the master, and makes lay advancement possible, even outside the professions or guilds. To consult different books it was no longer essential to be a wandering scholar. The interesting question is: how did recipes and recipe books change with the coming of print? Early printing was not so much concerned with printing newly composed texts, but rather with reproducing (and so preserving) existing ones in greater numbers than before. Print fixes the texts, which can confer authority. This is not necessarily a good thing – they are the same old material as before. We should not make the error of supposing that they were 'authoritative editions'. The printed texts were rarely checked for quality. Eisenstein considered that initially, in the early days of printing before the great blossoming of quality editing, textual corruption actually *increased*.[52]

More beneficial was the availability of more and different texts, allowing comparison, and thus criticism and exposure of inconsistencies or contradictions, on a scale as never before. (Eisenstein considered this the most important effect of print on learning.) In a similar way, the confusion and disorder of manuscript texts was often hidden by orality when manuscripts were studied and transmitted by reading aloud. The printed books, ingested in personal quiet study, exposed this disorganisation and lack of structure somewhat, and later printed books were notable for the far higher degree of organisation, chapter heads, indices etc. than were common in manuscripts.

Eisenstein asks: "What was the point of publishing vernacular manuals outlining procedures that were already familiar to all skilled practitioners of cer-

51. Early printed books were often simply replicas of these compendia. Early printers have been criticised for their retrogressive nature: but they were no more so than were scribal copyists.

52. Eisenstein *Revolution*, p. 74.

tain crafts?" and continues by concluding that "It is worth remembering, in any event, that the gap between shoproom practice and classroom theory was just becoming visible during the first century of printing and that many so-called practical handbooks and manuals contained impractical, even injurious, advice."[53] These observations and puzzles clearly apply with equal force and validity to *manuscript* recipe texts. Indeed, it seems to me that "the gap between shoproom practice and classroom theory" was already in place, but was perhaps widened somewhat during the first centuries of printing.

Literacy

Literacy may well have increased in the 14th-15th centuries. Statistics on mediaeval and early modern literacy are notoriously vague. Wide variation doubtless existed, e.g. between regions, or between town and country. Eisenstein concluded (giving supporting references) that "Thanks to unusual municipal schools ... the young [Italian] apprentice in architecture, painting or sculpture was also well ahead of his Northern counterparts in his mastery of the written word."[54] It does seem, even after accidents of survival are accounted for, that there was a great flourishing of 'how-to' books. Kieckhefer states:

> "In the fourteenth and fifteenth centuries ... European towns were flooded with popular writings on all topics. Medical writings taught people how to heal themselves ... books of charms ... [and] manuals for divination, long known in monastic and clerical circles, became common now among the laity. These and other materials, which had previously circulated for the most part in Latin, were increasingly available in the various vernacular languages."[55]

This seems true, but is perhaps an oversimplification. Who wrote these books? Who did they write them for? Were they mainly personal copies or compilations of extracts, or were they mainly professionally produced copies of standard texts? How available were they in fact – how many copies, if you like, per head of population, or per literate head? All this seems to need quantifying before making these kind of generalisations.

There is also a danger in concentrating on a literate elite. Literacy is not congruent with social status (consider the literate but lowly slaves, clerks and monks, or conversely the often illiterate squirearchy). The treatises of Ghiberti, Alberti, and Cennini were apparently originally written for aristocratic patrons, rather than for practicing artists, and it is often asserted that in the middle ages all but an elite few were illiterate. This, however, was not true, and it seems that by the fifteenth century it was even less true: in fifteenth century England,

53. Eisenstein *Revolution*, p. 34.
54. Eisenstein *Agent*, p. 252.
55. Kieckhefer *Magic*, pp. 63-64.

for example, manuscripts are being bought by merchants and bakers as well as lawyers or knights.[56] The versatility of some mediaeval and many Renaissance artists, like that of doctors, apothecaries (to certain guilds of which artists belonged – and surely literacy was required of apothecaries), architects and perhaps musicians, enabled them to break down somewhat the gap between the intellectual and the manual worlds, and stimulating dialogue between theoretical concepts and practical procedures. This makes the existence of literate artists – and hence written artists' recipe books – less surprising.

Magic recipes

Household recipes and especially medical recipes often contain an element of magic, e.g. charms, the time or manner of collecting plants or the employment of special words (but this could be religion), and ritual.[57] Artists' recipe books are, compared to these contemporary medical texts, largely, but not completely, free of magic. They are not, however, devoid of a spiritual or mystical element: there can be no doubt that certain recipes that appear at first to be for artists' materials, are in fact alchemic,[58] and Bucklow has written on the rôle of alchemic theories of matter in recipes for artists' materials, of which he gives several startling and convincing examples;[59] there is no space to talk about this now, but it is worth highlighting many surviving recipes for vermilion and Theophilus' recipe for 'Spanish gold' as examples.

Does *artes* literature become less magical with time? We should be careful about our use of the term 'magic'. 'Mediaeval' does not mean 'childlike and credulous', and concern with the occult was – at least for most 14th-15th century intellectuals – scientific, rather than the reverse. I would suggest that magic can perhaps be distinguished from science by the component of hope, by its results not being direct and inevitable, by its requirement for a magically prepared operator (it can not always simply be learned), and by its working being hidden, i.e. 'occult'.

Magic and learning are not mutually exclusive. Learning does not seem always to displace magic – at times (e.g., the present day) they increase simultaneously. This may always have been so. Cryptic and allegorical texts – alchemy – had their basis in the observation of certain crafts such as metalworking and glassmaking. Thorndike proposed that if "astrology and other occult sciences do not appear in a developed form until the Hellenistic period,

56. Eisenstein *Agent*, p. 62.

57. L. Thorndike, *A History of Magic and Experimental Science*, 8 vols. London, Macmillan & Co., and New York, Columbia University Press, 1923-58. See especially volume I, chapter 1.

58. Clarke *Art of all Colours*, pp. 35-39.

59. See the articles by Bucklow in *Zeitschrift für Kunsttechnologie und Konservierung* Volume 13 (1999) pp. 140-9; Volume 14, (2000) pp. 5-14; Volume 15 (2001) pp. 25-33, and Bucklow, this volume.

it is not because the earlier period was more enlightened, but because it was less learned."[60] Eisenstein noted that "In the age of scribes, for instance, magical arts were closely associated with mechanical crafts and mathematical wizardry. When 'technology went to press', so too did a vast backlog of occult lore, and few readers could discriminate between the two."[61] The 'occult' sciences in Europe are derived largely from Arabic sources arriving during the twelfth century, travelling alongside other 'scientific' learning.

For me, the greatest puzzle inherent in books of magical ' recipes' is that they don't work! What did people think when they failed? Why did they persist in copying and transmitting them? The mediaeval reader often had no way to verify or otherwise much of the material that we now consider incredible – he simply lacked access to direct experience in the form of alternative books. This led to a common acceptance of what we may consider fantastic – and here we might think of dragons and dragonsblood. Access to more books allowed comparison and exposed inconsistencies.[62] Similarly it was not until pharmacopoeiae were printed and revised in the *seventeenth* century that the more absurd recipes were recognised as such.[63]

Outstanding questions

The biggest questions remain: why and for whom were mediaeval recipe books written?

The idea of a target reader, or of imagining the reader, was largely alien to the mediaeval author.[64] The book dedicated to a patron, and the personal family housebook are perhaps the only mediaeval texts for which a clear idea of the intended reader was present in the mind of the compiler. (This is a generalisation with many exceptions.) It was not appreciated how different readers may have different requirements, and there was generally no idea of the end user.

60. Thorndike *History*, volume I, p. 22.

61. Eisenstein *Revolution*, p. 45.

62. This is a key argument throughout the books of Eisenstein; for its application to magic see *Revolution*, p. 129.

63. Eisenstein *Agent*, p. 539.

64. Unlike manuscript producers, who (with a few important exceptions) knew for whom a given text was being copied, printers had no concrete idea of readership, but had to rely on educated guesses and commercial acumen. Printers choose to produce books for the general reader in the hope of large sales. One class of texts they clearly *did* choose to print was a great number of 'how-to' books aimed at the general reader. Examples include "... 'how to' draw a picture, compose a madrigal, mix paints, bake clay, keep accounts, survey a field, handle all manner of tools and instruments, work mines and metals, move armies or obelisks, design buildings, bridges and machines" [Eisenstein *Agent*, p. 243]. The intended audience of these many printed sixteenth century *kunstbuchlein* has been debated hard. There is some evidence that they form a somewhat literary genre, due to indiscriminate compilation. See also Eisenstein *Revolution*, p. 34 and *Agent*, p. 554.

It is possible to imagine that mediaeval artists might have been literate. Recipes, especially involving quantities and mixtures, are not easily memorised, even using the formidable formal mediaeval techniques of memory;[65] this would increase the desirability of written records. (These go back to the Assyrians.) Were there such things as 'lodge books' preserved by the guilds of artists, wherein agreed, approved, standardised or approved recipes were recorded? Are these the manuscripts that survive? Eisenstein seems to have taken it for granted that "trade skills had been passed down by closed circles of initiates, ... [sometimes] written down and preserved in lodgebooks ..."[66] As far as I know, there is no evidence for this. (If such books were used, one has to wonder at the breaking of guild secrecy by printing them, and the disapproval of colleagues that this would entail. The prefaces to such early printed revelations are filled with self-justification, e.g. to serve art and posterity[67]).

It is possible that art technological handbooks were of interest to practitioners precisely because they provided information that was different to their own experience – i.e. a new look. Alternatively workshop practices may be recorded by interested non-practicing observers, (or rather, *also* to non-practitioners – we must not forget the Brussels manuscript which shows that even well known standard 'literary' texts were useful to a practicing illuminator). This interest goes back to Pliny (and arguably to the Assyrians too). Indeed, even the artist using a material may be only casually interested in the origins of his materials (e.g. dragon's blood), being in fact only interested in its quality, working properties and effect ("... the sensation of wonder would be reason enough for reading or listening, and wonder is a state that seldom notes carefully the bounds between fiction and fact"[68]). In the 14th-15th centuries, alongside a rise of lay literacy and a lay intelligentsia, came an increased interest and respect for crafts. Since the twelfth century there was already a tradition of scholarly interest in craft and technology. I (and others) would argue this already existed in Hellenistic Alexandria, where natural philosopher's observations and explanations of craft processes such as metallurgy and dyeing resulted, ultimately, in alchemy. Vitruvius, for example, was for a long time an important, widely copied and appreciated literary source (This becomes more prominent with early printing, with the publication of vernacular editions of e.g. Vitruvius).

It does seem that – in our present state of knowledge – we can place only very few of the manuscripts (with any confidence) as belonging in the work-

65. On which Carruthers' book cannot be too highly recommended: M. Carruthers (1990), *The Book of Memory: A Study of Memory in Medieval Culture*. Cambridge Studies in Medieval Literature 10. Cambridge: Cambridge University Press.

66. Eisenstein *Revolution*, p. 141.

67. Eisenstein *Agent* ..., p. 553, p. 558.

68. Kieckhefer *Magic*, p. 105.

shop. Where are the *working* manuscripts? Consider, for comparison, the 19th century workshop manuals from the English artists' colourman Winsor & Newton.[69] These were laid out and used as a working document, available for consultation and for emendation. Although there is a great difference in the appearance of mediaeval recipe books, with a spectrum between the rough simple plain small copies, to the fair copies with wide clean margins and coloured decorations, no mediaeval recipe books look like this. Several of the Winsor & Newton recipe manuscripts show heavy use and revision. No mediaeval artists' recipe book looks so 'used'. Certainly in the field of painting and illuminating we sadly lack such process-demonstrating manuscripts, apart from a couple of tiny fragments: when libraries were being broken up, it would be precisely such working manuscripts that would not be preserved – perhaps soiled, much-annotated, and neither decorated, nor *auctoritas*.[70]

So again we have to ask, w*hy* were they written? Were they useful or literary? To what extent did they reflect contemporary workshop practice? Were they *really* written for the use of mediaeval artists, and were they *really* intended to disseminate new techniques and practices?[71] If so, why might this have been desirable? Why *might* an artisan wish to inform his competitors, even if they were members of his own guild?!

The other contributors to the seminar on which the present volume is based have suggested a variety of *raisons d'être* for craft treatises:

– they may have been straightforward didactic texts

– they may have been used during training

– they may have been for the gentleman reader (general interest)

– they may be blind de-natured copying of texts *qua* texts, i.e. preservation for its own sake

– they may have been self-promotion or self-advertising

– they may have been associated with alchemic or medical theories of matter

Are the later (14th-15th century) texts more accurate with respect to workshop practices? I have said so before, but I am not so sure any more. A comparison may be made with early printed books which were no more accurate than the manuscripts that preceded them. I think that with the late technical manuscripts there are some that preserve old errors derived from old texts,

69. http://www-hki.fitzmuseum.cam.ac.uk/archives/wn/

70. Clarke *Codicological indicators*.

71. With the coming of print this is easier to determine: the difference between a printed treatise and a personal workshop notebook is obvious – when all books are in manuscript the distinction is not always so apparent.

some that re-work old texts (either to make them more literary or more practical depending on audience), and some that present new material derived from experience (albeit generally with a time delay – this delay perhaps arising from the time taken to migrate from a workshop text to the fair copy that survives today).

Only the preparation of a complete corpus of artists' technical recipe texts will help us untangle the relationships and development and transmission of mediaeval recipe texts. Approximately 12,000 pages of artists recipes survive from before *c.*1500 AD. It has been said that to produce a critical edition or stemma is impossible, yet this is exactly what must be attempted. We must, as for any other text, look at all textual variants, clues from the codicology, and consider the other contents of each manuscript as clues to the processes of transmission, and as clues to the likely *purpose* of each individual manuscript copy of the recipes.

In conclusion, we may be sure that – while clearly we must be careful, and not all the recipe manuscripts are reliable witnesses to the local workshop praxis at the time they were written – nevertheless they are a vital and fascinating source of information and insight into the mediaeval artists' workshop.

RECIPE BOOKS FOR ILLUMINATORS IN 15th CENTURY GERMANY AND NETHERLANDS – WORKSHOP PRACTICE AND ENCYCLOPEDIC AMBITION

Doris Oltrogge

Artists working in and for manuscripts seem extremely close to written documentation of their art. Therefore it will be not just chance that a considerable number of art technological texts deals with book illumination. But how close are these texts to the practice in an illuminator's workshop? The choice of subjects seems very heterogeneous, for example recipes for gilding are found elsewhere, instructions for the underdrawing only occasionally. Also the treatment of pigment production does not correspond exactly to their use in manuscript illumination. Often irreproducible recipes stand next to very practical information. Single recipes could be copied over very long periods, and even whole treatises like the 12th century *De diversis artibus* of Theophilus were still transcribed in the 15th century. Did these old fashioned texts remain part of the practical knowledge of the respective time or were they transformed into mere book knowledge? What was the purpose of the texts on techniques of illumination? To which extent and how were they really used in workshops? Which literary pretensions could they answer?

There is no single answer to these questions. The analysis of recipe books, the contexts in which they were transmitted, and their comparison with mediaeval workshop practice shows that these texts could have served different purposes and that the transition between practical use and scholarly learning may be fluent even in the same manuscript. Three case studies of 15th century German and Netherlandish recipe books shall exemplify this.

The *Liber illuministarum* from Tegernsee and the Trier Painters' Manual

The *Liber illuministarum*[1] and the Trier painters' manual[2] are typical representatives for two types of late mediaeval recipe books. The first was compiled over a certain period in the second half of the 15th century until about 1512 in the Bavarian monastery of Tegernsee. The latter was written in the second half of the 15th century probably in the monastery of Eberhardsklausen on the Mosel or at least in that area. The Liber illuministarum is a miscellaneous manuscript dealing with various aspects of practical and theoretical knowledge: mathematics, medicine, housekeeping, pyrotechnics, magic, chemistry / alchemy, metallurgy, tanning, dyeing, and painting, especially book illumination. Several scribes wrote this growing compilation, but the most important contributor was Konrad Sartori (in Tegernsee from 1480 until 1531), librarian, scribe and illuminator in Tegernsee. He added not only information on inks or illumination but was interested in all the topics mentioned above. Corresponding to the heterogeneous character and lengthy genesis of the Liber illuministarum there is no consistent order of the respective topics, medical recipes can be interpolated into sections of art technological instructions, than turn abruptely to magic or housekeeping items etc. But at least when text blocks were written continuously in a certain period a certain effort to put the things into an order can be observed.

The Trier manuscript on the other hand is confined to painting, only a few dyeing recipes have been added by a later hand on empty leaves. The subjects are clearly arranged: application, ornamentation and imitation of gold and silver, binding media, production and distempering of pigments and lakes, colour concords, copying techniques, preparation of the support and finally – in this context a somewhat strange topic – the production of artificial pearls.

Considering both manuscripts in general the Liber illuministarum is an accumulation of knowledge, encyclopedic in its range of interest, but not an encyclopedia in the sense of systematization[3] whereas the Trier manuscript is a sort of specialised encyclopedia in its form, not in its scope. Considering the recipes related to illumination, both manuscripts have much more in common, except for the copying techniques the same subjects are treated and these are the main topics also of most other late mediaeval recipe books. These shall be discussed here in the sequence of the Trier manuscript.

Gilding

Different types for the application and ornamentation of gold and silver, for the use of substitutes and lustres are major subjects in all recipe books dealing

1. A. Bartl, Ch. Krekel, M. Lautenschlager, D. Oltrogge, *Der "Liber illuministarum" aus Kloster Tegernsee. Edition, Übersetzung und Kommentar der kunsttechnologischen Rezepte*, Stuttgart, 2005.

2. Trier, Stadtbibliothek, Hs. 1957/1491, 8°; for an edition of the recipes see D. Oltrogge, *Datenbank mittelalterlicher und frühneuzeitlicher kunsttechnologischer Rezepte in handschriftlicher Überlieferung* (www.re.fh-koeln.de, Kunsttechnische Rezepte, s.v. Standort: Trier).

3. For type and structure of mediaeval encyclopedia see Ch. Meier, "Grundzüge der mittelalterlichen Enzyklopädik. Zu Inhalten, Formen und Funktionen einer problematischen Gattung," *Literatur und Laienbildung im Spätmittelalter und in der Reformationszeit*, ed. by L. Grenzmann, K. Stackmann, Stuttgart 1984, p. 467-500.

with painting or illumination. The Trier manuscript devotes 25, the Liber illuministarum even more than 160 instructions to this topic. Some recipes are similar in both manuscripts, these and some others were also copied in other places in Germany, Bohemia and Austria during the 15th and 16th centuries. Other recipes seem to be unique, for example that of Conrad (Sartori) himself in the Liber illuministarum (ch. 1131). The mode of language varies considerably between lengthy step by step descriptions and simple enumerations of ingredients and proportions. Sartori's recipe is one of the latter, notes from a workshop, written down as an aid to memory for oneself and only of little use for the unexperienced reader. Sartori recorded many other of these short notes. On the other hand he also transcribed one of the very descriptive instructions which was explicitely sent to him by a friend from the monastery of Undersdorf (ch. 279). But Sartori's interest was not restricted to own experience and contemporary workshop information, he also copied recipes from books, even the old fashioned verses on gold ink written by Heraclius (I, 7)[4] in the 10th or 11th century received his attention (ch. 153). It could not have been of much practical value to him since the text was corrupt already in the model. However this is an exception, the bulk of instructions was certainly practical information, and especially the short notes were most probably intended for personal use.

But what were the reasons for the huge number of recipes for gilding and gold substitutes in the Liber illuministarum and elsewhere? In late mediaeval German book illumination gold and gold colours played an important role. Often a range of golden effects is displayed on the same page: for example lustrous gilding with gold leaf on a cushion like gesso for the background, a less glossy, flat gold for the representation of golden objects like crowns or chalices, glittering mosaic gold for ornaments or architecture and glimmering golden highlights on blue, red or green garments. The lustrous gold could be ornamented with punches or with arabesques drawn with nearly transparent lakes or resins. Gold and silver were combined in several ways: applied as metal leaf on metal leaf, as golden coloured glazes on silver, etc. For all these effects different gessos or mordants for gold and silver leaf were needed, gold ink was used as well as the "substitute" mosaic gold, yellowish lakes and resins for a "golden" ornamentation on gold or silver.[5] Therefore one reason for the abundance of recipes related to this subject will be the wide spread use of different "gold" techniques in workshop practice. Another reason could be the technical difficulties of gilding. It is true that most of the job is exercise, on

4. A. Ilg, *Heraclius. Von den Farben und Künsten der Römer* (Quellenschriften für Kunstgeschichte und Kunsttechnik des Mittelalters und der Renaissance 4), Wien, 1874, Reprint Osnabrück, 1970, p. 35.

5. See for example the colour plates in U. Merkl, *Buchmalerei in Bayern in der ersten Hälfte des 16. Jahrhunderts. Spätblüte und Endzeit einer Gattung*, Regensburg, 1999, or R. Budde, R. Krischel (eds.), *Das Stundenbuch der Sophia von Bylant*, Köln, 2001.

the other hand the ingredients – for example fresh or putrid glair or glue, additives like sal ammoniac, myrrh etc. – in the gesso or mordant can be helpful to achieve certain effects of the gold or to facilitate polishing or punching so that even experienced illuminators will have been interested to get suggestions from their colleagues or to record their own successful experiences.

Beside these practical reasons we should not overlook that also more theoretical or symbolic thoughts could play a role in the treatment of the subject "gold." The sequence of the Trier recipes related to this topic – with which the manuscript conspicuously starts – deals mostly with very practical instructions for gessos, mordants, gold and silver inks, materials for ornamentation of gold, mosaic gold and bismute ink. However, it is interrupted twice by descriptions of symbolic alchemical processes of gold making, either from sulfur and mercury put into an egg and hatched by a hen or from egg yolk and sulfur.[6]

The gold brooding hen reappears in a number of other recipe books, often inserted into series of technical workshop notes on gilding, silvering and golden and silver colours. Gold was considered as the most valuable metal in many times and also in the late Middle Ages and Renaissance. Its inertness to nearly every chemical compound known at the time made it the highest possible aim of alchemical transmutation of metals. Material and symbolic value of gold should thus be considered as another reason for the special attention paid to this particular field of artistic practice and also as the reason to include non practical information.

Colour production: synthetic pigments and organic lakes

After gilding the Trier manuscript turns to pigments and their distempering. A major topic is the production of synthetic pigments: artificial blue pigments, verdigris, vermilion, red lead, lead white and, treated in the section of gilding, mosaic gold. Another important group of recipes is devoted to the making of red lakes from brazil wood, only some prescriptions deal with other organic colours: sap green, blue dyes from corn flowers and blue berries, yellow from the bark of apple trees.

These are common topics in recipe books related to painting.[7] If we consider the number of recipes for synthetic pigments in late mediaeval German texts the most important group deals with verdigris followed by artificial blue pigments and vermilion. Very popular are also instructions for making lead white, red lead and mosaic gold, whereas the preparation of burnt ochre or

6. On the relationship of these recipe types to mediaeval nature philosophy see S. Bucklow, "Paradigms and Pigment Recipes: Vermilion, Synthetic Yellows and the Nature of Egg," *Zeitschrift für Kunsttechnologie und Konservierung*, 13, 1999, p. 148-149.

7. Some of the synthetic pigments listed here are treated also in alchemical manuscripts, but in this paper only recipe books related to painting shall be considered.

lamp black are rarely described. If we compare this to the results of the analysis of contemporary book illumination we find most of the pigments mentioned in the recipes. Nearly in every manuscript and incunabula of the 15th century vermilion is the material for rubrication; vermilion, lead white, red lead, mosaic gold and black were used by almost every illuminator. Copper green pigments are also found everywhere, synthetic (verdigris) as well as mineral. There are only two striking differences between the texts and workshop practice: the ordinary yellow pigment in the 15th century is lead tin yellow, but only one pre-1500 Italian recipe is known for it.[8] On the other hand, up to now none of the artificial blues so often described in the recipes has been analysed with certainty – and some of these instructions even do not work.

Concerning the organic colours the predominance of instructions for making brazil lakes is characteristic for late mediaeval recipe books. Rather often also the production of sap green, of different yellow dyes or of blue and violet clothlet colours from corn or poppy flowers, from blue berries, elder berries or dwarf elder are described. Not present in the Trier manuscript but for example in the Liber illuministarum and some other recipe books are lakes made from the shearings of woolen cloth.[9]

The latter are difficult to identify in analysis because lakes could have been produced also directly from the same dyestuffs which were used for dyeing the wool (brazil wood, madder, kermes). The rather weak stability of anthocyans which are the dyestuffs of corn or poppy flowers and the different blue or red berries may be the reason why this group of dyes has been scarcely found in illumination up to now.[10] Yellow lakes and sap green, on the other hand, are common materials in 15th century illumination and hand coloring of prints, but certainly the most popular organic colours of the time were brazil lakes. Brazil wood was not very expensive,[11] and depending on the preparation – which is always rather uncomplicated – brazil lakes offer a broad variety of red, rose, brown and purple hues, a coloristic spectrum also found in many illuminations. There is evidence that brazil lakes were often produced by the painters themselves, for example the accounts of the scriptorium of the Westfalian monas-

8. Manuscript of Bologna, M. Merrifield, *Original Treatises on the Art of Painting*, Vol. 2, London, 1849, p. 529; see also A. Burmester, Ch. Krekel, "Von Dürers Farben," *Albrecht Dürer. Die Gemälde der Alten Pinakothek*, ed. by G. Goldberg, B. Heimberg, M. Schawe, Munich, 1998, p. 65.

9. On Italian recipes see A. Wallert, "'Cimatura de grana': identification of natural organic colorants and binding media in mediaeval manuscript illumination," *Zeitschrift für Kunsttechnologie und Konservierung*, 5, 1991, p. 74-83. For German recipes see Oltrogge, *Datenbank*, loc. cit, s.v. Farbtechnologie: Recyclingmethoden. For the Cologne model book see below.

10. For a possible anthocyan dye in the coloring of a 15th century woodcut see S. Fletcher, L. Glinsman, D. Oltrogge, "The Pigments on Hand-Colored Fifteenth Century Relief Prints from the Collections of the National Gallery of Art and the Germanisches National Museum," *The Woodcut in Fifteenth Century Europe*, ed. by P. Parshall, Washington, 2009, p. 288, p. 297, n. 50.

11. Burmester, Krekel, loc. cit., p. 71.

tery Werden (near Essen) record the purchase of mineral and ready made synthetic pigments but brazil wood was bought as raw material.[12] This allowed the illuminators to produce just the colour hue they needed, and it explains also why so many different recipes for brazil lakes have survived.

As mentioned before the synthetic pigments were bought ready made by the scriptorium of Werden. In the late Middle Ages synthetic colours seem to have been usually produced by specialists and sold in the pharmacy or by the grocer.[13] Even if in recipe books prescriptions for artificial colours stand next to instructions for distempering these colours we should be cautious to conclude that these techniques were practised by the same persons or workshops, or even by the compilers or owners of the respective recipe book. But of course, in some cases such multi-professional people existed, so we know that the painter Lucas Cranach also run a pharmacy, and thus a colour trade, in Wittenberg; nevertheless he had to employ trained people for the professional work of a pharmacist.[14]

If the scribe and illuminator Konrad Sartori who partly wrote the Liber illuministarum, or other people in the Tegernsee monastery produced synthetic pigments we do not know. But at least, Sartori was not interested in lead white or red lead, because he gives no recipes for them. The instructions for vermilion and mosaic gold in the Liber illuministarum are heterogeneous: some are practicable, others defective. This is not uncommon for recipes for artificial pigments that they can contain errors or traces of experiments which are not necessarily improvements. So in one of the vermilion recipes in the Trier manuscript the white excrements of a dog should be added – which makes no practical sense; in one of the recipes for the tin sulfide mosaic gold the absolutely necessary tin is omitted.

In general the written tradition of the technology of synthetic pigments seems to be less close to workshop practice than that of gilding technics or lake production. This is evident especially in the case of lead tin yellow and of the artificial blue pigments. Lead tin yellow was the most important yellow pigment of the late Middle Ages, in the 16th century it was produced at large scale in Venice and in the Netherlands, in the 15th century certainly in Venice and probably also in some German towns.[15] Except from the already mentioned

12. D. Oltrogge, "Pro lazurio auricalco et alii correquisitis pro illuminacione. The Werden accounts and some other sources on the trade of manuscript materials in the Lower Rhineland and Westfalia around 1500," *Trade in Artists' Materials. Markets and Commerce in Europe to 1700*, ed. by Jo Kirby, S. Nash, J. Cannon, London 2010, p. 191-192.

13. Burmester, Krekel, loc. cit., p. 81-82.

14. M. and D. Lücke, "Lucas Cranach in Wittenberg," *Lucas Cranach. Ein Maler-Unternehmer aus Franken*, ed. by C. Grimm *et al.*, Regensburg, 1994, p. 62.

15. Burmester, Krekel, loc. cit., p. 65-66; A. Burmester, Ch. Krekel, "Azurri oltramarini, lacche et altri colori fini: Auf der Suche nach der verlorenen Farbe," *Tintoretto. Der Gonzaga-Zyklus*, ed. by C. Syre, München, 2000, p. 197-198.

Italian recipe no 15th century description of its making is known. Obviously the producers succeeded in keeping the technique secret so that it was not as commonly known as the burning of red lead, of vermilion or of mosaic gold.

Artificial blue

Contrary to that there are countless recipes for artificial blue pigments which are all of questionable practical use; in the Trier manuscript we find seven instructions, in the Liber illuministarum eight. They describe four major techniques: blue produced from silver, from copper, from mercury and from tin. Prescriptions for silver and copper blue occur since the 11th century.[16] Depending on the procedure the copper blue recipes will result in copper compounds like copper acetate, copper chloride or copper calcium hydroxide,[17] but not in artificial azurite as has been supposed previously.[18] Only the copper calcium hydroxide, which is stable only in a very alkaline medium and therefore not suited for illumination, is a bright blue pigment, the other compounds are more green than blue. Even if we take into account that the perception of blue and green can be individually very different, they never have the deep blue quality of ultramarine praised in many recipes. But green copper compounds often show pleocroistic characteristics, which means that the colour changes between blue and green depending on the incidence of light. This effect can be observed especially in larger crystals. If they are finely ground to pigment grains they generally appear only green or blue green. Synthetic green copper pigments were widely used in painting since Antiquity, and we cannot decide if they were produced in the way described for one of these "blues" or according to one of the innumerable recipes for verdigris.

The silver blue is more complicated because a blue silver compound does not exist. Only if we assume that a silver copper alloy is used a blue green copper acetate can be produced, thus a similar pigment to that obtained from copper or brass.[19] But the silver influences the growth of the crystals in a different way so that more often large deep blue crystals originate on the surface of the silver leaf. The ground pigment would again be more green than blue. Certainly copper acetates analysed in manuscripts could have been produced from a silver alloy, but since the same result could be obtained in a much cheaper way directly from copper or brass it is questionable if the silver recipes had

16. M. V. Orna, M. J. D. Low, N. S. Baer, "Synthetic Blue Pigments: Ninth to Sixteenth Centuries. I. Literature," *Studies in Conservation*, 25, 1980, p. 53-63.

17. Ch. Krekel, K. Polborn, "Lime Blue – A Mediaeval Pigment for Wall Paintings?" *Studies in Conservation*, 48, 2003, p. 171-182.

18. R. J. Gettens, E. W. Fitzhugh, "Azurite and Blue Verditer," *Artists' Pigments. A Handbook of their Characteristics*, ed. by A. Roy, 2, New York, 1993, p. 31.

19. M. V. Orna, M. J. D. Low, M. M. Julian, "Synthetic Blue Pigments: Ninth to Sixteenth Centuries. II., Silver Blue," *Studies in Conservation*, 30, 1985, p. 155-160.

great practical importance. Nevertheless, they were frequently copied in the recipe books and often described as the most valuable blue. We may suspect that more the value of the silver than that of the product was the reason for the interest in this recipe.

The Trier manuscript pretends to produce a third type of artificial blue from tin which should either be burnt or put into an acetic atmosphere. Recipes for tin blue appear only occasionally in late mediaeval recipe books,[20] like in the Trier manuscript they are always variations to known processes of pigment production: the sublimation is influenced by recipes for making vermilion or mosaic gold, the reaction of the metal plate in an acetic atmosphere depends on the production of verdigris or lead white. Nevertheless, in the case of tin these methods are obsolete, blue tin compounds do not exist, the sublimation or better burning results in gray, the reaction with vinegar in white products.

Far more wide spread is the recipe for making a mercury blue[21] which is the third artificial blue in the Liber illuministarum. It should be obtained by burning mercury together with sulfur and sal ammoniac, a procedure clearly influenced by the production of vermilion. But whereas the sublimation of a mixture from mercury and sulfur actually results in the red pigment vermilion no blue pigment will be produced when sal ammoniac is added. Thompson suggested that the recipe originally meant vermilion whose Arabic name *al-zunğufr* was misunderstood in Latin as *azurium* (blue).[22] More likely is the relationship to alchemical ideas or to mediaeval nature philosophy. Bucklow referred to the Aristotelian concords between planets and colours where the planet mercury was associated with the colour blue.[23] The base materials mercury, sulfur and sal ammoniac were considered as most powerful alchemical agents, often used for all sorts of transmutation. Vermilion was produced from mercury and sulfur alone, mosaic gold from mercury, sulfur, sal ammoniac and tin, speculations on what happened when only three of the ingredients were mixed suggested themselves. The observation of experiments may have led to the idea of a possible blue product since the bluish black mercury sulfide metacinnabarite – with by-products as mercury chloride – results from the process described in the recipes.[24]

We find thus in the Liber illuministarum, the Trier painters' manual and in a number of other recipe books instructions for making artificial blue pigments

20. For example also in the Kunstbuch from Colmar (Bern, Burgerbibliothek, Cod. Hist. Helv. XII 45, p. 315-316) or in another Tegernsee manuscript, Munich, Bayerische Staatsbibliothek, clm 20174, fol. 181v, see Oltrogge, *Datenbank*, loc. cit, s.v. Farbtechnologie, Zinnblau.

21. D. V. Thompson, *The Materials and Techniques of Medieval Painting*, 2nd ed., New York, 1956, p. 155.

22. Thompson, loc. cit.

23. S. Bucklow, "Paradigms and Pigment Recipes: Silver and Mercury Blues," *Zeitschrift für Kunsttechnologie und Konservierung*, 15, 2001, p. 31-32.

24. This has been suggested by Bucklow, loc. cit. 2001, p. 33.

with little or no practical value. It is difficult to certainly identify the reasons for this, only some suggestions can be made.

One reason for the special interest in artificial blue could be the value of the blue mineral pigments. Mediaeval painters and illuminators used three major blue colour materials: lapis lazuli, azurite and indigo. Bright deep blue colour hues could be obtained only with good quality lapis lazuli or azurite. In the Middle Ages lapis lazuli was only mined in Badakstan (Afghanistan), it had to be imported via Persia and the Near East to Italy which made it an extremely expensive pigment.[25] Azurite was mined also in Europe, but only few good quality deposits were known. Furthermore the extraction of the pigment from the mineral was rather difficult so that in the 15th century the best quality azurite was the second-expensive pigment on the market after lapis lazuli.[26]

Thus, the beautiful azure blue pigments, so much desired by painters and customers, were of high material value. To imitate and substitute valuable materials and objects has always fascinated human intellect, and this may have attracted the compilers even to the practically not working recipes.

Distempering of pigments

It seems striking, that also the distempering of blue pigments is extremely often discussed in recipe books. The distempering is in general a common topic of books dealing with painting, in this context usually also mineral pigments are mentioned which do not occur elsewhere in the recipes like orpiment or ochre. But three materials stood always in the centre of interest: blue, vermilion and green pigments. For example in the Trier manuscript nine recipes are devoted to blue pigments, seven to synthetic and mineral copper greens and four to vermilion, but only one to lead white or to red lead respectively.

At first sight one could suppose that again the value of the mineral blue pigments was responsible for this interest. But in this case one should also consider the practical aspects. Usually the material of the blue is not specified, it is only called "lasur(ium)" or "blaw," names which could mean all sorts of blue colour: the minerals lapis lazuli or azurite, organic pigments or lakes made from indigo or other plants, or the artificial blue pigments.

In late mediaeval German book illumination azurite was the most important blue pigment, widely used despite its price, even the versals in manuscripts and incunabula were often written alternately with vermilion and azurite. Deep blue script or backgrounds should have more body than for example a light blue garment. This required different pigment qualities of azurite, coarser for the script, more finely ground for the azure blues, and the different pigment

25. Burmester, Krekel, *Dürer*, loc. cit., p. 87.

26. Burmester, Krekel, *Dürer*, loc. cit., p. 75-80, 99-100.

sizes required different binding media. Furthermore, the pigment grains of the two mineral blues lapis lazuli and azurite are usually rather uneven so that they do not adhere as easily in the medium as do the more regular and fine pigments vermilion or red lead. Mixtures of different media, gums, glues, glair or additives like myrrh or gum tragacanth could improve the adhesion. The choice of binding media and additives could influence also the colour hue. The coloristic perception of a pigment is always dependent on the refraction index of the pigment and its medium, but in the case of blue pigments changes are especially clearly perceptible even to the naked eye.[27] All this explains the interest of the compilers of recipe books in variations of binding media for blue pigments.

Also the attention paid to the distempering of vermilion can well be explained by workshop requirements. Vermilion was the most important colour material for rubrication, for this purpose it had to be written with a pen and often a shiny surface was desired. On the other hand, the illuminators preferred mat surfaces in the miniatures, where the pigment was painted with a brush. Again the different techniques and purposes needed different binding media, for example glair, perhaps with the addition of some yolk, for the shiny vermilion, gum for the mat surface.

Another reason is responsible for the interest in the distempering of copper green pigments. The mineral malachite becomes rather pale when it is ground finely. If it is put for some time into vinegar, the copper carbonate malachite reacts partly to copper acetate which gives the pigment a more brillant appearance.[28] Artificial copper green pigments on the other hand were from the beginning often mixtures of different copper compounds of varying green and blue green colours. The treatment with vinegar, tartar or different binding media could again influence the colour hue. The whole group of distempering recipes stands thus in extremely close relationship to the practice in the illuminators' workshops.

The encyclopedic idea

The conjunction of very practical technical instructions and irreproducible recipes without any value for every day's work looks strange to modern eyes and we could have doubts if the Trier manuscript was perhaps not only literary work of a scholar.[29] But the mixture of similar types of recipes in the Liber

27. Bartl *et al.*, loc. cit., p. 550, 602.

28. This product has been analysed in some Middle Rhenish manuscripts from the mid 15th century, see R. Fuchs, D. Oltrogge, "Untersuchungen rheinischer Buchmalerei des 15. Jahrhunderts," *Imprimatur*, N.F., 14, 1991, p. 62.

29. The question of the practical value has been lately discussed controversally in the case of vernacular medical literature, see O. Riha, "Das systemologische Defizit der Artesforschung," *Archiv für das Studium der neueren Sprachen und Literaturen*, 229, 1992, p. 255-276 and B. Schnell, "Die volkssprachliche Medizinliteratur des Mittelalters – Wissen für wen?," *Laienlektüre und Buchmarkt im späten Mittelalter*, ed. by T. Koch and R. Schlusemann, Frankfurt *et al.*, 1997, p. 129-145.

illuministarum written by a man who is reported to have been scribe and illuminator proves that practical usefulness was not the only criterium for an artist to compile recipes. Konrad Sartori and his companion compilers tried to amass as much knowledge as possible, personal workshop notes, information from colleagues or from books, ancient or new. The intellectual interest went far, art, alchemy, housekeeping, medicine, mathematics.

The Trier manuscript shows a more systematic approach. Obviously the compiler tried to give a most complete survey on the different topics of painting, i.e. gilding and all possible colours. The colours are represented by recipes for the production of synthetic pigments and lakes and for the tempering of artificial, organic and mineral colour materials. All available information is collected, the compiler's interest is not restricted to the practical needs of workshop practice, but includes all written, authoritative knowledge about the matter. The gold brooding hen or the mercury blue are topoi of alchemical literature and thus belong to the science of gold or blue as well as the common gilding of an initial or the tempering of the blue pigment azurite for painting or writing. This is clearly an encyclopedic choice. From this point of view it was not impossible to jump to other techniques devised by human ingenuity: the making of pearls written down as last recipe in Trier.

Encyclopedic structure could be achieved differently, and also the scope of encyclopedian interest could be different, only artificial pigments as in Trier or also the "nature" and refinement of the mineral blue pigments as for example in the late 14th century *Tractatus de coloribus.*[30] Other individual types of accumulation and systematisation of theoretical and practical knowledge are possible. But the absence or presence of encyclopedic ideas is not an argument for or against a close relationship to workshop practice, as we can see in Konrad Sartori's compilation. The compiler of Trier could have been also a practising illuminator as well as a scholar interested in the matter.

The case of the Model Books

Another group of interesting examples for the interrelation between recipe books, workshop practice and more or less theoretical knowledge can be found in three model books. Two of them, written about 1450 in Mayence and now preserved in Göttingen and Berlin, copy the same text.[31] A step by step

30. Munich, Bayerische Staatsbibliothek, clm 444; ed. M. F. Edgerton Jr., "A Mediaeval *Tractatus de coloribus*. Together with a Contribution to the Study of the Color-vocabulary of Latin," *Mediaeval Studies*, 25, 1963, p. 173-208.

31. Göttingen, Staats- und Universitätsbibliothek, Cod. Uffenbach 51 Cim. and Berlin, Staatliche Museen Preußischer Kulturbesitz, Kupferstichkabinett, Ms. 78 A 22; see: D. Oltrogge, R. Fuchs, S. Michon, "Laubwerk - zur Texttradition einer Anleitung für Buchmaler aus dem 15. Jh.," *Würzburger medizinhistorische Mitteilungen*, 7, 1989, p. 179-213. Facsimile: H. Lehmann-Haupt, The Göttingen Model Book, 2nd ed., Columbia, 1978.

description for the painting of three differently coloured acanthus leaves beginning with the metal point drawing and ending up with the highlights is accompanied by its respective illustrations. The same is repeated for a set of ornamental patterns which could be used for the backgrounds of miniatures or the grounds of initials. The recipes deal mainly with the tempering of the pigments necessary for the different models and with the preparation of the gold ground for the ornamental patterns. Only three recipes are concerned with pigment production: two for colours from brazil wood and one for mosaic gold. Patterns and painting technique are described on a basic, educational level. A small set of border and initial ornaments is presented, choosen out of a larger stock of patterns which were used in Mayence manuscripts and incunabula of the third quarter of the 15th century.[32] With their step by step description and execution of these formula, the model books look like exercise books of the apprentice in an illuminator's workshop, useful perhaps also for the amateur who occasionally wanted to decorate his newly bought incunabula himself. The recipes for making brazil lakes were most probably also intended for personal use, but one can doubt if the scribes and illuminators of the two model books were really experienced enough to try the rather complicated production of mosaic gold. More likely, this recipe was inserted for mere intellectual interest in the valuable, golden looking artificial pigment. And the final remark "so god wil" ("if god will"), a formula often found in alchemical manuscripts, alludes not only to the difficulties to practise this technique but also to the miraculous and ingenious work in general.

Contrary to the Göttingen and Berlin manuscripts, in which models and text form a unit, the Cologne Model Book is joined together from three separate parts: one quaternio of Netherlandish recipes, one ternio of Latin recipes, written by another hand, and a model sheet which was cut into two double leaves to fit into the booklet.[33] Nevertheless, some remarks to the models show that the painter did know the texts rather well so when he explicitly states that in model 19 (fol. 9v) he has painted vermilion with gum water because he did not want a shiny surface – according to the recipe on fol. 11r-11v this shiny vermilion should be painted with old glair. But in general there are only few coincidences between the recipes and the models. The topics of the recipes are again primarily the tempering of pigments, the making of gold grounds, furthermore the preparation of iron gall inks and of a few pigments: brazil lakes, lamp black, bistre, sap green and lakes from shear wool. Some of the colours

32. Fuchs, Oltrogge, loc. cit., p. 55-80.

33. Köln, Historisches Archiv der Stadt Köln (HAStK), 7010 293 (olim W. 8°293); see: A. Wallert, "Instructions for Manuscript Illumination in a 15th Century Netherlandish Technical Treatise," *Masters and Miniatures. Proceedings of the Congress on Medieval Manuscript Illumination in the Northern Netherlands* (Studies and Facsimiles of Netherlandish Illuminated Manuscripts, 3), ed. by K. van der Horst and J.-C. Klamt, Doornspijk, 1991, p. 447-456. A fourth part has been added about 1555, now HAStK 7010 293a.

occur also in the models as shown in the example of the vermilion, others are missing in the paintings, for example brazil lakes, whereas the *parisrot* lake used in the models has no recipe in either of the two recipe collections.

Another topic of the recipes in the Cologne manuscript are colour concords which prescribe coloristic combinations of pigments in the modelling, for example a red garment should be painted first with vermilion, on which the shades should be made with brazil lake, the highlights with red lead.[34] Colour concords were an important subject of recipe books at least since the late 11th century, and Theophilus devoted even more than a third of his book on painting to modelling.[35] As mentioned before, this 12th century text was copied still in the 15th century, and also other catalogues of colour concords were transcribed over a long period. This may seem strange since the coloristic appearance is dependent of style and therefore more subject to changes than the production of for example lead white. Therefore we may assume that in many cases the lists of colour concords became mere book knowledge. But this must not have been always the case. In the Cologne models these formula become a matter of experiment: the painter tried variations of mixtures, changing the proportions of the ingredients, or he played with different colours for the modelling. Even if he did not use any of the concords described in the Latin respectively Netherlandish texts in the existing model sheet, they were most likely an interesting source of ideas to him.

The recipes for distempering pigments, for making organic lakes, lamp black or inks could have been also useful as references to him. Purely encyclopedic material is not found in the two small recipe collections. However, the compiler of the Latin recipes had some idea of a systematic, encyclopedic classification, so when he puts colours together according to their value, not their material: *Item duplex est color rubeus Minium et sinobrium*[36] or *Item purpureus Color est duplex scilicet Rosule et tornessoel*[37] and so on. Again theoretical concepts and practical workshop use are intertwined.

Summary

Late mediaeval recipe books on illumination or painting matters can contain various sorts of information. Main topics are gilding, distempering of pig-

34. Cologne Model Book, fol. 7v; H.J. Leloux, "Nordoostmiddelnederlands in Keulen. Een Keuls manuskript met laatmiddeleeuwse recepten voor verf en inkt voor het schrijven en verluchten van boeken," *Driemaandelijkse bladen voor taal en volksleven in het oosten van de Nederland*, 29, 1977, p. 27.

35. Ch. R. Dodwell, *Theophilus. De Diversis Artibus. On Divers Arts*, London, 1961, p. 5-16.

36. Köln, HAStK, 7010 293, fol. 11r. "There are two kinds of red colour, minium and vermilion."

37. loc. cit., fol. 11v. "Than purple, which is of two kinds, [brazil] rose and tournesol."

ments, production of organic lakes and inks, and in these fields many agreements with contemporary illumination can be stated. It seems that most of this information came directly from workshop practice and that it was also often thought to be used as reference by illuminators. This does not mean that every technique reported in a recipe book was actually used in the respective workshop.

In the case of synthetic pigments the interrelation between the workshop practice and the texts is less strong, defect or nonpractical recipes stand beside practical ones, instructions for an important pigment like lead tin yellow are nearly completely missing. There is evidence that the compilers of the recipe books often had only second hand information on the production of synthetic pigments. Even if recipe books could be used as reference by illuminators this was not necessarily their only purpose. They could be also conceived as encyclopedia on the whole matter of painting or just as a note book for a scholar, be he scribe, illuminator or just a man interested in techniques of all sort.

For the learned illuminator and painter pragmatic texts and encyclopedic interests were no contradiction. How wide spread such intellectual occupations – or even literacy – were among the lay artists, for example the "Briefmaler" who painted and colored cheap woodcuts in Nürnberg and other towns, remains an open question.

The *Mappae Clavicula* Treatise of the *Codex Matritensis* 19 and the Transmission of Art Technology in the Middle Ages

Stefanos Kroustallis

Introduction

Medieval compilations of technical instructions constitute the main historical reference regarding materials and techniques used in artistic praxis. Over 400 treatises are known, most of them dating from the 8th century to the 15th century (Clarke, 2001). Despite the miscellaneous character and the diversity of length of these texts, it is possible to distinguish one specific group of recipe books, the *Compositiones-Mappae clavicula* family, according to the techniques described, the date of their copy and the type of manuscripts they form part.

The treatise known as *Compositiones ad tingenda* of the late 8th or early 9th century (Cod. 490, Biblioteca Capitolare Feliniana di Lucca, Italy) is the oldest known compilation of technical instructions in Latin related to the artistic practice in Western Europe. The *incipit* of the treatise is quite explicit about a content focused mainly on sumptuary arts: *compositiones ad tingenda musiva, pelles et allia, ad deaurandum ferrum, ad mineralia, ad chrysographiam, ad glutina quaedam conficienda, aliaque artium documenta.* Of a very similar content – they share half of their recipes and many others are close related – is another treatise known as *Mappae clavicula*, a compilation of recipes developed and circulated already since the 9th century,[1] although the oldest extensive text that has come down to us dates back to the 10th century (Ms 17, Bibliothèque Humaniste de Sélestat).

The close relation between both treatises points out that both are part of a specific group of medieval compilations of instruction of art technology that

1. A work called *Mappae clavicula* was listed in the library catalogue of 821-822 of the Benedictine Monastery at Reichenau and a fragment of the early ninth century survives from the Klosterneuburg (Eamon, 1994: 32-33; Smith and Hawthorne, 1974: 4).

we can call *Compositiones-Mappae clavicula* (or simply, *Mappae clavicula*) family.[2] The main features of these treatises are: an early date of production (from 8th to 10th century); the copy and dissemination through the network of monastic scriptoria; a content dedicated to sumptuary arts; and a direct link with Hellenistic and Byzantine artistic tradition and praxis. The continuity in the copy of this group of treatises of art technology in monastic scriptoria is surprising and indicates the importance attached to this text within the scheme of the transmission of technical knowledge and praxis in Western Europe.[3]

The Lucca manuscript has no title, unlike *Mappae clavicula.* But the later title is quite puzzling. The use of the Latin term *clavicula* has been considered as a symbolic reference to the use of a small key that would help to open up the doors of knowledge. This explanation agrees with the metaphor used in the prologue of the same treatise. Moreover, it can be linked to the alchemical corpus where technical recipes were "little kees" to accomplish ceremonial and initiatory rituals (Eamon, 1996: 34). On the other hand, the term *mappae* generated confusion, as it means cloth or chart. Phillips in his transcription of the Phillipps-Corning manuscript has not made any mention of this term, as he translated the title as "Little key of drawing or painting" (1846: 183). Since that moment, the title "Little key to painting" becomes quite common, although the content of the treatise has little to do with painting techniques.[4] Smith and Hawthorne stated that the title is enigmatically redundant as its literal translation should be "A little key to the chart" in the sense of a map, like the *mappaemundi* of the known world; that is why they used the title "a Little Key to the word of Medieval Techniques" in order to "catch the overtones of the two Latin words of the title and to reflect the contents more accurately" (1974: 14). Wilson suggested that the puzzling title of the treatise is a corruption of the original Greek *Baphes kleis* (A key to dying), Latinized initially as *Baphae clavicula* and then altered by scribal error to *Mappae clavicula* (quoted in Eamon, 1996: 373). Halleux proposed a much more attractive explanation (1989: 28): the Latin term *mappae* is a misunderstanding and mistranslation of the Greek word *cheirokmeton* (χειρόκμητον, something made by hands) that was misread as *cheiromaktron* (χειρόμακτρον, a hand towel) (Halleux, 1989: 28). In fact, the term *cheirokmeta* was also used as a title to Greek practical alchemical treatises (Berthelot, 1885: 157, 177). The term *mappae* could also be related to a signal-cloth, in particular the cloth with which the signal for starting was given to racers in the circus (Kroustallis, 2005: 241). In any case

2. Smith and Hawthorne suggested that the treatises belong to two different traditions: the *Compositiones* at the south of the Alps and the *Mappae clavicula* noth of them (1974: 4). However, I don't find any reason for such differentiation, mostly when they share a very similar content.

3. So far, more than eighty treatises of the *Compositiones-Mappae clavicula* compilations are known, the later being the example of the 17th century manuscript *Voss. Lat. octavo* 33, in the library of the Rijksuniversiteit in Leiden.

4. For example, Brunello, 1973: 125 or Eamon, 1966: 32.

the last two proposals fit better with the probable original title of the treatise, *Mappae clavicula de efficiendo auro volumen 1* as it appears at the catalogue of the Reichenau library: "a little key to the manufacture of gold" or "starting keys to the manufacture of gold." Consequently, I think that Halleux and Meyvaert were right in their proposal that the *Mappae clavicula* derives from a Greek alchemical text (1987: 7-15),[5] but I would only apply such an assertion only to the prologue and the first 50-56 recipes, the ones that agree with the index of the treatise and, moreover, they were copied as a block in Ms. 17 of Selestat and in Ms. 19 BNE.

The *Codex Matritensis* 19

The Ms 19 of the National Library of Spain (from now on Ms. 19 BNE) is an organized miscellaneous codex, structured around a corpus of treatises of *computus*, together with a group of texts related to the natural sciences, such as astrology, medicine, geology or geography. Two of these treatises (Bede's *De temporum ratione* and Aratus' *Phaenomena*) are splendidly illuminated. The *Mappae clavicula* recipe treatise was copied at the final folios of the manuscript (ff. 194r-198v) and at present contains only 90 recipes, because the manuscript ends abruptly, due to the damage and loss of its last pages.

The Ms. 19 BNE does not provide any information about the date or the place where it was made; even more, it is not known when or how the National Library acquired it.[6] The manuscript is written in Caroline minuscule in transition towards Gothic forms, a feature that can date the script in the first half of the 12th century.[7] The quality of the script is exceptionally good with clear strokes and very uniform.[8] Meyvaert saw in some abbreviations a slight influence of the Beneventan script (Mayvaert, 1966: 349-377). The standardization of Caroline minuscule and its diffusion throughout Europe between the 9th and 12th century is a major obstacle not only to date manuscripts but also to

5. Authors also emend Smith and Hawthorne translation and offer several alternative readings like "hermetic books" (*in hermetic libris*) and not "since I possess many books (*in hec meis*) of the prologue" (Halleux and Mayvaert, 1987: 14-15).

6. The first mention of the manuscript appears in 1769 in Iriarte's work *Regiae Bibliothecae Matritenis, Codices Graeci Mss*, v. 1: 206.

7. Codicological and palaeographical studies of recipe-containing manuscripts is a major drawback in several editions of medieval treatises on art technology. An accurate examination of all codicological and palaeographical aspects can offer not only important tools to date and locate its production; also allows a better understanding of the social context in which these recipes were copied, discern the authors' intentions, establish relations between several of these technical texts or even relate it to workshops practises and artistic influences. The importance of such studies was highlighted by Kroustallis (2008) and especially by Clarke (2009) as an indicator of the practical use of some treatises.

8. The script becomes only less careful at the final part of the manuscript that contains the recipes of art technology. Cordoliani saw the hand of only four copyists and he quoted that the script is so uniform that copyists should belong to the same scriptorium (Cordoliani, 1951: 6).

attribute them to a particular *scriptorium*, especially when they lack any kind of data like the case of Ms 19 BNE.

In 1912 the palaeographer John M. Burnam started the publication of his *Palaeographia Iberica* where he studied the manuscript, although he only advanced a hypothesis about its origin in the transcription of the *Mappae clavicula* treatise he published in the same year. There, he proposed the monastery of Santa Maria of Ripoll in Catalonia as the place of its making and he set the date to the year 1130 approximately (Burnam, 1912: 6). Although his hypothesis was not based on solid palaeographical, codicological or even historic arguments, the attribution of the manuscript to the *scriptorium* of Santa Maria of Ripoll was generally accepted.[9] Other researchers, however, rejected this attribution[10] gaining more strength an Italian provenance (Domínguez Bordona, 1933: 233-234; Mundó, 1961: 395). Zinner and Jones went further and suggested the Italian *scriptorium* of Montecasino (Zinner, 1925; Jones, 1943: 159-161). This thesis has been very well argued by García Avilés, who related the content of the Ms. 19 BNE with the production of the *scriptorium* Cassinense (García Avilés, 2001: 116-120). The *computus* treatises presents some extraordinary similarities with two later manuscripts, both of Italian origin but not attributable to a particular *scriptorium*. The first of these is the Ms. lat. 7418 of the Bibliothèque Nationale de France (13th-14th century) and the second is the Ms. Strozzi 46 of the Biblioteca Medicea Laurenziana in Florence (early 14th century). These similarities of content are so many that can suggest even the copy of both manuscripts directly from the Ms. 19 BNE (Cordoliani, 1951: 13). Another argument in favour of an Italian origin of Ms. 19 BNE is the treatise *Phaenomena* of Aratus as it is one of the most important examples of the Italian line in the transmission of this text.[11] Finally, two specific texts can shed light to this question as they can relate MS. 19 BNE with the production of the *scriptorium* of the monastery of Montecasino. The first is the *De annis a principio*, a *computus* and chronology treatise in poem written by Paul the deacon, an important figure of the Lombard Court and monk at Montecasino at the beginning of the 8th century. At present, this poem is known only by the Madrid codex and suggests a direct line of copy with the production of the Cassinense *scriptorium*. More conclusive is the colophon of the *Martirologium Bedae*,[12] where we can read that it was written by *Erchenperto sancti Benedicti de Castro Casino Monaco* (f. 48 r, col. 2) a prominent Cassinense monk and scribe.

9. Millás Vallicrosa defended strongly Burnam's proposal and added some linguistic arguments in favour of the Catalan origin of the manuscript (Millás Vallicrosa, 1931: 237-238).

10. For example, Karl Strecker proposed a possible relationship with the monastery of San Martin of Tours (Strecker, 1914: 674-682).

11. According to García Avilés, the *Aratea* of the Madrid codex comes with the *Scholia strozziana*, the typical comments of the Italian family of this text. (García Avilés, 2001: 121-123).

12. The oldest known text of this work is still conserved in the library of the Montecasino monastery (Ms. 439, 10th c.).

The first conclusion we can draw is that the Ms 19 BNE is an organized miscellaneous manuscript that copies much older texts produced at the monastic *scriptorium* of Montecasino, during the 10th and 11th centuries. But it is not just a compilation of texts from the library of the monastery but an exact copy of a manuscript produced at Montecasino probably in the 11th century. The script is Caroline minuscule in transition and not Beneventan, as we could expect if the Ms. 19 was written at Montecasino. This fact suggests that the manuscript was produced in a place where the tradition of the Caroline minuscule script was predominant and that the scribe (or scribes) was very skilful and exercised in this script. But Meyvaert's observation of a Beneventan influence in the execution of some abbreviations indicates that the Madrid codex copies an original in Beneventan script. The survey I carried out of the script of the early twelfth century codices from the monastery of Santa Maria of Ripoll didn't allow me to establish any kind of relationship with the Caroline minuscule used in Ms. 19 BNE. However, the script looks quite similar with examples of Caroline minuscule used in manuscripts in eastern France *scriptoria* during the 12th century.[13] The establishment of St. Benedict's rule during the 10th and 11th centuries in several monasteries in the south of France could be a possible explanation for the copy of a manuscript containing interesting treatises for the life in a monastery, like *computus* and excerpts of encyclopaedic and scientific works, moreover if its origin was an important Benedictine centre as Montecasino.

Another interesting aspect of Ms. 19 BNE is that not only copies older material of Cassinense origin, but also that it is a copy carried out with the utmost conscientiousness regarding the original. This is quite exceptional and the reasons are to be found in the importance granted to the manuscript, based both on the authority of the authors of the treatises (Aratus, Beda, Isidore, etc.) and the authority of being antique material, thus valuable (like the *Mappae clavicula* compilation). Two examples can sustain such hypothesis. First, some of the *argumenta* of the *Computus Graecorum et Latinorum* indicate that they were calculated for the year 904 (Ms. 19 BNE, fols. 75-80) and the copyist of the treatise did not update. Second, the illuminations of the manuscript copy faithfully models of the late Roman antiquity, filtered through the Byzantine culture. It is worth to mention that Bede's *De temporum ratione* illuminations, where is represented a system of counting through various positions of the body and fingers (Ms. 19 BNE, fols. 2v-4v) are very rare, and the only manuscripts that contain full-length figures are the Madrid codex, the already mentioned Ms. lat. 7418 of the Bibliothèque Nationale de France (Italy, 13th-14th century) and the manuscript Ms. 3 of the library of the monastery of the Santissima Trinità di Cava de' Tirreni (Italy, 11th century). It is worthy to mention

13. Check, for example the scripts of datable and attributable manuscripts from eastern France in Samarán and Marichal, 1974, t. V.

that the three manuscripts, beside the similarities of the *computus* treatises and the illuminations, they share another thing in common like a *Mappae clavicula* recipe collection (Tosatti, 2007: 31; Johnson, 1937: 86). This is quite important as is shows how the recipe treatises on art technology were copied and in what type of manuscript they formed part. In our case, it seems that during the 10th or the 11th century was formed and copied at the *scriptorium* of Montecasino a collection of recipes on art technology (Speciale, 1990: 347). This collection has been integrated in an important manuscript, like one that organizes time and activities (religious, intellectual and manual) in a monastic community. In this sense it should be understood the faithful copy of the original manuscript (text and images) and its transmission through the Ms. 3 of Cava de' Tirreni (11th c.), Ms. 19 BNE (12th c.) and Ms. lat. 7418 BNF (13th-14th c.)

A comparative study of the *Mappae clavicula* from Ms. 19 BNE

As we have already mentioned, the *Mappae clavicula* recipe treatise occupies the final folios of Ms. 19 BNE (ff. 194r-198v) and it is incomplete due to the damage and loss of the last pages. The treatise contains only 90 technical instructions that describe several artistic procedures related mainly to sumptuary arts, like gilding, fabrication of gold and silver objects, alloys and imitations of gold and silver, niello and enamel decorations, imitations of gems and glass pastes, argirorgaphy and chrysography, fabrication of artificial pigments and dying recipes.

A comparative study of the recipes of Ms. 19 BNE with similar recipes from other treatises of art technology provided the necessary keys to interpret the information they contain and highlight some specific aspects in the transmission of this technical knowledge. In other words, distinguish the different type of texts; establish possible influences among them; find what artistic techniques interested to copy; come across of possible actualizations of the recipes; understand the process of elaboration of this kind of compilations; clarify dark passages, correct mistakes and textual omissions in the process of copying; or even discern if there was any relation with coetaneous artistic praxis.

From the 90 recipes of the *Mappae clavicula* treatise of the Madrid codex, sixteen matches with the Lucca manuscript or *Compositiones variae* (8th c.); twenty-five with the *Mappae clavicula* of the Sélestat manuscript (10th c.); forty-eight with the *Mappae clavicula* of the Phillips-Corning manuscript (XIIth c.); eleven with the so-called *Liber Sacerdotum* (14th c.); nine with the *Greek papyrus X* of Leiden (3nd-4th c.);[14] and seven with Greek alchemical texts from *Mose's chemistry*[15] (7th c.).

14. As reference texts for the recipes of these treatises the following publications were used: Hedfors. 1932, 2003; Smith and Hawthorne, 1974; Berthelot, 1893; and Halleux, 1981.

The relationship of the recipes from Ms. 19 BNE with the recipes from the *Compositiones-Mappae clavicula* family is obvious. What is quite interesting is the fact that the formulation of the Madrid recipes is closer to the oldest recipe books like the Lucca and the Sélestat manuscripts.[16] Moreover, there are cases where the text from Ms. 19 BNE presents the same mistakes as the ones found in both manuscripts, a fact that suggests an original-copy relationship, at least in the case of these specific recipes.[17] Consequently, the compilation of recipes of Ms. 19 BNE was copied in a *scriptorium* where *Compositiones ad tingenda* and Ms. 17 of Sélestat coexisted or in a *scriptorium* where there was already a compilation based on both treatises and copies later Ms. 19 BNE.

The comparative study of the recipes also made possible to observe a tendency to copy recipes in groups of identical technical content, creating a more or less common structure, applicable to all *Mappae clavicula* family: first copied the recipes of gold;[18] then the recipes of silver; alloys and imitations of gold and silver; decorative techniques; preparation of raw materials; and finally, the dying recipes. Of course they are not coherent and homogeneus treatises, but even in both indexes of the Ms. 17 Sélestat and the Phillipps-Corning manuscript[19] it is possible to see an attempt to organize the recipes into coherent categories. The fact that both original compilation and later additions were made in form of blocks of recipes, arises the question whether the sources of these new additions were oral or written. An oral source[20] process could explain the interpolation of later material, duplications or differences between similar recipes and, moreover, could explain why copyist didn't notice it and why there was no editing process. The use of written sources

15. Berthelot has catalogued the treatise *Mose's chemistry* as alchemical (Berthelot, 1888: 300-319). However, a closer look at the recipes shows that it closer to a practical manual dedicated to the preparation of raw materials (metals and minerals) and gold and silver works. Even the author in the introduction wanted to disassociate the text with any hint of relationship with alchemical practice, noting that it was God that taught mechanic arts to men.

16. It is curious to notice also that the influence of the Greek language is stronger in the Lucca manuscript, the Ms. 17 of Sélestat and the Ms. 19 BNE: for example, in the spelling of some Latinized Greek terms it used the letter *– y* where the Greek word is written with the letter upsilon (-υ), like *elydrium, cyprum, mysi, hyrcini, lydium,* while in Ms. Phillips-Corning are always written with he letter – i. Finally, another common feature among the above mentioned three manuscripts is the conservation of the aspirant letter – h in words like *helydrium* or *hyrcini.*

17. For example, the Ms. 19 BNE and *Compositiones* write *terbentina* instead of *terebentinam* (Hedfors. 1932: 30); *refricdare* instead of *refrigidare* (Hedfors. 1932: 10-11); and the same with the Ms. 17 Sélestat: *flabum* instead of *flavum* (f. 9r), *amonii conopice* instead of *amonii canopice* (f. 10r), and *substraveris* instead of *substernis* (f. 10r).

18. Bear in mind that this was probably the nucleus of the original *Mappae clavicula* treatise, as we mentioned earlier.

19. The table of contents of the Phillipps-Corning has 209 numbered recipes with their title, although the work contains 382 recipes and the table agrees with the work itself as far as number 56; 66 chapters bear original numbers and the rest were numbered by Phillipps (Smith and Hawthorne, 1974: 9).

20. Byzantine or Byzantine influenced artists and craftsmen that transmitting oraly recipes that were translated at the same moment in Latin. Moreover, we have also the case of a recipe written in Greek with Latin characters, so an expert could translate it later.

would mean the circulation and accessibility of a large number of technical recipe books in the network of monastic scriptoria; and that the reason why they were not conserved is because they were practical workshop manuals that the continuous use and a manufacture with less durable materials, did not facilitate their conservation.[21] Of course, both processes could be possible.

Another important aspect is the literal copy of recipes from the *Greek papyrus X* of Leiden and the *Mose's chemistry*, that make obvious not only the continuity in the transmission of technological knowledge, but also that these recipes were copied directly from Greek texts and translated at the same moment into Latin.[22] This process of direct copy from Greek texts can also be seen in the use of the expression *quod greci vocant* to explain the Latinized name of ingredients, the continuous use of Greek terms or even he transcription in Latin characters of a recipe in Greek, like in *Compositiones ad tingenda* (Caffaro, 2003: 160). As they were compiled and translated from earlier Greek sources it is logic to be closely related to Greco-Latin encyclopaedic knowledge from authors like Dioscorides, Theophrastus or Plinio, but this does not necessarily means that they are obsolete texts with technical connection only with a remote classical past rather than current medieval techniques, like Smith and Hawthorne – among others – quoted (1974: 15).

An argument against the generalized opinion that the *Mappae clavicula* treatises have no relation with early medieval workshop practise (Halleux and Mayvaert, 1987: 25; Smith and Hawthorne, 1974: 15) is the fact that these compilations were live texts that were evolved (or actualized) by the voluntary intervention of the copyists that, at least, had a minimum knowledge of the technical instructions described. There are several cases that an older (and obsolete) terminology is replaced with terms of the coetaneous craft practice at the time of the copy. For example, the terms of Greek origin *chrisographia* and *argirographia* of the older treatises (like *Compositiones*, Ms. 17 Sélestat or the Ms. 19 BNE) are replaced by *auri inscriptio* and *argentis litteris scribere* in the 12th century (Phillips-Corning manuscript). Another example of this preference to use coetaneous technical terms is the process to decorate with niello: Ms. 19 BNE uses the verb *infestare*, the Ms. Phillips-Corning the verb *inpistare* (Phillipps, 1847: 202), while in the *Liber sacerdotum* is used the expression *ad faciendum niellum*, closer to current term niello (Berthelot, 1893: 198). The intentional intervention of the scribe can be seen also in the case of the instruments employed, like in a recipe of chrysography of the Madrid codex which is identical to a recipe from the Ms. 17 of Sélestat, except the instrument to burnish the gold leaf: the scribe of the Madrid texts advises to use the stone hematite (*poli cemathite*) while the author of the Sélestat text recommends the

21. Regarding medieval workshop manuals and their identification see Clarke, 2009.

22. An oral transmission through a Byzantine artist should be ruled out due to the fidelity of the translation between the Greek text and the Latin version.

use of a tooth (*dente splendifica*, f. 13r). The metric system employed in the common recipes of the four manuscripts reveals that it was also updated with the times. Thus, the frequent use of *drachmas* in *Compositiones* and Ms. 19 BNE was replaced by the use of ounces in the Ms. 17 of Sélestat and the Phillips-Corning manuscripts (Phillipps, 1847: 193, recipe n° 10 and Ms. 17 of Sélestat f. 6r. Phillipps, 1847: 201, recipe n° 52); while in other cases the Phillips-Corning manuscript replaced *drachmas* by a proportional system (Phillipps, 1847: 200, recipe n° 43).

Finally, work simultaneously with several similar recipes made possible to create a 'reference text' for many recipes. This text can be a valuable instrument in the cases of uncertain readings or textual omissions, inconveniences that can easily lead to erroneous interpretations and valuations. In most of the cases, omissions affect actions to carry out (considered, probably, less important or obvious by the copyist), while textual corruption affects mainly the ingredients of the recipes. To highlight the help of such 'reference text' we will point out some differences between the new edition of Ms. 19 BNE (Kroustallis, 2005) with the Burnam's transcription (Burnam, 1912) as well as with similar recipes of the edition of *Mappae clavicula* of Smith and Hawthorne (1974). In the first case, Burnam suggested an Arab-Hispanic linguistic influence in the text, together with the use of Syrian terms, close to an alchemical jargon. But such an interpretation is due to Burnam's misunderstanding of the text. For example, he consideres of a Syrian origin the term *flacaminas* and translates it as "black," when actually is a copy error for *fac laminas*, i.e. make sheets; he translates *luco* "almond" when, in fact, means "varnish;" or he relates the term *zumbri* with the Arabic measure *azumbre*, when it is a copy error for *cupri*. i.e. copper. In the second case, Smith and Hawthorne read *fac portionem ferream* in the recipe n° 6 of their edition of the *Mappae clavicula* and translate "you make a potion containing iron." However, the Madrid texts reads *fac caccabelum in quo hec universa trita mittis ferreum*, i.e. "prepare an iron pot, where you can mix all [ingredients] well pulverized," a translation more adequate to the technical procedure described. Again, in the recipe n° 58 Smith and Hawthorne didn't manage to explain the phrase *uno veda*, when it is really a copy error for *unguenda*, as we can see in the text of Ms. 19 BNE. Finally, Smith and Hawthorne consider that the recipe n° 50 was technically useless as there is no mention of lead in it, when it is an omission in the copy as in the Madrid text we can read precisely *tolle ex inde plumbum*.

The *Mappae clavicula* and the transmission of craft knowledge

We have already mentioned that the continuity in the copy of the *Mappae clavicula* group of treatises is surprising and the large number of extant copies suggests that they were considered as important books. However, as Smith and Hawthorne quoted, "there was no theory neither philosophy and it is rather surprising that the work achieved such popularity and longevity in the philosophical inclined Middle Ages" (1974: 19). Of course, the reason for their copy is not their literary character (Eamon, 1966: 33) or because they were consider as "a fountain into witch the brethren dipped to purify their Latin" (Smith and Hawthorne, 1974: 20), as they would probably get a contrary result.[23] Neither is credible the argument that these treatises that have become confused and almost all useless as instruction manuals were still copied by monks "perhaps with no interest at all in the material" due to an esoteric purpose (Ball, 2008: 92). We have already seen some arguments against such hypothesis. But there is also a codicological one. The type of the early *Mappae clavicula*-containing manuscripts[24] and the quality of their execution (like the Ms. 19 BNE) do not match with a manuscript containing literary exercises for Latin learners or a compilation of technical recipes that made no sense for the copyists and readers. On the other hand, the same argument can be used against their use as workshop manuals, mostly when the state of conservation of the manuscripts suggests a copy for a library than a practical use in a workshop.[25]

I believe that the keys to understand the causes of the sudden rise since the 8th century of the compilation of the *Compositiones-Mappae clavicula* treatises and their subsequent success are: first, the progressive loss of technical knowhow in the Western Europe as a cause of the crisis of the third century, especially with the techniques related to sumptuary arts and to the manufacture of luxury objects; and second, the need that felt West to revive, precisely, these arts, through the systematization of technical knowledge in the form of recipe collections. In this way, the copy of the treatises of art technology opens in the West during the 8th and 9th century with the first texts of the *Compositiones-Mappae clavicula* family. Its production is geographically placed in the area between the northeast of France, south of Germany and central and north Italy. Both the date and the location of these recipe collections are clear indicators of the relationship with the Carolingian culture and its successor, the Ottonian.[26] In both cases, we have a solemn and majestic art, employed by the political and ecclesiastical hierarchy to achieve splendour similar to the one of his main rival (political and spiritual), Byzantium.[27] The recipes of art technology of the *Compositiones-Mappae clavicula* group constitute an example of this artistic and aesthetic expression of the *Renovatio imperii* of Charlemagne

23. The language of the recipes is vulgar Latin, mixed with a craft jargon and a lot of Latinized Greek terms.

24. The *Liber Pontificalis* contains the *Compositiones ad tingenda* treatise and the Ms. 17 from Sélestat contains Vitruvius' *De architectura*, Faventinus' compemdium and the *Mappae clavicula* treatise.

25. On this subject see the codicological indicators proposed by Mark Clarke, 2009.

26. The relationship between the Carolingian culture and the appearance of the first technical treatises of the Middle Ages was highlighted by Le Goff (1971) and Bischoff (1971).

27. Migrating Byzantine craftsmen probably helped to enter technical literature in Western Europe in form of recipe books. Bear in mind that in the 8th century the iconoclastic movement caused the exile for several artists.

and the reforms of the Ottonians and, probably they were considered as an important instrument to carry them out. Monasteries with their workshops were the centres of this artistic and cultural activity and it is no surprise that the recipe collections of this group were also an exclusive product of the monastic *scriptoria* of that period.

This feeling of a lost knowledge, along with comments indicating the bad situation of artistic techniques, we can find in various texts from the 10th to 12th century, as well as in the same recipe treatises (Kroustallis, 2008: 36-37). For example, the author of the *De coloribus et artibus Romanorum*, commonly known as the *Treatise of Heraclius* (10th c.) in the proem of his work remembered with longing the ancient Roman splendour and lamented the situation of the artistic production at his time. Moreover, he added that the only solution left to gain knowledge of the antique techniques was to investigate the manufacture of objects (old ones or imported) and to experiment in order to achieve his purposes (Merrifield, 1967, v. 2: 189). Bishop Einhard (770-840), the director of the Imperial workshops of Charlemagne, act in the same way as he collected ancient works to study them and find out their manufacture technique (Hinks, 1962: 108-110); or Bernward, Bishop of Hildesheim (993-1022) who also got trained as goldsmith and illuminator, spend a lot of time studying the manufacture of all foreign gifts sent to King Henry II to understand their making process (Mayr-Harting, 1999: 44). The biography of Desiderius, Abbot of Montecasino (1058-1086), also conveys this feeling that a splendid art was lost together with the artistic connection with the past. His biographer, Leo of Ostia, praised all the artistic works carried out by the Abbot at the monastery because "Western Christendom had lost the talent of these arts for more than five hundred years and now, thanks to his efforts, and with God's help, he was able to recover them and protect them against a new death in Italy" (Dodwell, 1993: 226). The method the Abbot Desiderius employed consisted of stretching the political and artistic ties with Byzantium, inviting Byzantine artists to work in the Basilica of the monastery or commissioning works to the workshops of Constantinople. Moreover, he planed a long-term policy by training his own monks to several artistic techniques, mostly the ones related to the manufacture of objects of sumptuary arts. Although our source does not mention anything about the copy of technical treatises, it is imaginable that Desiderius project would not have been possible without the copy of the technical instructions as, precisely, the *Mappae clavicula* recipes of the original Casinense manuscript that copied the Ms. 19 of the National Library of Spain.

Another important issue to deal with is the relation of the *Mappae clavicula* treatises with alchemy. Undoubtedly, technological progress in art and crafts owe much to alchemy (alloys, dyes, artificial pigments, etc.). But there is a moment around the 3rd century when craft technology followed its own line of development, away from practical alchemy and its eagerness of transmuting base metals into gold. I believe that the reason of this division is the growing importance of industrial arts in the Late Roman Empire and the desire of artists and craftsmen to get disengaged from a practice regarded as fraudulent. Diocletian's edict of 296 that ordered the burning of all alchemical writings in Egypt was a clear attempt to avoid disrupting economy and prevent fraud related to the preparation of false gold and silver.[28] An interesting anecdote is found in *Historia Langobardorum* of Paul the Deacon (8th c.): at the end of the 6th century the Saxons, allies of the Longobards, wanted to return from the North of Italy to their homeland; but along the way, they cheated and ruined a lot of people because they used gold-coloured bonze to buy supplies (Paulus Diaconus, 1974: 99). Technically there is no difference between a craft recipe for a gold alloy with copper and an alchemical recipe to transmute copper into gold. In fact, the difference is in the use or, in other words, the difference between fake and imitation, concepts that were quite clear in the medieval mind.[29] This estrangement between craft and alchemical praxis can be noticed also in the compilation of recipes on art technology.[30] A good example of this moment of separation is the Greek Papyri of Leiden and Stockholm (3rd-4th c.), considered as one of these alchemical manuals, saved from burning. The relation of the alchemical recipes of the Greek Papyrus X of Leyden with some metallurgical recipes of the *Compositiones-Mappae clavicula* group led several researchers to think that also the latter would have been alchemical texts or of alchemical origin and purpose.[31] But, as mentioned above, there is a difference in the intent of the technical instructions. In the Greek Papyrus X of Leiden counterfeit is obvious: one recipe is explains that goldsmiths cannot detect fraud and another warns that silversmiths could notice the difference (Halleux, 1981). On the contrary, craft treatises use a clear terminology – absent of a dark language appropriate only for the initiated –, the technical process is well explained and metrology is well described.[32] Even when a rec-

28. The sources that cite this episode are collected in Clarke, 2001, p. 15, note 59.

29. For example, St. Isidore of Seville (*Etym.*, II, 267 and 382) quotes that imitation was not a mere copy material, but an extension of the activity of nature: man watched what were the laws that allowed nature to achieve certain results and following these laws are tried, through their work, to obtain similar results. And St. Augustine argues against any kind of falsehood in his work *De mendacio* (Bruyne, 1987: 59).

30. The two different lines of development of alchemical and craft recipe manuals was fist proposed by Arie Walert (1990: 156).

31. For example, Burnam quoted that that the Ms 19 BNE was "alchemical, as well as chemical and practical" (1912: 5). Halleux suggests that the primitive nucleus of the compilation *Mappae Clavicula* was an alchemical text written in the 4th century and translated into Latin around the 5th century (Halleux, 1990: 174). Finally, Ball states that these recipe collections is an echo of alchemical treatises whose "theoretical component has atrophied and what is left is a strange mixture" (Ball, 2008: 92).

32. Obscure passages have to do more with copy errors (misreading, omissions, etc.) than an intentional action. For example, there is only one only case in Ms 19 BNE where alchemical terms are used but immediately it is explains that *batrachium* is copper and *chrisanthimum* is the gold that painters use (the same recipe can be found in the Selestat manuscript (f. 8r) and in the Phillipps-Corning manuscript (Phillipps, 1847: 194).

ipe for a gold amalgam "that never fails" calls for maintaining the secrecy of the proceedings it seems closer to craft confidentiality than intent to fraud (Smith and Hawthorne, 1974: 31, recipe nº 11; 32; recipe nº 14). More explicit on this is the final comment from a recipe of a gilding imitation, where the author quotes that it was better to keep it secret; otherwise he ran the risk of being considered dishonest (Smith and Hawthorne, 1974: 35, recipe nº 52). Obviously, this is a completely different ethics from the one we have just seen in the Greek Papyrus X of Leyden. In short, the *Mappae clavicula* recipe compilation of art and craft technology is an example of the vernacular science, a system of practical knowledge used by artists and craftsmen to understand matter and nature and carry out their work.[33] And, although this system of practical knowledge shared a common theoretical framework with alchemical treatises, it was transmitted only for craftsmanship purposes, long before the transfer of alchemical and Greco-Roman scientific knowledge from the East to Western Europe through the Latin translations of Arabic texts during the 12th century.

As we have already mentioned, the recipes of the *Mappae clavicula* family are mainly focused on sumptuary arts. Consequently, and contrary to the most widespread thesis, its relations with painting techniques is secondary. This misconception was caused, possibly, by the emphasis placed on the studies of the history of painting[34] and by the arbitrary division between "major or fine arts" and "minor arts" that, to some extent, has distorted our perspective to these texts. But to be more precise, the *Compositiones-Mappae clavicula* group don't deal with general instructions for art and craft procedures. The emphasis in these recipe collections is given on how to prepare raw materials, elaborated ingredients and different mixtures (alloys, metallic solders, artificial pigments, enamel, niello, glass pastes, varnishes, or imitations of pearls and gems). In that sense, it is easy to explain why they don't contain any theoretical discussion; why there are no step-by-step instructions regarding the most outstanding medieval art techniques; the absence of any coetaneous technological novelty an art technology;[35] and, finally the presence of a strong traditionalism in craft terminology. In this way should be interpreted expressions like "you will have it for use," "make whatever work you want" or "make use of it as you wish" at the end of several recipes (Smith and Hawthorne, 1974: 30, recipe nº 6; 34, recipe nº 36; 35, recipe nº 48; 41, recipe nº 97; 46, recipe nº 132).

33. I would like to stress that reconstructions of the *Mappae clavicula* recipes and laboratory analysis can confirm the practical character of the treatise and its relation to coetaneous art techniques. I will quote as an example, the reconstructions of green and blue copper pigments, the processes of coating metallic surfaces or the preparation of niello (Scott, 2002: 225, 237 and 282-286).

34. For example, Smith and Hawthorne quote that "the Mappae clavicula seems to have been preserved mainly because of the apparent connection of its recipes with the art of the painter" (1974: 15-16).

35. Unlike later treatises, like Theophilus' *Schedula* or the Montpelier treatise (Clarke, 2011).

As a conclusion I would like to point out that traditionally, medieval recipe books on art technology were regarded as simple texts of technical instructions related to the artistic or handicraft work. This idea contributed to their undervaluation in the studies of medieval art, augmented by the apparent lack of any theoretical section.[36] However, in the same recipes it is possible to glimpse the medieval aesthetics and mentality, like sumptuousness, the concept of light and beauty, an art that stresses the inherent quality of the materials, an attempt to transform modest into luxurious and to embellish everything or the mimetic function of art. In this sense it would be possible to consider these recipe collection as the theory of art praxis, a set of essential precepts and technical rules for an artist or artisan to carry out his work: art not only could be learned and taught, but also could encoded and compiled in the form of recipes.

Bibliography

BERTHELOT, M (1885): *Les origines de l'alchimie*. Paris: Steinheil.

BERTHELOT, M (1888): *Collection des anciens alchimistes grecs*, v. 2 y v. 3. Paris: Steinheil.

BERTHELOT, M. (1893): *La chimie au Moyen Âge*, v. 1. Paris: Imprimerie National.

BISCHOFF, B. (1971): "Die Überlieferung der Technischen Literatur," in *Artigianato e tecnica nella società dell' Alto Medioevo occidentale, Spoleto, 2-8 Aprile 1970*, t. II. Spoleto: Centro Italiano di Studi sull' alto Medioevo: 267-296 (*Settimane di Studio del Centro Italiano di Studi sull' alto Medioevo*, XVIII).

BRUNELLO, F. (1973): *The Art of Dyeing in the History of Mankind*: Vicenza: Neri Pozza Editore.

BURNAM, J.M. (1912-1920): *Palaeographia Iberica*, Fascicule II. Paris: Libraire Ancienne Honoré Champion.

BURNAM, J.M. (1912): *Recipes from Codex Matritensis A16 (ahora 19). Palaeographical Edition from a Black on White Facsímile*. Cincinnati: University of Cincinnati Press (*University of Cincinnati Studies*, series II, vol. VIII, part 1).

CAFFARO, A. (2003): *Scrivere in oro. Ricettari medievali d'arte e artigianato (secoli IX-XI). Codici di Lucca e Ivrea.* Napoli: Liguori Editore.

CLARKE, M. (2001): *The Art of all Colours. Medieval recipe books for painters and illuminators*. London: Archetype Publications.

CLARKE, M. (2009): "Codicological indicators of practical medieval artists' recipes," in Hermens, E.; Townsend, J.H. (eds.), *Sources and Serendipity*. London: Archetype: 8-17.

CORDOLIANI, A. (1951): "Un manuscrit de comput ecclesiastique mal connu de la Bibliothèque Nationale de Madrid," in *Revista de Archivos Bibliotecas y Museos*, LVII: 5-35.

CORDOLIANI, A. (1955): "Un autre manuscrit de comput ecclesiastique mal connu de la Bibliothèque Nationale de Madrid," in *Revista de Bibliotecas Archivos y Museos*, LXI: 435-481.

36. The only exception is the treatise of Theophilus, or *Schedula diversarum artium*, with the three prefaces.

DOMÍNGUEZ BORDONA, J. (1933): *Manuscritos con pinturas*, t. I. Madrid: Centro de Estudios Históricos.

EAMON, W. (1994): *Science and the Secrets of Nature: Books of Secrets in Medieval and Early Modern Culture*. Princeton: Princeton University Press.

EAMON, W. (ed.) (1982): *Studies on Medieval Fachliteratur*. Brussels: UFSAL (*Scripta: Medieval and Renaissance Texts and Studies*, vol. 6).

HALLEUX, R. (1981): *Les Alchimistes grecs, Tome I: Papyrus de Leyden, Papyrus de Stockholm, Recettes*. Paris: Les Belles Lettres.

HALLEUX, R.; MEYVAERT, P. (1987): "Les origines de la Mappae clavicula," in *Archives d'Histoire Doctrinale et Littéraire du Moyen Âge*, 62: 7-58.

HALLEUX, R. (1989): "Recettes d'artisan, recettes d'alchimiste," in Jansen-Sieben, R. (ed.), *Les Artes mechanicae en Europe médiévale*, Bruxelles: Archives et Bibliothèques de Belgique (no. spécial 34)

HALLEUX, R. (1990): "Pigments et colorants dans la Mappae Clavicula," in *Pigments et Colorants de l'Antiquité et du Moyen Âge, Colloque International du Centre National de la Recherche Scientifique*. Paris: Éditions du CNRS: 173-180.

HEDFORS, H. (1932): *Compositiones ad tingenda musiva*. Uppsala: Almquist & Wiksells Boktryckeri-A.-B.

HINKS, R. (1962): *Carolingian Art*. Michigan: The University of Michigan Press.

GARCÍA AVILÉS, A. (2001): *El tiempo y los astros. Arte, ciencia y religión en la Alta Edad Media*. Murcia: Universidad de Murcia.

IRIARTE, J. (1769): *Regiae Bibliothecae Matritensis. Codices Graeci Mss*, vol I. Madrid: Antonio Pérez de Soto (imr.).

JONES, C.W. (1943): *Bedae opera de temporibus*. Cambridge: Medieval Academy of America.

JOHNSON, R.P. (1937): "Some Continental Manuscripts of the Mappae Clavicula," *Speculum*, XII: 84-103.

KROUSTALLIS, S. (2005): *Edición crítica y estudio de un trabajo de tecnología artística medieval: el Codex Matritensis 19*. Unpublished PhD thesis, Complutense University of Madrid, Faculty of Geography and History.

KROUSTALLIS, S. (2008): "Los recetarios medievales de tecnología artística" in Kroustallis, S., Townsend, J.H., Cenalmor, E., Bruquetas, R., Stijnman, A., San Andrés Moya, M. (eds.), *Art Technology. Sources and Methods*. London: Archetype: 35-41.

LE GOFF, J. (1971): "Travail, techniques et artisans dans les systemes de valeur du haut Moyen Âge (V-X siècles)" in *Artigianato e tecnica nella società dell' Alto Medioevo occidentale, Spoleto, 2-8 Aprile 1970*, t. II. Spoleto: Centro Italiano di Studi sull' alto Medioevo: 239-266 (*Settimane di Studio del Centro Italiano di Studi sull' alto Medioevo*, XVIII).

LEVEY, M. (1962): *Medieval Arabic Bookmaking and its Relation to Early Chemistry and Pharmacology*, Philadelphia: American Philosophical Society, (*Transactions of the American Philosophical Society*, new series, vol. 52, part 4).

MAYR-HARTING, H. (1999): *Ottonian Book Illumination: an Historical Study*. London: Harvey Miller.

MERRIFIELD, M.P. (1967): *Original Treatises dating from the XIIth to XVIIIth Centuries on the Arts of Painting*. New York: Dover Publications.

MILLAS VALLICROSA, J. M. (1931): *Assaig d'historia de les idees fisiques i matemàtiques a la Catalunya medieval*, v. I. Barcelona: Institució Patxot.

MILLAS VALLICROSA, J.M. (1959): "Sobre el Manuscrito 19 de la Biblioteca Nacional de Madrid," *Revista de Archivos Bibliotecas y Museos*, LXVII, 1: 119-126.

MUNDÓ, A. (1961): "Códices isidorianos de Ripoll," *Isidoriana, Estudios sobre S. Isidoro de Sevilla en el XIV centenario de su nacimiento*, 36: 389-400.

PAULUS DIACONUS (1974): *History of the Lombards. Paul the Deacon.* Foulke, W.D. (trans.) and Peters, E. (ed.). Philadelphia: University of Pennsylvania Press.

PHILLIPPS, T. (1847): "Letter from Sir Thomas Phillips ...," in *Archaeologia*, 34, pp. 183-244.

SAMARÁN, C.; MARICHAL, R. (1974): *Catalogue des manuscrits en écriture latine, portant des indications de date, de lieu ou du copiste*, t. V, Paris, Éditions du Centre National de la Recherche Scientifique.

SCOTT, D.A. (2002): *Copper and Bronze in Art: Corrosion, Colorants, Conservation.* Los Angeles: Getty Trust Publications; Getty Conservation Institute.

SMITH, C.S.; HAWTHORNE, J. G. (1974): *Mappae Clavicula: A Little Key to the World of Medieval Techniques'.* Philadelphia: American Philosophical Society (Transactions of the American Philosophical Society, new series, vol. 64, part 4).

SPECIALE, L. (1990): "Indicazioni di colori in un disegno cassinese dell' XI secolo: il foglio dell' Ascensione nell' Omiliario Cas. 99 (Archivio della Badia, ms. 99, p. 409)," *Pigments et colorants de l'Antiquité et du Moyen Age. Teinture, peinture, enluminure, études historiques et physico-chimiques. Collôque International du CNRS*, Paris: CNRS: 339-350.

STRECKER, K. (1914): "Poetarum latinorum medii aevi," *Monumenta Germaniae Historica*, t. IV, pars. 11: 674-682.

THOMPSON, R. C. (1937-1938): "A Survey of the Chemistry of Assyria in the Seventh Century B.C.," *Ambix*, I (1): 3-16

TOSATTI, B.S. (2007): *Trattati medievali di tecniche artistiche.* Milano: Jaca Book.

WALERT, A. (1990): "Alchemy and Medieval Art Technology," in Von Martels, Z.R.W.M. (ed.), *Alchemy Revisited: Proceedings of the International Conference at the University of Groningen*, 17-19 April 1989. Leiden: Brill, 154-161.

ZINNER, E. (1925): *Verzeichnis der astronomischen Handschriften des deutschen Kulturgebietes.* Munich: C. H. Beck.

Local Traditions and the use of local products in a XV Century English Manuscript

Cheryl Porter

The recent (2005) identification of the use of Genista tinctoria L. (Dyer's broom) and (possibly) Reseda luteola (weld) for the making of yellow lakes in a fifteenth century English manuscript, has provoked much interest and serves to bring to our attention that cheap local products could be, and were used for even the most prestigious of works. The discovery is surprising, partly because organic yellows fade faster than other colours, and partly because there is not much yellow, either organic or inorganic, used in manuscript painting from the late 12th century – various forms of gold being the preferred choice. Vegetable yellows are not noted in trade records, for the simple reasons that they are cheap, and that dyers and painters would certainly have had local sources. In 1582, Richard Hakluyt recorded that "yellowes and greenes are colours of small prices in this realme by reason that Olde (weld) and Greenweed ... be natural here and in great plenty."[1] Historically, we have few recipes in England for their use in manuscript (or other) painting until the late 16th and 17th centuries, where they were curiously known as "pinkes."[2]

In the 14th and 15th centuries, there are a number of recipes for the making of organic lake colours – mostly red, but also yellows. However, it is one thing to have the recipes, and another to prove that artists are using them. In order to be quite sure, it is necessary to analyse the colours used, and so far, for a number of very good reasons, little analytical work has been done on the colours used to paint in manuscripts.[3]

1. C. Bolton, 1935-40, "Contributions on the History of Dyeing," *The Dyer and Textile Printer*, lxxiv, p. 66.

2. J. Kirby, 2000, "Sir Nathaniel Bacon's Pinke," *Dyes in History and Archaeology*, 19, pp. 37-50.

3. This is not the place to rehearse the arguments for/against the taking of samples, but it should be noted that most of our samples were taken from offset and detached fragments of paint – though in fact, before conservation, the condition of much of the manuscript was so poor, that sample – taking was not an issue.

Now on permanent loan to the Hallward Library, Nottingham University, Manuscript 250, the Wollaton Antiphonal, was made in the East of England, probably around Norwich-Lincolnshire about 1420-30. The scope, size and illustration programme are exceptional, with more than four hundred (calf) parchment folios, twelve full borders, fifteen partial borders, large-scale coats of arms in borders and fifteen historiated initials. The cost of such a large and prestigious manuscript must have been enormous.[4] It was made for Sir Thomas Chaworth of Wiverton, after his marriage in 1411 to Isabella of Aylesbury. A number of decorative traits link it to Lincolnshire or Eastern England: the Calendar connects it to East Anglia and particularly the diocese of Norwich. Our results from analysis of the colours used to paint the miniatures in the manuscript add weight to this locality.

The importance of this manuscript is reflected also in the choice and the abundance of the costly pigments, expensive lake colours and the gold and silver used throughout. Those inorganic colours identified in the manuscript illuminations[5] include silver leaf, gold leaf and powder, vermilion, verdegris and other copper salts (sulphate and/or chloride – we don't know if these were used originally, or are a conversion product of the verdegris), lead tin yellow, azurite, ultramarine, a layering of ultramarine over azurite (a difficult technique to execute) and admixtures of azurite and ultramarine.[6] Of particular interest has been the identification of the organic colours used to paint in this manuscript, including indigo/woad blue (with tannin),[7] a large amount of the costly kermes (and maybe madder),[8] sap green, from the buckthorn (Rhamnus cathartica L.), Reseda luteola L (weld), Genista tinctoria L (Dyer's broom) and a purple anthocyanine colouring matter for the pen flourishing – used as a juice, with no alum detected in its make-up.[9] This is the first time that the sap green, weld and broom have been identified in manuscript painting. The colours are often

4. Kathleen L. Scott, 1996, *Later Gothic Manuscripts 1390-1490* Catalogue and Indexes. Harvey Miller Publishers.

5. Paint samples were analysed using polarising microscopy. Other analysis was done at the National Gallery in London: Fourier Transform Infrared Microscopy (FTIR) by Catherine Higgitt, energy-dispersive X-ray analysis in the scanning electron microscope (SEM-EDX) by Marika Spring and High Performance Liquid Chromatography (with a diode array detector – HPLC-DAD) by Jo Kirby.

6. The technique of mixing and the layering of ultramarine and azurite is not uncommon in manuscripts from the 12th century.

7. Analysis of organic blues using HPLC by Jan Wouters and his team at the Koninklijk Instituut voor het Kunstpatrimonium/Institut Royal du Patrimoine Artistique, Jubelpark 1/Parc du Cinquantenaire 1, B-1000 Brussels, Belgium. There is no way to tell if this was blue from woad, grown locally, or the imported indigo. Both were available.

8. Some small amount of madder component(s) seems to be present. We cannot be sure if the madder was mixed in with the kermes, or is a separate pigment. Many of these organic lake colours were made from textile shearings. These (usually woollen) textiles were dyed first with madder and then top-dyed with kermes. This practice was especially prevalent in England and the Netherlands.

9. It is not a buckthorn or a folium (turnsole).

mixed together, for example, azurite and ultramarine, copper greens with lead tin yellow and yellow organic lake pigment, indigo/woad and yellow lake and most interesting – mixtures of broom and weld.[10]

Some questions arising from the analysis of the colours used in this manuscript were:

– Why and how is this a reflection of local practice and tradition?
– Is there something about the use of these colours that makes them "English?"
– Is the use/preparation of these colours different in any way, or completely in keeping with contemporary practices/recipes?

Yellow lakes

The identification of Genista tinctoria L. (broom) and Reseda luteola L. (weld) is of special interest. Until well into the 15th century, metal gold was the most important yellow used to paint manuscripts. Vegetable and pigment yellows were mostly used to modify reds and greens, or to imitate gold but, as Thompson writes, they were rarely used to represent yellow things. Yellow anyway, was used very sparingly, since the medieval artist was wary that strong yellows could disturb the balance of a painting. Of course, it may also be relevant that until the invention of lead tin yellow, the main yellow pigment *orpiment* was poisonous and extremely reactive and unstable.

From the 14th century, it seems from contemporary recipes, that one of the most important gold substitutes was vegetable yellow, especially buckthorn (Persian Berries and Rhamnus catharticus) and weld.

Merrifield and others explain that these vegetable yellow pigments were of two kinds: those that were precipitated onto a white base – such as the different kinds of yellow lakes, and those used simply as a juice – such as saffron.[11] Analysis has proven this to be the case with all of the yellow lakes in the Wollaton Antiphonal. In keeping with contemporary recipes, they are all made with alum, and then precipitated onto either chalk or lead white, and in one case, also lead tin yellow. Our analysis shows that these yellow lakes are all made with Genista tinctoria L. (broom), and sometimes Reseda luteola L. (weld) is added to the broom. Surprisingly, the green tested from folio 241 (three priests and choir book) is sap green, from the Rhamnus cathartica (buckthorn). It was made with alum and mixed with broom and possibly also weld. What is interesting is that the ripe berries were used to make the green but that

10. These mixtures are increasingly common in manuscript painting from the 14th century.

11. M. P. Merrifield, 1999, *Medieval Renaissance Treatises on the Arts of Painting*. Dover Publications, NY, p. clxiv.

the unripe berries were *not* used in the making of any of the yellow lakes that have been tested.[12]

Genista tinctoria L.

(Dyer's greenweed, genusta, genestrum, Wdewise, brom, Dyer's broom) is commonly found in open parkland, grassy fields and embankments, on the edges of woods (it doesn't like shade) and especially on dry, sandy and gravel soil, as well as boulder clay. It is widespread throughout Europe and England. Recipes stipulate that only the flowering branches are used for dyeing. We have a few recipes for its use as a dye plant in Germany, France and England since medieval times. The earliest English record for dyeing with broom comes from a late c.12th manuscript from the British Library, MS Cotton Titus D.XX1V ff. 131r-2v. It was made at the Cistercian abbey, of the BVM Nottinghamshire, and includes epitaphs, poems, hymns and didactic exercises, as well as a series of recipes on colour making:

> "To make yellow, take (wood) ash lye and pick dyer's broom and boil it in lye. Then remove the broom and dye the cloth and it will be yellow."[13]

In Britain, Dyer's broom is generally associated with England and sandy and chalky soil. It is rarely found in Scotland, except occasionally in the south.[14] It is rare in Ireland also. Broom was known, and certainly very much used, by English dyers and evidence of its use has been found at a number of archaeological sites in England: most notably, the site at Beverly, York, dated c.9th-11th and at the Dundas wharf site in Bristol, dated c.13th-14th. In both sites, the samples were found in waste from a dye bath.[15] Dyer's broom was an important article of trade between Ireland and England as early as 1367[16] and the old English Guilds in the neighbourhood of Winchester used this plant for dyeing wool as early as 1400. Both silk and wool could be dyed with it. Although we have no particular recording of broom being cultivated to any extent as a dyer's plant – either in Britain or the Continent, there are a number

12. The sample tested high for *kaempferol,* indicating that the green was made from *ripe* buckthorn berries).

13. T. Hunt, 1995, "Early Anglo-Norman Recipes for Colours," *Journal of the Warburg and Courtauld Institutes,* Volume 58, p. 206. It is interesting to note that this recipe is in a Carthusian manuscript. Carthusians were averse to the use of gold in their manuscripts since it was deemed to be incompatible with their austere way of life. In fact, it is not uncommon to find recipes in these compilations, and it can be supposed that they served as a sort of reference book – things to know for teaching apprentices, for taking to a new "house," or simply to aid the memory.

14. *New Atlas of British and Irish Flora,* 2002, Defra Publications, London, p. 145.

15. Penelope Walton Rogers, verbal communication and recorded in D. Cardon, 2003, *Le monde des teintures naturelles,* Germany, p. 154.

16. C. D. Mell, 1932, "A Brief Historical Sketch of Dyer's Broom," *The Textile Colorist,* 54, p. 27.

of references connecting it to the dyeing industry in Suffolk and Norfolk, in the east of England.

– 15th century records show that in Suffolk and Norfolk, the wild plant was being collected and sold to dyers for 8 pence per hundredweight of the green plant.[17]

– Earlier, the *Staverton Park and Estate* accounts of 1268-1306, record that in 1294-5, the sum of 8 shillings was achieved for the sale of a fence enclosing 40 acres of broom. Further records show, in 1300-01, "nothing for broom" and 1301-2, "6/- sale of broom." Though it is unclear which "broom" is being referred to, it is worth noting that Staverton Estate is near Ipswich and exactly within those areas where Dyer's broom was grown and sold to the dyers.[18]

Later recipes for broom yellow lakes

There are very few recipes for dyeing with broom and even fewer for painting with it.[19] Those that we do have are from much later; mostly the 16th and 17th centuries, and broom (and weld) has been identified in paintings from this time. However, even at this later time, a number of the recipes describe the process that we recognise from our 15th century English manuscript. The simplest recipes describe the boiling of the flowering branches in water and adding eggshells or ground chalk mixed with a little water, then adding powdered alum. English recipes of 1606 and 1612 describe "woodewaxen" (dyer's broom), picked "in the latter end of May" and adding alum and then precipitating onto chalk.[20]

Reseda luteola L. (Weld)

Also known as dyer's weed, dyer's rocket, green weed, green weld, wolde, herba luteolae, and herba resedae, it is said to have been grown in England from at least Roman times, though the earliest reference to it is from Chaucer's *Former Age* (1374), where he writes: "No madyr, welde, or wod no litestere ne knewh."

17. C. D. Mell, *ibid.*, p. 26.

18. I owe thanks to Dr Oliver Rackham of Corpus Christi College Cambridge for drawing this to my attention.

19. There is evidence of the use of broom as a textile dye on a fabric from Florence late 14th, early 15th century. For later dyeing records see the Plictho, by Gioanventura Rosetti for dyeing Italian cloth, and C. D. Mell, *op. cit.*, p. 28 for English dyeing of inferior woollen goods with broom.

20. For a full description of these later yellow lakes and their use in panel and cloth painting, see J. Kirby, *op. cit.*

Since the beginning of the 14th century, weld was systematically cultivated in many parts of Continental Europe and in England, especially in Essex and Norfolk. It remained an important industry around dyeing centres until soon after the middle of the 19th century.[21] All of the plant is used for colour-making, except for the roots. It is harvested when it flowers in mid-summer and grows easily on disturbed soils on waysides, in moist ground as well as dry – even on the poorest soil, and especially where the subsoil is chalk or lime. In Britain this chalk can be found down the east side of England and swinging across the South to the west. (Medieval maps show a chalk pit in the very centre of the town of Norwich). It was imported into Scotland, though it does now grow wild in many parts.

Recipes for weld lake-making

There are a number of recipes from medieval times documenting its use for painting in books. Thompson notes recipes for weld lakes from the 14th century and claims that weld lakes were preferred to be somewhat opaque. For this the juice was precipitated onto lead white, calcium or egg shells.[22] The 14th century *Arte Illuminandi* has a recipe for precipitating the extract of *erba luza* (weld) onto rock alum and then incorporating this into lead white or marble dust.[23] At the end of the 14th century, Cennini noted that weld (arzica) was chiefly used for miniature painting; especially in Florence, but gave no recipe for its manufacture, except to note that it is made alchemically.[24] In the 15th century Bolognese manuscript, a pound of weld is cut finely and boiled in enough water to cover it, until the water is "half wasted." To this is added 2oz finely ground travertine (calcareous white stone from Pisa-Tivoli) or 2oz lead white and half oz roche alum ground very fine. These were then slowly boiled, stirred, removed from the fire when nearly dry and the remaining water poured off. This arzica (lake) should then be laid on a new brick hollow in the middle and "let it settle." It was then put onto a small, well-polished board to dry.[25] British Library, Sloane MS 2584, f.5r, a 14th century English recipe for making a weld lake boils weld with lye and then adds alum. The lake thus produced was used without filtering and was not precipitated onto chalk or lead white.

21. C. D. Mell, "A Brief Historical Account of Weld," in the *Textile Colorist*, 54 (1932), p. 335-7 and D. Cardon, *Le monde des teintures naturelles*, Berlin (Germany) 2003, p. 150.

22. D. V. Thompson, 1956, *The Materials and Techniques of Medieval Painting*, Dover Publications, Inc, New York, NY, p. 188.

23. F. Brunello, 1992, *De Arte Illuminandi*, Neri Pozza Editore, Italy, p. 200.

24. Cennino d'Andrea Cennini, 1960, *Il Libro dell'Arte*, Dover Publications, Inc, New York, p. 30.

25. M. P. Merrifield, *op. cit.*, pp. 482-5.

An East-Anglian Manuscript?

The identification of the use of Dyer's broom and weld, adds weight to art historian's locating of this manuscript around the East Anglia – Lincolnshire region. The Calendar of the manuscript has been specifically connected to Norwich. From the 14th century, Norwich was a wealthy city, second in size only to London. 130 different trades – including manuscript illuminators – were listed, and the city is much noted in documents relating to the production of wool, the production of manuscripts, the richness of its dyers, its well-developed and profitable trade links with the Low Countries, even the use of broom and weld in the local dyeing.[26] For yellows, the choice of which plant as a source of colorant was influenced by local availability. In many cases, the dyes used for textile dyeing were also used for painting in manuscripts. Painter's and illuminators' treatises often specified: "the herb that the dyers use," and both weld and especially broom were very connected with the east of England

Conclusions

The answers to the questions we originally posed are clear: the preparation of the yellow lakes in this manuscript is in keeping with the recipes from this time – whether from Mediterranean sources or from north of the Alps. The colouring matter is extracted by boiling in water, alum is added and it is then precipitated onto a white base to create an opaque or semi-opaque lake colour. The use of weld and (especially) Dyer's broom is in very much in keeping with local traditions, and the use of locally-grown products ear-marks the manuscript to be not only English, but perhaps, more specifically, connects it to East Anglia and maybe even the great medieval city of Norwich.

Towards the end of the middle ages, yellows came to be used more frequently in manuscript painting. Gold – traditionally associated with heaven, monarchy, power – was now itself used in a different form in manuscripts. It became increasingly rare to see whole skies and drapery covered in gold leaf. Instead, gold is used as a paint, to highlight pigmented surfaces, almost completely replacing lead white highlights. New types of yellows became available – both organic and inorganic – and by using these yellows, rather than gold, the artist was able to create images that looked more natural and were also rather more "painterly." The relatively new technology of making lake pigments from plants and animals changed the way that the medieval book artist painted in that the materials used were chosen not for the cost or prestige of

26. Weld is mentioned in Norwich customs accounts, as well as woad, madder and alum in the late 14th century. The National Archives, E122/149/28 and E122/150/11 for 1475-6. C. D. Mell also mentions Norwich as a centre for the cultivation of weld and its use in woollen manufacture. See p. 336.

the materials themselves, but for their naturalistic look and their suitability for the subject. The humble vegetable yellows were a part of this.

I owe thanks to The Historic Manuscripts Trust for funding the conservation and research on the manuscript; to Dorothy Johnston of Nottingham University Hallward Library and her staff for their continued support; to Jo Kirby, Catherine Higgitt and Marika Spring of the Conservation Science Department of the National Gallery in London, and Jan Wouters of the Koninklijk Instituut voor het Kunstpatrimonium, Brussels. My thanks also to Jenny Dean for working with me in picking and making yellow lakes, with weld and broom from her garden.

O livro de como se fazem as cores or a Medieval Portuguese Text on the Colours for Illumination: A Review

Luís U. Afonso/António J. Cruz/Débora Matos

Introduction

O livro de como se fazem as cores (LCFC), that is, *The book on how to make colours*, is a Portuguese technical text with relevance for the knowledge of the materials used in medieval painting, especially in manuscript illumination. It is known through a miscellaneous Hebrew manuscript kept at the *Biblioteca Palatina de Parma* (Ms. Parma 1959) and was first published in 1928 (Blondheim).

Several editions and translations of LCFC were produced and several opinions on the authorship and date were advanced (see Cruz & Afonso, 2008), but the first detailed study about the work was only published in 1999 (Strolovitch). This study, which focused on language issues, was developed and incorporated into a doctoral thesis (Strolovitch, 2005), but until 2005, when two of us presented a communication to the Córdoba Symposium, neither the technological aspects raised by LCFC had been studied nor its historical and artistic framework had been significantly developed. In that communication, in addition to the necessary review of the literature, the date of the text was discussed based on the Portuguese historical context and the vocabulary used in connection with materials and techniques. We thus concluded that it was more pertinent to consider the text as a fifteenth century work, as some claimed, rather than a thirteenth century text, as was argued by others based on other texts contained in the same manuscript volume. Finally, we also discussed the characteristics and influences of the text, namely its Hebraic, Spanish and alchemic influences. Shortly after the symposium, we made significant progress on the main aspects discussed in our presentation, and due to the publication delay of the proceedings a more detailed study with new data was published in a journal (Cruz & Afonso, 2008). Therefore, it is no longer relevant to publish again the text that we had prepared for the proceedings of the symposium.

LCFC has been the subject of several studies since then, ours and by others, and significant progress has been made on some of the issues we had initially raised. Hence, it now seems appropriate to review the current state of research on LCFC. Among the new studies is a Master dissertation (Matos, 2011) supervised by one of us, somehow as a major development of the communication presented at the symposium, which partly carried out the literature review that we intend to do here. Therefore, this text incorporates parts of that dissertation and it comes signed by its author. Considering the theme of the symposium, this present review focuses on the fundamental issues related to the date and authorship of LCFC, its sources and technological aspects. Only studies published after 2005 have been considered, as earlier publications have already been identified in our first paper (Cruz & Afonso, 2008). Likewise, the historic and artistic contexts are not treated here either, since they were revised and developed in the cited dissertation (Matos, 2011).

The manuscript and its date

LCFC occupies folia 1 to 20 of the Ms. Parma 1959, a manuscript volume mostly written in Hebrew with a total of 221 folia. According to Débora Matos (2011), who directly examined the manuscript, this volume contains 20 philological units, although a smaller number of texts is mentioned in other studies. Traditionally, it was considered that LCFC was composed at Loulé, south of Portugal, around 1262, by Abraham Ibn Hayyim due to an ornamented signature with that name at the end of LCFC (fol. 20r). The same name, together with the place and the date (in an abbreviated form without the century), also appears in the colophon of another work (f. 195v), the *List of the oddly shaped letters found in an accurate Torah scroll*.

It has been stressed the homogeneity of the volume in terms of format, justification, number of lines, composition of books and decoration, which accounts for the association between the two texts, although in some cases the variety of calligraphy has been mentioned. Luís Afonso identified nine hands and concluded that the two texts were written or copied by two different writers (Afonso, 2010). Consequently, although it can not be ruled out the initial possibility of Ibn Hayyim being the author of the two mentioned texts, he could not have written them by his own hand and no relationship can be established between the monogram in the first book and the colophon in the other – and therefore ceases to exist any justification for the traditional dating and authorship attribution. However, Matos (2011), who analysed the palaeography of the entire volume, suggests that the handwriting differences between the two texts are not significant and their apparent difference must be understood otherwise. Both texts are written in a careful semi-cursive script, and even though LCFC seems to be more formal, she considers that there is no sufficient evidence to assume that they were copied by two different hands. She argues that the apparent differences should be understood in terms of modes of writing in Hebrew: semi-cursive script is used in texts dealing with para-textual elements relating to the copy and decoration of Hebrew sacred texts (with some justified exceptions) and for liturgical texts, whereas *halakhic* texts are written in a more current semi-cursive script. Consequently, she concludes that there is a direct relation between the nature of the texts included in the Ms. Parma 1959 and their modes of writing, and they seem to have been copied by the same person (Matos, 2011).

Moreover, Matos established that the ink used in both the ornamented signature at the end of LCFC and the main text is the same, currently a brownish ink, and not inks with different colour as it had been stated previously, which clearly indicates the simultaneous writing of both the signature and the text. In view of the entire manuscript volume Matos considers that Abraham Ibn Hayyim must not be the author of LCFC but rather its copyist, similarly to the majority of the texts included in the volume (Matos, 2011).

It was already known that the paper of the manuscript had unspecified watermarks in use between 1423 and 1488, but in the meanwhile new data about the paper arose. Inês Villela-Petit (2011)[1] reported that according to the information given to her four watermarks were found: a crown, a horn, a pair of scissors and a cross. The former corresponds to a model that was common in Europe precisely between 1423 and 1488, with strong similarities to watermarks in papers of Italian origin, more specifically from Piedmont, dated from between 1459 and 1473. The author also pointed the pair of scissors to be an exclusively Italian watermark, most likely from Genoa. None of these watermarks was detected in LCFC, only in some of the other texts that comprise the volume (Villela-Petit, 2011). Matos (2011), through the direct analysis of the manuscript, was able to confirm the crown (Fig. 1) and the pair of scissors (Fig. 2), yet the watermark that had been described as a horn was an ox head instead (Fig. 3). Also, in the place of cross appeared to be a column (Fig. 4). These differences partly result from the fact that some watermarks are cut and incomplete. Moreover, Matos, who noticed a greater number of folia with watermarks, detected one additional watermark – a ring with three stones (Fig. 5). She was equally able to establish that two folia of LCFC have watermarks (a crown in folia 1 and 2). Considering the distribution of the watermarks across the volume, Matos believes that the first 37 folia, corresponding to the first eight texts, were written in the same paper with a crown watermark, whereas from fol. 84 onwards the paper used is the one with the scissors watermark. All the other watermarks are found in one specific text in the middle of the volume, and their variety corresponds to the textual divisions (Matos, 2011).

1. Although published in 2011, this paper corresponds to a communication presented in a symposium held in 1999.

Fig. 1: Crown watermark on fol. 37 (Matos, 2011).

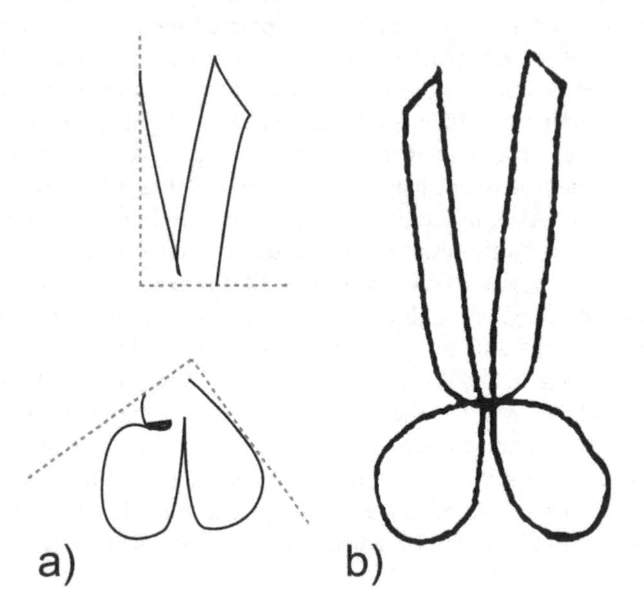

Fig. 2: a) Fragments of the scissors watermark on foll. 51 and 52; b) a similar watermark from Genoa. Redrawn from Matos (2011).

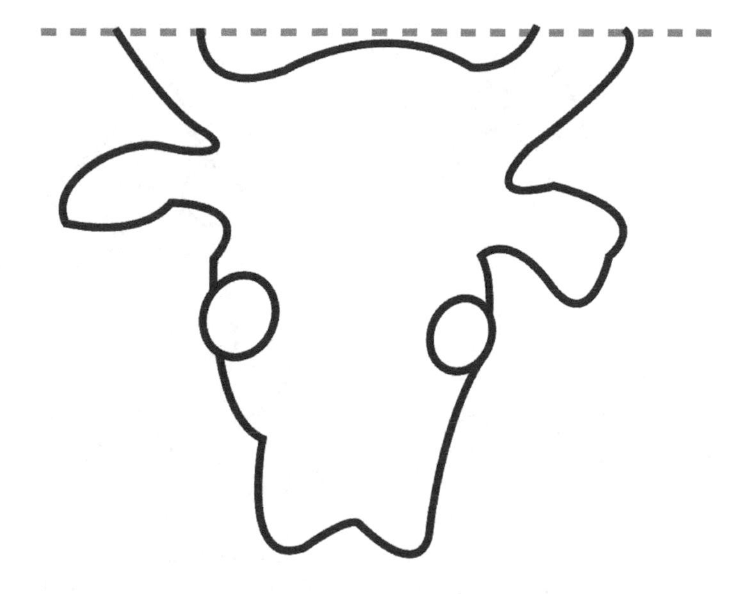

Fig. 3: Ox head watermark on fol. 44 (Matos, 2011).

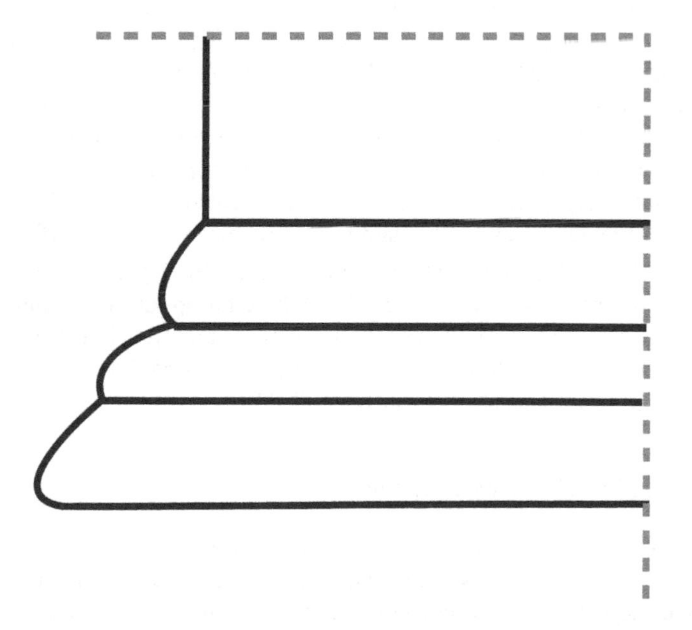

Fig. 4: Column watermark on foll. 73 and 74 (Matos, 2011).

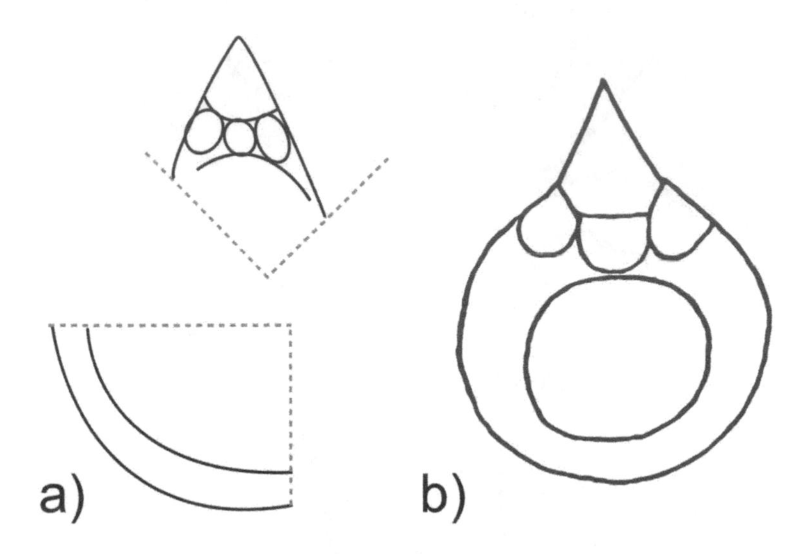

Fig. 5: a) Fragments of the ring watermark on foll. 45 and 41;
b) a similar watermark from Cologne. Redrawn from Matos (2011).

Regarding the date and origin of the watermarks, Matos (2011) agrees with Villela-Petit regarding the crown watermark, and found parallels of the several watermarks in papers from Italy, as is the case of the scissors (Genoa, 1448-1463) and the ox head (Genoa, 1462), and from Germany (Cologne, 1457) and the Low Countries (Utrecht, 1462) in the case of the ring. Based in these findings, Matos suggests that the manuscript was copied somewhere between 1423 and 1463 and that the year of 62 (without the century) on the colophon on the *List of the oddly shaped letters* most likely corresponds to the year 1462 – which, ultimately, may be the date of the manuscript (Matos, 2011).

In addition to the previously mentioned researches, the text of LCFC was printed two other times: based on his PhD dissertation, Devon Strolovitch published the transliteration (Strolovitch, 2010a) and an English translation (Strolovitch, 2010b). Moreover, there is a forthcoming French translation by Anne-Marie Quint and Michel Larroche (2009).

The text: philological issues and date

Aiming to clarify the doubts about the date of LCFC, António João Cruz and Luís Afonso (2008) explored the history of the vocabulary used to name materials, equipment and procedures related with the colours preparation or its use, as well as the history of the materials themselves. With this approach, they intended to avoid the formal changes (namely those related to the spelling) that could have occurred during any possible copy in order to discuss the date of the original text.

With respect to the vocabulary, they used a set of 122 words, many corresponding to materials or objects of common use, of which, according to the documentary corpus used by the best Portuguese dictionary with historical data, only 50 percent were known in a written form in the thirteenth century or before, while 20 percent had their first appearance in fifteenth century or later. Of the ten words documented only after the fifteenth century, nine correspond to terms of restricted use, and considering the probable reduced number of technical texts in the corpus, it is not unexpected that those words might already have been in use before. Therefore, they concluded that the words employed by LCFC suggest for the text a date not earlier than the fifteenth century (Cruz & Afonso, 2008).

Regarding the materials used in the recipes comprised in the LCFC, Cruz and Afonso pointed out that the use of paper for pigment storage, as mentioned in the text, in a situation where the paper did not have any advantage compared to other storage materials, could only be justified at a time when paper was a common material in Portugal, i.e. only in the fifteenth century or later. They equally noted that mosaic gold, a pigment for which two recipes are presented in LCFC, is not mentioned in any of the known technical treatises dated from the thirteenth century, being the oldest references from the fourteenth century. Similarly, they considered the references to gum Arabic and to turnsole (*catasol*) as more probable in the fourteenth and fifteenth centuries than in the thirteenth century. Consequently, they concluded that the materials mentioned in LCFC suggest a date not earlier than the fourteenth century for the original text (Cruz & Afonso, 2008).

In relation to the vocabulary, Ivo Castro (2010) drew attention to the fact that the date recorded in a dictionary for the first usage of a word is not the date on which that word appeared in that language, since before recorded in written form the word circulated orally, and it does not necessarily correspond to the oldest document but only to the oldest document known to the dictionary's author. Consequently, the dating done by Cruz and Afonso based on the set of 122 words should be regarded with reservations, and in any case should be considered as provisional. Castro also showed that the represented orthography through Hebrew script of many words of LCFC is characteristic of Old Portuguese (thirteenth and fourteenth centuries) and that these words did not circulate with those forms in the fifteenth century, except in the case of the copy of previously existing texts. In conclusion, Castro, considers that the fifteenth century paper, and the language of LCFC shows that its known text is a copy of older sources (Castro, 2010).

Regardless of the language issues, Afonso (2010) investigated the possibility of dating the text offered by the reference to the gold coins (*dobras*, *escudos* and *florins*) used for gilding. According to his research about the circulation of coins in medieval Portugal, the text cannot be earlier than 1266, when *escudos* were minted for the first time in Europe, but it is much more likely to have been written somewhere between early fourteenth century and mid-fifteenth century, the period when the three coins had significant circulation. However, Afonso considered that the LCFC comprises three parts which correspond to different sources (see below) and, hence, this dating does not apply to the whole book but only to the first part (where the three coins are mentioned) (Afonso, 2010).

The structure of the text and its sources

As a result of the analysis carried out on LCFC, Cruz and Afonso (2008) concluded that the text is a collection of recipes, organised in one introduction plus 45 chapters (including the missing chapters 22 and 23) compiled from several sources, and not an original coherent book, and for that reason they avoided the use of the term 'treatise' for it. The authors mentioned the lack of organisation, the employment of different systems of nomenclature for similar materials, the duplication of procedures and the use of different materials in identical situations. Furthermore, the diversity of significant influences found by them, namely alchemic, Castilian and Arabic, was also interpreted as an indication of the variety of sources.

Afonso (2010) took this analysis further and identified two incipits in the text, one in the beginning and another on chapter 25, which are an indication of the juxtaposition of at least two sources. Taking into account the recipes in terms of contents (materials, procedures, aims, etc.), as well as in terms of language (syntax, words, etc.), he concluded that the first 24 chapters can be split into two parts. Therefore, he considered the whole LCFC composed of three parts, each originated from different sources: part I (introduction and chapters 1 to 15), dealing with the preparation of pigments and with how to temper them, is characterised by a language with strong elements of an alchemic tradition; part II (chapters 16 to 24), dealing with recipes for dying objects made of wood and bone, has a stronger presence of Portuguese words of Arabic origin; and part III (chapters 25 to 45), consisting mainly of recipes for the mixture of pigments and dyes, is where the Castilianisms are more frequent (Afonso, 2010).

A similar division was proposed by Villela-Petit (2011): a first part (introduction and chapters 1 to 15) only about pigments; a second part (chapters 24 to 45) less ordered and with recipes about binders, glue, varnish, pigment mixtures and indications for use; and a very heterogeneous set of six recipes (chapters 16 to 21) related several activities. Nevertheless place the beginning of the second part in Chapter 24, she noted, as Afonso, that the first words of chapter 25 resemble a prologue.

Matos (2011) also addressed this issue related to the constituent parts of LCFC, as she found a mark (Fig. 6) that is repeated in various parts of the book. This led her to conclude that this may be a signal of textual division. Hence Matos proposes an alternative division of the text into four parts: part one (introduction and chapters 1 to 16), with some alchemic language; part two (chapters 17 to 23), the most coherent one, dealing exclusively with the dyeing of wood and bone, with a strong Arabic influence; part three (chapters 24 to 40), which deals mostly with colour tempering, shades and variations, and mixtures and mordants; part four (chapters 41 to 45), with no specific subject (Matos, 2011).

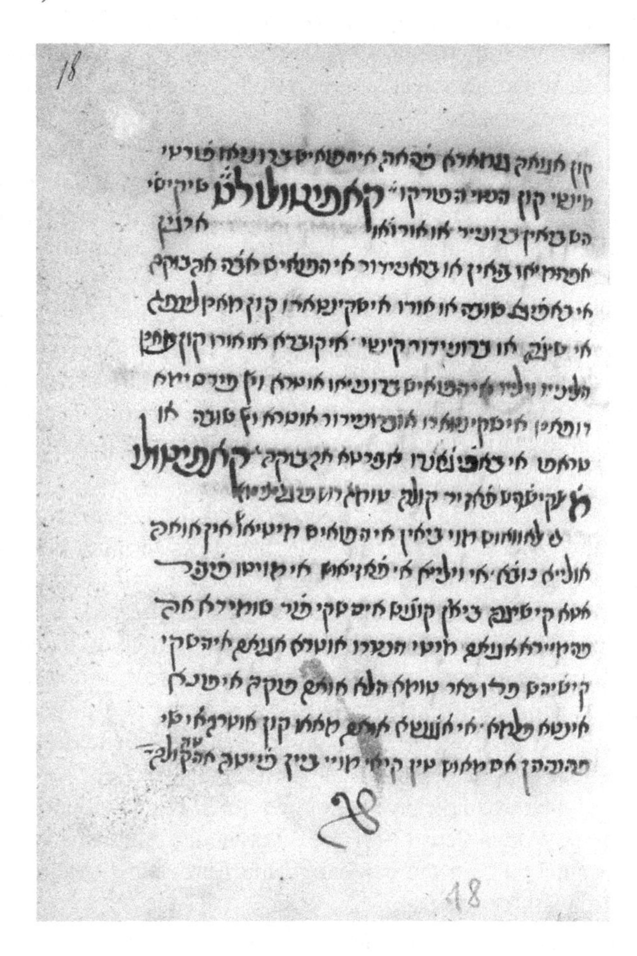

Fig. 6: Sign at the end of chapter 40 (Matos, 2011).

Regarding the specific sources compiled in LCFC, little progress has been made, although some affinities have been pointed out. Cruz (2010) searched for similarities for it in the *Mappae Clavicula*, a work referred in Portuguese medieval libraries, which is a source of many other manuscripts, but he found no connection. Yet he detected some similarities between LCFC and Jehan Le Begue treatise's *Experimenta de Coloribus* (1431), the *H490 Montpellier Manuscript* (1460-1480) and, to a lesser degree, the *Bolognese Manuscript* (first half of the fifteenth century). These similarities are manifested through the procedures' details on how to make some pigments, namely mosaic gold and silver blue, and through thc names attributed to certain pigments or pigments' raw materials. The recipes in question from the manuscript of Le Begue are part of a set of recipes collected by him in Italy, especially from a manuscript which, in turn, reproduced recipes of another manuscript of Fra. Dionysius of Milan. The second source is a miscellany, which particularly collects documents of medical and botanical nature, probably compiled by Juan de Celaya, master or bachelor of arts, of Salamanca. The last work, also known as *Segreti per Colori*, is a well-organised compilation of recipes, in Italian and Latin, which gives an account of common practices in Bologna. The contact points detected do not reflect the direct use of these works by LCFC, but only the existence of common sources, not identified, which eventually may be relatively remote (Cruz, 2010).

With a similar methodology, Villela-Petit (2011) found some lexical similarities between the pictorial conventions for the representation of clothing and fabric expressed in LCFC and the same conventions expressed in *De Coloribus et Mixtionibus*, a text dated from the twelfth century, but often copied up to the fourteenth century. However, there are also significant differences between the two texts.

With another methodology, that of comparing the general characteristics of LCFC with the general characteristics of the works forming the corpus of the medieval technical texts, Mark Clarke (2010) concluded that LCFC is a characteristic work of the fourteenth or fifteenth century because it is a systematic vernacular composition and not merely a compilation or translation of old material, but rather a new, locally composed, coherent, usable treatise, comprising in part some re-worked pre-existing material and in part newly composed material. Moreover, he considered that the manuscript may reflect fifteenth century practices, although it might also reflect somewhat earlier practices, and that it may have belonged to a Jewish illuminator (Clarke, 2010). Also for Villela-Petit (2011), the vernacular language is an important argument to put LCFC in the context of the European literature of the fourteenth and fifteenth centuries.

This conclusion agrees with the contextualisation of LCFC in the Portuguese history which led Cruz and Afonso (2008) to consider that a technical text on colours for illumination would have been useful in Portugal during the second half of the fifteenth century, coinciding with a large production of Hebrew manuscripts but not before, particularly in the thirteenth century.

Historical accurate reconstructions

Recently, there has been a strong interest in testing recipes from ancient texts through historical accurate reconstructions. In some cases, the obtained materials, besides being chemical and physical characterised, have been compared with those found in some contemporary works of recipes.

The recipe for vermillion recorded in LCFC is one of the tested recipes, which was done by Catarina Miguel and collaborators. From a theoretical point of view, the instructions stated in LCFC for the synthesis of that pigment from cinnabar and sulphur appeared to be one of the best found in a medieval text, although some important details are missing from the recipe. In practice, as a result of the complex mechanism involved in the synthesis, some of the experiments were carried out without success, but other experiments allowed to obtain vermillion that matched the reds employed in medieval manuscripts of Lorvão – an important Romanesque monastery centre of manuscript production and illumination in Portugal (Melo & Miguel, 2011, in press; Miguel *et al.* 2011, 2012).

In another successful experiment, a dark saturated green colour with a glassy appearance observed in the same manuscripts of Lorvão, a colour identified as bottle-green, was obtained from copper and vinegar, following instructions contained in LCFC (Miguel *et al.*, 2009). The preparation of red lead from lead white, as is described in LCFC, also provided a paint that likewise matched some colours of the Lorvão manuscripts (Miranda *et al.*, 2010). Some of the materials obtained in these successful experiments were precisely used as reference materials in a detailed analytical study of Lorvão manuscripts (Claro, 2009).

However, in other cases, the desired materials were not obtained or the materials have failed to reproduce the features observed in Portuguese medieval manuscripts, namely its colours. This happened, for example, with the "fine carmine" mentioned in chapters 13 and 14 of LCFC (Miranda *et al.*, 2010). The recipe for another lac, the brazilwood, presented in chapter 9 was also reproduced (Miguel *et al.*, 2009), but there seems to be no details available about the success of these experiments.

Regarding the other colours mentioned in LCFC, the work seems to be in progress and some preliminary remarks about the recipes or the processes that will likely be involved in the recipes were already presented (Miguel *et al.*, 2009).

Conclusion

If before the Córdoba symposium LCFC was generally considered a thirteenth century work or, at least, such date was plausible, the studies published meanwhile clearly suggest that this is a fifteenth century text, compiled from a variety of sources, some considerably older. As Clarke (2010) points out, if this manuscript is not the original compilation, it is a fifteenth century copy of a slightly earlier original. Be that as it may, LCFC is a valuable source for the knowledge of the medieval painting in the Iberian Peninsula, and it should be particularly considered in the light of Hebrew manuscript illumination in Portugal.

Bibliography

AFONSO, LUÍS U., "New developments in the study of O livro de como se fazem as cores das tintas," in L. U. Afonso (ed.), *The Materials of the Image. As Matérias da Imagem*, Lisboa, Cátedra de Estudos Sefarditas "Alberto Benveniste" da Universidade de Lisboa, 2010, pp. 3-27.

BLONDHEIM, D. S., "An Old Portuguese work on manuscript illumination," *Jewish Quarterly Review*, 19, 1928, pp. 97-135.

CASTRO, IVO DE, "Notas sobre a língua do Livro de como se fazen as cores (ms. Parma 1959)," in L. U. Afonso (ed.), *The Materials of the Image. As Matérias da Imagem*, Lisboa, Cátedra de Estudos Sefarditas "Alberto Benveniste" da Universidade de Lisboa, 2010, pp. 87-96.

CLARKE, MARK, "The context of the O livro de como se fazem as cores: late mediaeval artists' recipes books (14th-15th centuries)," in L. U. Afonso (ed.), *The Materials of the Image. As Matérias da Imagem*, Lisboa, Cátedra de Estudos Sefarditas "Alberto Benveniste" da Universidade de Lisboa, 2010, pp. 45-73.

CLARO, ANA LUÍSA DO VALE FONSECA, *An Interdisciplinary Approach to the Study of Colour in Portuguese Manuscript Illuminations*, PhD dissertation, Lisboa, Faculdade de Ciências e Tecnologia da Universidade Nova de Lisboa, 2009.

CRUZ, ANTÓNIO JOÃO, "Em busca da origem das cores de 'O Livro de Como se Fazem as Cores': sobre as fontes de um receituário português medieval de materiais e técnicas de pintura," in L. U. Afonso (ed.), *The Materials of the Image. As Matérias da Imagem*, Lisboa, Cátedra de Estudos Sefarditas "Alberto Benveniste" da Universidade de Lisboa, 2010, pp. 75-85.

CRUZ, ANTÓNIO JOÃO; AFONSO, LUÍS URBANO, "On the date and contents of a Portuguese medieval technical book on illumination: O livro de como se fazem as cores," *Medieval History Journal*, 11 (1), 2008, pp. 1-28.

MATOS, DÉBORA MARQUES DE, *The Ms. Parma 1959 in the Context of Portuguese Hebrew Illumination*, MA Dissertation, Lisboa, Faculdade de Letras da Universidade de Lisboa, 2011.

MELO, MARIA JOÃO; MIGUEL, CATARINA, "À volta de um vermelho. Apresentação de edição d'O livro de como se fazem as cores, sob o olhar da ciência e tecnologia," *Revista de História da Arte-Série W*, 1, 2011, pp. 290-293.

MELO, MARIA JOÃO; MIGUEL, CATARINA, "The making of vermilion in medieval Europe – historically accurate reconstructions from 'The book on how to make colours'," in S. Kroustallis and M. Del Egido (ed.), Fatto d'Archimia: History and Identification of Artificial Pigments, Madrid, IPCE, in press.

MIGUEL, CATARINA; CLARO, ANA; LOPES, JOÃO A.; MELO, MARIA JOÃO, "Copper pigments in medieval times: green, blue, greenish-blue or bluish-green?," in E. Hermens, J. H. Townsend (ed.), *Sources and Serendipity. Testimonies of Artists' Practice*, London, Archetype Publications, 2009, pp. 33-38.

MIGUEL, CATARINA; MIRANDA, ADELAIDE; LOPES, JOÃO A.; MELO, MARIA JOÃO; CLARKe, MARK, "A study in scarlet – vermilion red and colour paint formulations in medieval illuminations," in J. Bridgland (ed.), *Preprints of the ICOM-CC 16th Triennial Conference*, Lisbon, ICOM, 2011, paper 0107.

MIGUEL, CATARINA; MIRANDA, ADELAIDE; OLIVEIRA, MARISA; MELO, MARIA JOÃO; CLARKE, MARK, " 'The book on how one makes colours of all shades in order to illuminate books' revisited," in S. E. Green, J. H. Townsend (ed.), *The Artist's Process: Technology and Interpretation*, London, Archetype Publications, 2012, pp. 60-66.

MIRANDA, ADELAIDE; LEMOS, ANA; MIGUEL, CATARINA; MELO, MARIA JOÃO, "On Wings of Blue: the history, materials and techniques of the Book of Birds in Portuguese scriptoria," in L. U. Afonso (ed.), *The Materials of the Image. As Matérias da Imagem*, Lisboa, Cátedra de Estudos Sefarditas "Alberto Benveniste" da Universidade de Lisboa, 2010, pp. 171-184.

QUINT, ANNE-MARIE, personal comunication, 19-10-2009.

STROLOVITCH, DEVON L., "Selections from a Portuguese Treatise in Hebrew Script: Livro de Como se Fazen as Cores," *Cornell Working Papers in Linguistics*, 17, 1999, pp. 185-196.

STROLOVITCH, DEVON L., *Old Portuguese in Hebrew Script: Convention, Contact, and Convivência*, PhD Dissertation, Cornell, Cornell University, 2005.

STROLOVITCH, DEVON L., "O libro de komo se fazen as cores das tintas todas (Transliteration)," in L. U. Afonso (ed.), *The Materials of the Image. As Matérias da Imagem*, Lisboa, Cátedra de Estudos Sefarditas "Alberto Benveniste" da Universidade de Lisboa, 2010a, pp. 213-223.

STROLOVITCH, DEVON L., "O libro de komo se fazen as cores das tintas todas (Translation)," in L. U. Afonso (ed.), *The Materials of the Image. As Matérias da Imagem*, Lisboa, Cátedra de Estudos Sefarditas "Alberto Benveniste" da Universidade de Lisboa, 2010b, pp. 225-236.

VILLELA-PETIT, INÊS, "Les Recettes pour l'enluminure. Do Livro judaico-português de como se fazem as cores," *Medievalista*, 9, 2011. http://www2.fcsh.unl.pt/iem/medievalista/MEDIEVALISTA9/petit9005.html

Recipes for *Iniziali Fligranate* in Manuscripts: A Separate Tradition

Arie Wallert

The Biblioteca Comunale in Siena holds a very interesting manuscript. The major part of the manuscript, Ms I.II.19, consists of a rather common religious text, but the last part of this manuscript contains a collection of recipes for the illuminator's craft. This section concludes with the remark that both the confessional, and the recipes to make more colours were *Scripto per mano di me, Ambruogio di Ser Pietro da Siena, nel Castello di Chiusure. Cominiato a di 13 d'aprile 1462 e finito a di 18 giugno anno sopradecto.*[1] It is rather exceptional to find a collection of recipes with the name of the scribe, a date, and a precise geographical indication.[2]

The Castello di Chiusure belonged since the middle of the thirteenth century to the Tolomei, one of the most important families in Siena. It was on this family property, that Bernardo Tolomei founded in 1333, together with Ambrogio Piccolomini and Patrizio Patrizi the Monasterium Monte Oliveto de Chiusure.[3]

More than hundred years later, in 1457 the Abbazia di Monte Oliveto started an ambitious project for the making of lavishly illuminated Choral Books, liturgical books that served to the core activity in the Benedictine monastic existence: the *officium divinum.*[4] Ritual prayer at regular times each

1. On fols. 1r-97v is a: 'Libro della confessione in volgare per fratrem Antoninum de Florentia ordinis predicatorum'. The last, tenth quire of the manuscript has on fols. 99r-106r, the: 'Ricepte da fare piu colori'. Both texts are written in the same clear hand. Organised as 1-1/0/1-1/0. f. 102. 29 rr.: 28<82>30] * [35<143>39] J.M.M. Hermans and G.C. Huisman, *De Descriptione Codicum*, Groningen 1981, pp. 34bis-34sexies. Although no simili were found, the watermarks in this manuscript resemble nrs 7386-7389 in C.M. Briquet *Les filigranes, Dictionnaire historique des filigranes de papier*, Paris, 1907, which is consistent with the date and origin from the Tuscan area.

2. Manuscript found by reference in: U. Forni, *Manuale del Pittore Restauratore*, Florence, 1866, p. 86-87. F. Brunello, *De Arte Illuminandi; e altri trattati sulla tecnica della miniatura medievale*, Vicenza, 1975, p. 97, note 2.

3. P.M. Lugano, *Origine e primordi dell' Ordine di Monte Oliveto*, Florence, 1904.

4. T. Leccisotti, "Il Missale monasticum secundum morem et ritum Casinensis congregationis alias sanctae Justinae," *Miscellanea Giovanni Mercati*, vol V: Storia ecclesiastica-diritto, Rome, 1946, pp. 367-375.

day of the year. The contents of these prayers were different each day depending on the position of that day in the liturgical year; the liturgical year being based on two parallel running cycles: the *Temporale* and the *Sanctorale*. Liturgical books for the Olivetan monks were the missale, graduale, antifonarium, and psalterium, each having its own specific texts and musical notations. Since prayers were sung by all the monks together from the same text, the books that they used had to be legible from some distance. Choral books are big books.

The illumination in them has a significant role; not just for embellishment but also to clarify the hierarchy in the structure of the texts. Illuminations served to indicate important passages and mark the places were new moments in the cycle of prayers would begin. Each verse of the psalms usually has a capital initial in a geometric pattern with fine penwork in different colours, (because of the intricate patterns also described as filigrane initials: iniziali filigranate). The initial of the whole psalm is usually larger and done in brushwork and gold. Different times in the liturgical year are often indicated by a large initial depicting the subject of the psalm. For the making of these illuminations, the Olivetan community attracted the most important specialised artists of the time: Liberale da Verona, Gerolamo da Cremona, Venturino Mercati, Sano di Pietro, Francesco di Giorgio.[5] Choral books were not just big books; they were very beautiful big books.

It has been convincingly demonstrated that the recipes in Ambruogio's manuscript have been used for the making of illuminations in the Monte Oliveto choral books. In the choral books materials, production processes and techniques have been identified that match precisely with descriptions in Ambruogios's recipes.[6]

Not only the Olivetan monastery needed choral books. In fifteenth century Tuscany there was generally an enormous upsurge in the production of these liturgical implements.[7] Since all these projects have similar needs, it did not come as a complete surprise that more recipe books for illuminators can be found. A collection of recipes closely related to those of Ambruogio begins

5. After their work for Monte Oliveto Maggiore, many of these illuminators moved on to Siena, to produce the stunningly beautiful choral books of Siena Cathedral. H.J. Eberhardt, "Sull' attività senese di Liberale da Verona, Girolamo da Cremona, Venturino da Milano, Giovanni da Udine e prete Carlo da Venecia," *Atti del II Congresso di Storia della Miniatura Italiana, Cortona 1982*, Florence, 1985, pp. 415-434.

6. A. Wallert, *Kookboeken en Koorboeken*, Groningen, 1991. This has been done with various scientific methods, such as polarised light microscopy (PLM), scanning electron microscopy with energy dispersive spectrometry (SEM-EDX), x-ray diffraction (XRD), microchemical analyses (MCA), thin layer chromatography (TLC), fluorescence spectrometry (FS), and direct temperature resolved mass spectrometry (DTMS).

7. *Codici Liturgici Miniati dei Benedettini in Toscana*, cat., coordinamento delle ricerche Maria Grazia Ciardi Duprè dal Poggetto (Centro d'Incontro della Certosa di Firenze), Florence, 1982.

with: *Incipiunt experientie magni scriptoris f. Bartolomeo de Senis, ordinis cartusiensis.*[8]

Both recipe collections were written in a monastic environment in Sienese territory, in approximately the same time, by a scribe of Sienese origin. The recipes in both manuscripts deal with the practice of manuscript illumination and describe various aspects of the craft, such as the making of inks, the preparation of parchment, the application of metal foil and the preparation and application of inorganic pigments and various translucent organic colorants.

Recipe texts often were compilations of traditional alchemical texts of sometimes Arabic origin, such as the texts of Geber (Jabir ibn Hayyan), or from the *Liber Sacerdotum*, or from western texts from Arnoldus de Villanova with recipes from the *Mappae Calvicula*, from Eraclius, from Theophilus or from the *Compositiones*. Of course, in none of these compilations the original source for the recipes is indicated.

The ninth century *Compositiones ad tingenda musiva* contains recipes that stem from fourth century Alexandrian papyri. The earliest extant manuscripts of that other famous body of recipes; the so called *Mappae Clavicula*, is a tenth century manuscript in Sélestat (Sélestat, Ms 17). This text already contains almost all of the recipes of the earlier *Compositiones*. The second earliest source of the *Mappae Clavicula*, a twelfth century text, known as the Philips-Corning manuscript is found to contain most of the *Compositiones* recipes and the Sélestat recipes. Transcripts of transcripts of transcripts of *Mappae Clavicula* recipes in numerous combinations can be found in more than eighty manuscripts dating from the ninth to fifteenth centuries.[9]

A good example of the meeting of recipe texts from several sources can be found in the manuscript of yet another monk entitled: *Liber Magistri Petri de Sancto Audemaro de Coloribus Faciendis.*[10] This Pierre de Saint Omer (in Dutch: Sint Omaars) came from the Flemish part of France as is indicated from his name, his use of language and the names of various pigments he described. Various recipes in his *Liber de Coloribus* are literal repetitions from the *Mappae Clavicula*, from Theophilus' *Schedula* and a few other sources. For one of these sources we may think of a late fourteenth-century *Liber de Coloribus*,

8. Siena Biblioteca Comunale, Ms L.XI.41, fols. 34v-38v. Explicit: 'Finite le ricepte di donno Bartolomeo da Siena dell' Ordine di Certosa, et quale e grandissimo miniature e scriptore, le quail ricepte forno tucte provate'. Bastarda, Organised as 1-1/0/1-1/0. f. 36. 32 rr.: 14<105>18]* [25<145>29], see Hermans and Huisman, *op. cit.*, (note 1), pp. 34bis-34sexies. Also this manuscript could be placed in Sienese territory. On fol. 27r the typically Sienese saints Ansanus and Savinus are mentioned. Below the calendar a date of 1437 has been written. Manuscript found by reference in: A. Lisini, *De la Prattica di Comporre finestre a Vetri Colorati, Trattatello del secolo XV, edito per la prima volta*, (Miscellanea Senese), Siena, 1885, p. 12.

9. R.P. Johnson, "Some manuscripts of the *Mappae Clavicula*," *Speculum*, 10, 1935, pp. 72-81.; R.P. Johnson, "Some Continental Manuscripts of the *Mappae Clavicula*," *Speculum*, 12, 1937, pp. 84-103. W. Eamon, *Science and the Secrets of Nature*, Princeton, New Jersey, 1994, pp. 30-35.

10. Merrifield, *Original Treatises on the Arts of Painting*, London, 1849, vol. I, pp. 112-165.

now living in the British Library.[11] And also this source is in its turn a compilation of recipes from other sources. Isolated individual handbooks hardly ever occur.[12]

For a short moment it would seem that the famous Strasburg manuscript would constitute such an individual "pure" source. This manuscript was known as the oldest known handbook on painting in the German area. Ernst Berger published in 1897 its most interesting recipes after a transcript ordered by Eastlake some fifty years earlier.[13] The original manuscript itself had gone lost as result of the German-French war, and the ensuing fire of the Strasburg library in 1870. Until recently no precedents for the Strasburg- text could be found. Recent textual analyses and comparison with recipes from Southern Germany in Amberg (Cod. 77), Munich (Cgm. 821) and in the Leiden University Library (Ms Voss. Chym. Oct. 6) showed that also these recipes stand within a tradition and can be defined as a group.[14]

The relationships between such technical texts can be defined by comparison of individual recipes. This is also the case with the texts of Ambruogio di Ser Piero and Bartolomeo da Siena. Both texts appeared to have striking similarities. The end of Ambruogio's recipe IX describing the improvement on a poor quality azurite, reads quite similar to that of Bartolomeo's recipe XI.[15] Bartolomeo's recipe XIV: 'El modo da dividare et assotigliare l' azurro de la Magnia', is an almost literal repetition of Ambruogio's recipe X. Relationships with other treatises could also be found. A good example is in Ambrogio's recipe XV, not just describing the same procedure as in Bartolomeo's recipe XXIV, but can also be found back in unpublished manuscripts in the Biblioteca Casanatense in Rome (Ms 1793, fol. 13 v) and in the Biblioteca Nazionale in Florence (Ms Palatina 916, fol. 110r). The same recipe could also be found back in the well-known Bolognese manuscript *Segreti per Colori*, published by Merrifield.[16] Recipe 80 from the Bolognese *Segreti* recurs almost literally in Ambruogio's recipe XXIII. And the same recipe can also be found in Ms 1793 in the Biblioteca Casanatense (fol. 12r) and in an equally unpublished manuscript in the Bodleian Library in Oxford (Ms Canon. Misc. 128, fol. 21r).

11. D.V. Thompson, "Liber de coloribus illuminatorum siue pictorum from Sloane Ms. No. 1754," *Speculum*, I, 1926, pp. 280-307.

12. Rare exceptions are a recipe text in the Biblioteca Capitolare in Ivrea (Ms. 54, LXXXVII), and one in the Leiden university library (MS. Voss. Lat. 4° 33). P. Giacosa, 'Un Ricettario del secolo XI, esistente nell' Archivio Capitolare d' Ivrea" , *Memorie della R. Accademia delle Scienze di Torino*, 37, 1886, pp. 3-22., B. Bischoff, 'Farbrezepte', *Anecdota Novissima ; Texte des vierten bis sechzehnten Jahrhunderts*, Stuttgart, 1984, pp. 219-222.

13. E. Berger, *Quellen und Technik der Fresko-, Oel- und Tempera-Malerei des Mittelalters*, vol. III in the series *Beiträge zur Entwicklungsgeschichte der Maltechnik*, 1897, pp. 143-176.

14. E. Ploss, "Das Amberger Malerbüchlein, Zur Verwamtschaft der spätmittelhochdeustche Farbrezepte," *Festschrift für Hermann Heimpel*, Göttingen, 1971, pp. 693-702. A. Wallert, 'Aquae Conficiendae ad Temperandos omnes Colores; Manuscript in the Leiden University Library Voss. Chym. Oct 6', *Technologia Artis*, 3, 1993, pp. 134-146.

15. Numbering of recipes as done in their critical edition and translation (Dutch) in A.Wallert, *op. cit.*, note 6.

16. Merrifield, *op. cit.*, (note 10), vol. II, p. 469.

Where Ambruogio writes: *Tolli fiori di guado e amido di grano e impasta in sieme con l'orina e aceto, e fane un migliaccio ...*; the manuscript in Oxford reads: *Tolli li fiore di guado a modo de grano, impastalo insieme cum urina et cum aceto et fa una pasta a modo de una fugaza ....* The Casanatense manuscript stays slightly closer to Ambruogio's recipe: *Tolli fiore di guato amido di grano. Impasta insieme con urina e con aceto e fane uno miglaco ....* Of these recipes the *Segreti* shows the strongest deviation: *Ahvve fiore de guato et amido bene candido e impasta insiemj cum orina preparatae stillata per filtro et cum aceto bianco forte tanto de luno quanto del altro e fane uno migliacio ....*[17] Ambruogio's recipe XII is literally repeated in Ms 1793 of the Biblioteca Casanatense (fol. 12v-13r), wile there are strong technical similarities with Ms α.T.7.3 of the Biblioteca Estense in Modena (fol. 10v). The description in the Bolognese *Segreti* on how to recognize good ultramarine on the basis of particle size shows strong similarities with the *Experientie*. Recipe 263 in the chapter *Ad lapides anullorum componendos* of the Bolognese manuscript recurs almost literally in Ambruogio's recipe XVI. Both recipes appear to be translations in Italian from Pieter van St. Omaars Latin text: *De Coloribus faciendis*, and have in some aspects resemblance to a recipe in Ms Canon Misc.128 (fol. 13rv). The specifications for the quality of oak galls and vitriol for the making of ink in Ambruogio's prescription XVIII must stem from the same source as the Bolognese recipes 389 and 390: *A cognosciare la bona galla*, and *A cognosciare la bona vitriolo*; and were also found in recipe III in Bartolomeo's manuscript.[18] A recipe in Ambruogio's text describes the making of a liquid to remove ink – letters or spots – without scraping: *Tolle sal nitro e vetriuolo romano di ciascuno un libra, e sia chiaro, poi tolle una spogna e bagna la di questa aqua. E mecte la in sulla carta e leveranne le lecture.* There is not very much deviation in the Manuscript in the Biblioteca Casanatense: *Tolli sal nitro una lb. vetruolo una lb. a falo stilare e poi togli una spungia e bagnia la in detta aqua. E frega in sulla charta a levare le lettere.* Recipe XXI that Bartolomeo gives for the same procedure clearly stems from the same source: *Piglia sale nitro et vetriuolo romano, tanto del'uno quanto dell'altro et fallo distillare. Poi tolle una spogna pichola et intignela in quella aqua distillate e fragala in sul quelle lettere che voi levare ....*

There is a whole series of other similarities. Ambruogio's recipe XX could be connected to a recipe in Florence Biblioteca Riccardiana Ms 1243, fol. 26v; Ambruogio's recipe XXII with that of Bartolomeo XXV and Modena Biblioteca Estense Ms α.T.7.3, fol. 5r; Ambruogio's recipe XXIV with Rome Biblioteca Casanatense Ms 1793 fol. 10v and with Oxford Bodleian Library Ms Canon. Misc. 128, fol. 2v. Bartolomeo's recipe XXII shows strong resemblance to a recipe in Florence Ms Palatina 916, fol. 111r.

17. *Idem*, p. 417.
18. *Idem*, p. 389-390.

The relationships between these sources are complicated and problematic. Occasionally, we find similar prescriptions, but in manuscripts that otherwise do not have very much in common. Merrifield pointed for the Bolgnese *Segreti* to a use of words that could be characteristic for several, mainly Lombardic, Northern Italian dialects; but also words from the Venetian dialect occur. Both manuscripts in the Biblioteca Comunale in Siena, the *Ricepte* from 1462 and the *Experientie* from 1437 share orthography and use of words and, based on other texts in the manuscripts and names of the scribes, appear to be of Sienese origin. Similar diversity holds for the other manuscripts that are related tot eh Sienese group. Ms Palatina in the Biblioteca Nazionale of Florence dates from 1460, according to an inscription on fol. 6v: *1460 a dì 31 di maggio, Questo libro fu di Johani di Simone Biffoli e fu iscripo a Ciptà di Castello*; and manuscript 1793 of the Biblioteca Casanatense is dated 1422 and is signed on the name of a Simone di Monte Dante. The precise origin and exact dating of Oxford Bodleian Library Ms Canon. Misc. 128, Florence Biblioteca Riccardiana Ms 1243, and Florence Biblioteca Laurenziana Ms Ashburnhamia 349 cannot be given. Not anything more precise than: "Italian; fifteenth century." Modena, Biblioteca Estense, Ms α.T.7.3, has on fol. 104 the inscription *Yo Joannes de Fossato scripsi hoc lybro anno 1441*, and may come from the Emilia.

Thompsons remark that Ambruogio's texts could be a "work of original composition based immediately upon workshop practice," suggesting them to consist of unique notes that Ambruogio directly would have made in the studio wile practicing his craft, does not seem to hold ground.[19] These are not lab notes. There is no immediacy here. Rather, the relations between individual recipes are so close that both collections must stem from the same origin. Both stem from another, earlier, unknown text.

Yet they are not related to the well-established *Mappae Clavicula* tradition, texts from Theophilus' *Schedula de Diversarum artium*, the Eraclius texts, or for that matter, texts that stem from the *Liber de Coloribus*.

Ambruogio's and Bartolomeo's texts rather appear to be related to a very specific group of technical treatises comprising a number of little-known, mostly unpublished recipe texts devoted to the illuminator's craft.[20] Both texts deal exclusively with the recipes that are of interest for the maker of iniziali filigranate. Recipes that do not seem useful for that particular purpose have

19. D.V. Thompson, "The 'Ricepte daffare più colori' of Ambrio di Ser Pietro da Siena," *Archeion*, 15, 1933, p. 342.

20. In Florence, Ms Magliabecchiana CL XV 8 bis; Ms Magliabecchiana CL XVI 80 bis; Ms Palatina850; and Ms Palatina 951 of the Biblioteca Nazionale, Ms R. 1246 and Ms R. 1247 of the Biblioteca Riccardiana, as well as Ms Ashburnhamia 190 and Ms Ashburnhamia 349 of the Biblioteca Medicea Laurenziana provide useful information and sound basis for comparison of techniques. In Rome this type of information can be found in Biblioteca Casanatense Ms 1793 and Ms 2265, in London in the British Library, Ms Sloane 122, Ms Sloane 345, Ms Sloane 2135 and Ms Sloane 416 (so called Venetian manuscript), in Oxford, in Bodleian Library Ms Raoul D. 251, and in Modena, Biblioteca Estense, in Ms γ.O.6.15, and Ms α.O.9.13.

been left out. Clearly, these recipes were collected with a specific purpose in mind. These recipes have by scientific examination been demonstrated to be used in the production of penwork initials for liturgical books.[21] Some of the recipes in the two Sienese manuscripts recur in slightly different wording in other manuscripts. The picture that appears from comparison with these different manuscripts is very complex and so diffuse, that a real stemma could not be made. Relationships of passages in the two manuscripts with certain manuscripts in the Roman Biblioteca Casanatense, the Modena Biblioteca Estense, the Bolognese Biblioteca Universitaria, the Oxford Bodleian Library, and the Florentine Biblioteca Nazionale, however, do indicate the existence of a distinctly separate tradition. The *Ricepte daffare piu colori* and the *Experientie*, though remote from the origin, seem to form the most recognisable and uncontaminated core of that separate tradition.

21. *Wallert, op. cit.*, note 6, pp. 124-166.

Verdigris ink. Compilation and interpretation of recipes extracted from medieval manuscripts

Natalia Sancho Cubino/Margarita San Andrés Moya

Introduction

Verdigris is one of the green pigments that has been most widely used in the art world, although this use has been shared with other ends, amongst those the dyeing of certain materials, the elaboration of inks and some uses of a therapeutic nature. It originally had a medicinal use; whilst Dioscorides states how it was a very useful eyewash and a remedy for gums and calluses,[1] Pliny explains how it was used to cure various eye problems, mouth ulcers, fistulas and lesions caused by leprosy.[2] This therapeutic application lasted over the centuries. Specifically in Spain, and within this medical context, it can be seen in some authors. In 1616, Sorapan de Rieros, in his *Medicina Española contenida en proverbios vulgares* [Spanish Medicine in Popular Proverbs], states how it could be mixed with egg, sugar and alum in order to make an ointment to "clarify the sight."[3]

Depending on the time and place of utilization, verdigris has had various names.[4] Some of the most widely-used Latin terms have been *viridis cupri*, *viridis eris*, *viride rami*, *aerugo*, *viride salsum*, *viride Grecum*, *viride Hispanicum* or, simply, *viride*. In Spanish, it is preferably called *cardenillo* or *verdete*, also being known as *verdet* in some regions of the country, such as Valencia, as Covarrubias tells us in the 17th century in his *Tesoro de la lengua castellana*

1. P. Dioscorides: *Acerca de la materia medicinal y de los venenos mortíferos*, (original title *De materia medica*, 1st C., trans. into Spanish from Greek and com. by Andrés de Laguna, Salamanca, 1566, current edition with pres. by Agustín Albarracín), Ed. Arte y Bibliofilia, Madrid, 1983, Chapter LI, p. 375.

2. K.C. Bailey: *The Elder Pliny's Chapters on Chemical Subjects*, Parts I, II, Edward Arnold & Co, London, 1929, Part I, XXXIV, 110, 114, 115, 116, pp. 40-43.

3. I. Sorapan de Rieros: *Medicina Española contenida en Proverbios vulgares de nuestra Lengua*, Martín Fernández Zambrano, Granada, 1st ed. 1616.

4. M. San Andrés, N. Sancho, S. Santos, J.M. de la Roja: "Verdigrís. Terminología y Recetas de Preparación," in *Fatto d'Archimia. Historia e Identificación de Pigmentos Artificiales*, Ed. Secretaría General Técnica. Subdirección General de Publicaciones, Información y Documentación, Ministerio de Cultura, Madrid, 2011, pp. 199-236.

o española [Treasure of the Spanish Language].[5] Gual Camarena also uses the terms *verdeth, uerdet* or *verdete* as synonyms of *cardenillo.*[6]

Other words commonly used today are *verderame*, in Italian, *vert de gris*, in French, and *Grünspan*, in German.

Verdigris is a pigment of synthetic origin, whose process of obtaining was known since Antiquity; there are many recipes for making it and they appear in many texts through the ages, going from Classical Antiquity to the early 20th century.[7] These documentary sources are: texts on natural history and medical subjects, alchemy texts, recipe books, books of secrets and art treatises in general. The ingredients mentioned can vary from some recipes to others, although the most commonly recommended ones are copper and vinegar.[8] In some cases, instead of vinegar, the use of other by-products of wine-making is specified, such as: wine marc, wine lees and tartar. Other ingredients may be added, for example: soap, honey and common salt. Instead of vinegar, some texts also make mention of the use of urine, on occasions mixed with vinegar, adding other substances such as sal ammoniac (ammonium chloride – NH_4Cl –).

This great variety of recipes also leads to certain differences concerning the chemical composition of the pigments obtained[9] and therefore also regarding their physical properties, amongst others, their colour.[10]

5. S. de Covarrubias: *Tesoro de la lengua castellana o española*, (ed. by Felipe C. R. Maldonado, rev. by Manuel Camarero), Ed. Castalia, Madrid, 2nd ed. 1995, p. 272.

6. M. Gual Camarena: *Vocabulario del comercio medieval*, Ed. El Albir, Barcelona, 1976, p. 448.

7. C. Lombardi, M. Mariconda, S. Salvatori, F. R. Minieri, U. Santamaria, R. Valori-Oiazza: "Experimental study on the production and use of artificial blue and green pigments (copper acetates and carbonates) from Antiquity to the sixteenth century," *Art et Chimie. La Couleur*, CNRS editions, Paris, 2000, pp. 31-37; D. Scott, Y. Taniguchi, E. Koseto: "The verisimilitude of verdigris: a review of the copper carboxylates," *Reviews in Conservation*, 2, 2001, pp. 73-91; S. Santos Gómez, M. San Andrés Moya, J. L. Baldonedo Rodríguez, A. Rodríquez Muñoz, J. M. de la Roja, V. García Baonza: "Procesos de obtención del verdigris. Revisión y reproducción de antiguas recetas. Primeros resultados," *I Congreso del GEIIC. Conservación del Patrimonio. Evolución y Nuevas Perspectivas,* Valencia, 2002, pp. 383-388; S. Santos Gómez, M. San Andrés Moya, J. L. Baldonedo Rodríguez, A. Rodríquez Muñoz: "Recetas de preparación del verdigris. Resultados preliminares de la obtención de la variedad conocida como *viride salsum*," *Pátina*, 12, 2003, pp. 41-52; M. San Andrés *et al., op. cit.*, 2011.

8. N. Sancho Cubino: *Revisión y reproducción de los antiguos métodos de obtención de los pigmentos de cobre.* Research project presented for the Advanced Studies Diploma [Diploma de Estudios Avanzados-DEA]. Directed by M. San Andrés Moya. Faculty of Fine Arts, Department of Painting (Painting-Restoration), Complutense University of Madrid, 2004.

9. M. San Andrés, J. M. de la Roja, S. Santos, N. Sancho: "Patrones de identificación del verdigrís. Elaboración a partir de la reproducción de recetas antiguas," in *Fatto d'Archimia. Historia e Identificación de Pigmentos Artificiales*, Ed. Secretaría General Técnica. Subdirección General de Publicaciones, Información y Documentación, Ministerio de Cultura, Madrid, 2011, pp. 237-260.

10. N. Sancho, S. Santos, J. M. de la Roja, M. San Andrés: "Variación cromática del verdigris en función de su composición química," *Optica Pura y Aplicada*, 37, 2004, pp. 119-123; J. M. de la Roja, M. San Andrés, N. Sancho, S. Santos: "Variations in the colorimetric characteristics of verdigris films depending on the process used to produce the pigment and the type of binding agent used in applying it," *Color Research and Application*, 32, (5), 2007, pp. 414-423; M. San Andrés, J. M. de la Roja, N. Sancho, V. G. Baonza: "Verdigris, a Pigment with Different Hues. Relation between Chemical Composition and Colour," *Lasmac 2009 & IMRC09. Proceedings of the 2nd Latin-American Symposium on Physical and Chemical Methods in Archaeology, Art and Cultural Heritage Conservation*, Ed. Universidad Nacional Autónoma de México (UNAM), Universidad Autónoma de Campeche (UAC), Instituto Nacional de Antropología e Historia (INAH), México, 2010, pp. 4-9.

This pigment is made by the corrosion of copper or its alloys (brass or bronze) due to the reaction of the acetic acid content of vinegar (4-5 %). As a result of this process, copper acetates are formed with different degrees of hydration and/or basicity. The way vinegar reacts on the copper sheet: due to immersion of the sheet in vinegar or the action of its vapours on the surface of the sheet, influences the morphology of the particles and their degree of crystallization[11]. It is also very important the time of actuation of vinegar (15 days or several months).[12] However, when other additives are present, the composition of the pigment is notably affected. For example, when one of the ingredients used is a chloride (common salt – NaCl – or sal ammoniac – NH_4Cl –), apart from hydrated copper acetates, chlorides such as tolbachite ($CuCl_2$) and atacamite [$Cu_2Cl(OH)_3$] are formed.[13]

Nevertheless, in the old texts in which information appears regarding the materials used as pigments, the terms used for verdigris, in general, do not refer to the different varieties obtained according to the recipe used. The *Ms. of Heraclius,*[14] *Treatise of Theophilus,*[15] *Pietro di Sant Audemar Manuscript,*[16] and *Tractatus de coloribus from Munich*[17] are an exception. In these texts the varieties *viridis color cum sale* and *viride salsum* are mentioned, whose name suggests its synthesis from salt (common salt or sal ammoniac). Just as reflected in the treatises, this variety was quite frequent. To this respect, in many of the recipes connected with the preparation of verdigris, the use of common salt or of other types of salt is prescribed;[18] examples of this can be found in the text by Dioscorides (1st C.), which specifies the use of common

11. S. Santos Gómez, *et al., op. cit.*, 2002.

12. M. San Andrés, J. M. de la Roja, V. G. Baonza, N. Sancho: "Verdigris pigment: a mixture of compounds. Input from Raman spectroscopy," *Journal of Raman Spectroscopy*, 41, (11), 2010, pp. 1178-1186.

13. S. Santos Gómez, *et al., op. cit.*, 2003; J. M. de la Roja, M. San Andrés, N. Sancho, S. Santos, V. G. Baonza: "Varieties of verdigris containing copper(II) hydroxychlorides and other compounds: preliminary results," *Lacona IX. Lasers in the Conservation of Artworks* (London, 7-11 September, 2011), London, 2012, (in press).

14. Heraclius: "De coloribus et artibus romanorum" (*Manuscript of Heraclius*), 10th-12th C., in M. P. Merrifield, *Medieval and Renaissance Treatises on the Arts of Painting: Original Texts with English Translations*. Vol. 1. Dover Publications, Inc., New York, 1999, rec. XXXVII, pp. 236-237.

15. J. G. Hawthorne, C. S. Smith: *"On Divers Arts" of Theophilus. The Foremost Medieval Treatise on Painting, Glassmaking and Metalwork (11th-12th C.)* (original title *Schedula Diversarium Artium*), translated into English and notes by J. G. Hawthorne and C. S. Smith, Dover Publications, New York, 1979, Chap.1, p. 15.

16. Pietro di Sant Audemar: "Liber magistri Petri de Sancto Audemaro de coloribus faciendis," 13th-14th C., in M. P. Merrifield, *Medieval and Renaissance Treatises on the Arts of Painting: Original Texts with English Translations*. Vol. 1. Dover Publications, Inc., New York, 1999, rec. 150, pp. 116-117.

17. D.V.Thompson: "More medieval color-making: Tractatus de coloribus from Munich, Staatsbibliothek, Ms. Latin 444," *Isis*, 24, 1935-1936, pp. 382-396.

18. S. Santos Gómez, *et al., op. cit.*, 2003.

salt (NaCl)[19] or the 17th century treatise by Pierre Lebrun, which mentions the utilization of sal ammoniac (NH_4Cl).[20]

The abundant information regarding the preparation of verdigris contained in the old texts is directly related with its frequent utilisation in the application of polychromes and in the preparation of green-coloured inks and dyes, although the latter uses are maybe not as well known.

These types of inks have been widely used throughout history for representing areas of a bluish-green colour; in this sense, they are frequently found in the illumination of books and for colouring maps and plans.[21]

It is necessary to point out that, due to the abundant use of this ink throughout history, its recipes of preparation appear in texts covering a vast chronological period up until the 19th century. However, for this paper, the bibliographical review carried out has focused on all those texts corresponding to the Middle Ages in which recipes appear for the elaboration of pigments, colourings and artistic materials in general.[22] The types of sources consulted have been: recipe books, books of secrets and art treatises.[23] Starting from this first review, we have chosen those in which way recipes for preparing verdigris ink specifically appear.

Medieval texts on artistic materials

In many of the manuscripts corresponding to the Middle Ages, especially the Early Middle Ages, a significant influence of some texts from Classical Antiquity can be found and, in some cases, recipes appear which are identical to those mentioned in classical texts. However, unlike the latter, those corresponding to medieval times usually focus on a specific topic. Therefore, those that refer to artistic materials only talk about this subject, although, in general, do so in a rather disorderly way. This type of works could be included in the

19. P. Dioscorides, *op. cit.*, pp. 374-375.

20. P. Le Brun: "Recueil des essaies des merveilles de la peinture" (*Brussels Manuscript*, 1635) in M. P. Merrifield, *Medieval and Renaissance Treatises on the Arts of Painting: Original Texts with English Translations*. Vol. 2. Dover Publications, Inc. New York, 1999, Chap. VII, rec. 10, pp. 806-807.

21. M. Giménez Prades, M. San Andrés Moya, J. M. de la Roja: "El color y su significado en los documentos cartográficos del Cuerpo de Ingenieros Militares del s. XVIII," *Ge-Conservación*, 0, 2009, pp. 142-160.

22. S. Bordini: *Materia e imagen. Fuentes bibliográficas de las técnicas de la pintura* (original title *Materia e imagine. Fonti sulle tecniche della pittura*, Ed. Del Serbal, Barcelona, 1995; M. Clarke: *The Art of All Colours. Mediaeval Recipe Books for Painters and Illuminators*, Archetype Publications, London, 2001.

23. M.I. Báez Aglio, M. San Andrés Moya: "La práctica de la pintura a través de las antiguas fuentes documentales," *PH, Boletín del Instituto Andaluz del Patrimonio Histórico*, 34, 2001, pp. 64-77; S. Muñoz Viñas: "Original written sources for the history of mediaeval painting techniques and materials: A list of published texts," *Studies in Conservation*, 43, 1998, pp. 114-124.

group of the so-called recipe books, since they are structured like a compilation of recipes.

Within the period corresponding to the Early Middle Ages (4th-11th centuries) the manuscript known as *Lucca Manuscript*[24] has been revised, dating from the late 8th/early 9th centuries. Here recipes related with the preparation of mineral and vegetable colours are collected for their use in miniatures. The number O 12-16 is referred to verdigris pigment, which is named as *iarin*.

The texts included in the period covering the High/Late Middle Ages are: *Manuscript of Heraclius* (10th-13th C.), *De Clarea* (11th C.), *The Manuscript of Theophilus the Monk* (11th-12th C.), *Mappae Clavicula* (12th C.), *Asrarul Khat*, also known as *The Secrets of Calligraphy* (1102), *El Livro de cómo se facen as cores* (1262), *The Book of Secrets* of Albertus Magnus (13th C.), *The Innsbruck Manuscript* (1330), *The Pietro di Sant Audemar Manuscript* (13th C.), the anonymous texts *De coloribus naturalia exscripta et collecta* (13th-14th C.), *Tractatus de Coloribus Illuminarum Sive Pictorum* (14th C.), *De Arte Iluminandi* (2nd half of 14th C.) and, lastly, *Il libro dell'arte* by Cenino Cennini (14th C.).

The Manuscript of Heraclius[25] (10th-13th C.) consists of three books, the first two are written in verse and the third in prose. They deal, in a general way, with the preparation of the materials that are indispensable for carrying out the miniature technique; it describes the elaboration of colours with vegetable dyes and the making of parchments, paintbrushes, inks and glues. In recipes XXXVIII and XXXIX explains how to obtain verdigris, which is named as *viridis color cum sale* and *viridem cupri,* respectively, and the recipe XI is related to verdigris ink (*viride colore*).

The manuscript known as *De Clarea* or *Anonymus Bernensis*[26] (11th C.) also deals with the materials required for making miniatures. However, it does not have any recipe related to verdigris.

The text *Schedula Diversarium Artium* (11th-12th C.), also known as the *Manuscript of Theophilus the Monk*, is the first medieval art technical encyclopaedia.[27] It tells us of how work was carried out in the convents. It consists of three books; the first one deals with subjects on miniature-painting, painting on board, wall painting and also contains several recipes for making colours. The second book is about the use of glass as a pictorial support, and the third deals

24. Anonymus: "Composittiones ad tingenda musiva, pelles et alia, ad deaurandum ferrum, ad mineralia, ad chisographiam, ad glutina, quedam conficienda, aliquae artium documenta, ante anno nonagentos scripta," 8th C. (*Lucca Manuscript*), Codex Lucensis 490 of the Lucca Chapter Library, translated and notes in Swedish by H. Hedfors, University of Uppsala, 1932.

25. Heraclius: "De coloribus et artibus romanorum ...," *op. cit.*, pp. 238-239, 194-195.

26. D. V. Thompon: "The *De Clarea* of the so-called *Anonymus Bernensis*," *Technical Studies in the Field of the Fine Arts*, I, 1932, pp. 8-19, 69-81.

27. Theophilus: "Schedula diversarium artium," 11th C., The various arts, translated from Latin with introduction and notes by C. R. Dodwell. Thomas Nelson & Sons LTD. London, 1961.

with questions related to ivory, precious stones and metallurgical procedures. In this text are explained procedures to obtain *viride salsum* (Chap. 35) and *viride hispanicum* (Chap. 36).

The anonymous text *Mappae Clavicula* (12th C.), is, in fact, a compilation of previous writings (10th-11th C.), as well as being similar to Classical texts such as: Vitruvius (*Ten Books of Architecture*-1st C.), Pliny (*Natural History*-1st C.), Dioscorides (*On Medical Matters and Deadly Poisons*-1st C.). It gathers recipes on colours, dyes, varnishes and glues, amongst other substances. This manuscript describes several procedures to obtain verdigris. In the paper of Smith and Hawthorne[28] this pigment is named as Byzantine green (rec. v), Rouen green (rec. vi), green pigment (rec. 96) and verdigris (rec. 106). The recipe 92-B describes how to prepare green writing, so it is related to verdigris ink.

The manuscript of Fadlu'llah Ansari Wal Faruqi (1102), called *Asrarul Khat* and also known as *The Secrets of Calligraphy*, explains the techniques of the art of calligraphy.[29] It contains highly valuable information about the material required for writing; the preparation of the inks, vermilion, lapis lazuli and verdigris ink (*zangar*), as well as the necessary tools for doing calligraphy.

The Jew of Portuguese origin Abraham Ben Judah Ibn Hayyim is the author of *Livro de cómo se facen as cores* (1262) which is contained in *Ms de Rossi 945,* held at the Palatina Library of Parma.[30] Part of it is in Hebrew writing and describes recipes for preparing the substances to be used in painting miniatures. The author describes how to prepare verdigris (*azinhavre*) in Chapters XI and XII.

The Book of Secrets of Albertus Magnus, whose original title is *De Rebus metallicis et mineralibus* (13th C.),[31] contains recipes for making some pigments, for example, minium and ceruse, but there is not any one referring to verdigris.

The manuscript *Liber magistri Petri de Sancto Audemaro de coloribus faciendis* by the monk Pietro di Sant Audemar (13th-14th C.), is a compilation of recipes for making pigments, dyes, glues and gilding; all related to miniature-making. Some of these recipes come from the *Lucca Manuscript, Manuscript of Heraclius* and from the first book of *Teophilus the Monk.* In this manuscript there are several recipes related to verdigris pigment: rec. 150 (*viridem colorem de sale*), rec. 155 (*viride eris Grecum/viride commune*), rec. 156

28. C. S. Smith, J. G. Hawthorne: "Mappae Clavicula (12th C.). "A little key to the world of medieval techniques," *The American Philosophical Society*, New Series, 64, (4), 1974, pp. 1-122

29. Y.K. Bukhari: "Pigments," *Marg, Bombay*, 16, 1963, pp. ii-iii.

30. S. Blondheim: "An old Portuguese work on manuscript illumination," *The Jewish Quarterly Review*, XIX, 1928, pp. 97-135.; S. Muñoz Viñas ..., *op. cit.*, 1998.

31. Albertus Magnus: "De Rebus metallicis et mineralibus," 13th C. *The Book of Secrets of Albertus Magnus*, edited by Michael R. Best and Frank H. Brightman, Oxford in Claredon Press, 1973.

(*viride rothomagensian*) and rec. 159 (*viridi*).[32] There are also recipes for verdigris ink; these are, rec. 153 (*aqua viridi colore*) and rec. 157 and 160 (*viridi eris*).[33]

The anonymous manuscript *De coloribus naturalia exscripta et collecta* (13th-14th C.) is contained in *Ms Amplonius Quarto 189*. It is a short text that describes the making and practical use of pigments, especially for paintings and manuscripts.[34] This text contains several paragraphs related to verdigris; these are number [10] (*viride colore*), [12] (*viridem colorem*), [14] (*viridi hispanico*) and 18 (*viride*). Paragraph [19] is referring to verdigris ink (*viridi*).

The anonymous text known as *The Innsbruck Manuscript* (1330) is a compendium of recipes for inks and colourings, both vegetable and mineral.[35] This text explains how to make a green dye, named *grünspat*, from verdigris pigment.

Another text which shows many recipes referring to verdigris is the anonymus manuscript *Tractatus de Coloribus*, also known as *Ms from Munich* (14th C.). Presently is in the Bayerische Staatbibliotek of Munich (Ms. Latin 444). According to Thomson this is the extended version of the *Tractatus qualiter artificiales color fieri possit*, *Ms. Latino 6749* of the National Library of Paris.[36] This text contains around of forty paragraphs deal with the making of pigments and colours. Paragraphs related to verdigris pigment are: [28] (*viride grecum*), [28a] (*viride salsum*), [29] (*viridi salso*), [30] (*viride hyspanicum*), [31] (*viride romanicum*), [33] (*viride grecum*), and [34] and [35]; in this last, the pigment is not specifically named. Paragraph [6] describes the dyeing of several materials and mentions the use of *viride grecum*, and paragraph [32] is deal with verdigris ink (*viridis*).

Three texts belonging to the end of the 14th century have been consulted; the first of them is the anonymous *Liber de coloribus illuminatorum siue pictorum*. The original is in the British Museum, classified as the *Sloane Manuscript 1754*. It contains recipes from previous treatises and, which, in general, are dedicated to the materials and the technique of illumination.[37] This manuscript describes the making of verdigris pigment, named as *viride Greco* and *viride rotomagense*, in paragraph [VI]. It also deals with verdigris ink in para-

32. Pietro di Sant Audemar: "Liber magistri Petri de Sancto Audemaro de coloribus faciendis," *op. cit.*, pp. 116-117, 124-127.

33. *Ibidem*, pp. 122-123, 126-129.

34. D. V. Thompson (1935): "De coloribus, naturalia excripta et collecta, from Erfurt, Stadtbücherei, Ms. Amplonius Quarto 189 (XIII-XIV Century)," *Technical Studies in the Field of Fine Arts*, III, pp. 133-145.

35. Anonymous: "The Innsbruck Manuscript," 1330. On http://costume.dm.net/dyes/innsbruck [Consulted on 30th March 2012].

36. D. V. Thomson (1935-1936): "More medieval color-making: Tractatus de coloribus from Munich, Staatsbibliothek, Ms. Latin 444," *Isis*, 24, pp. 382-396.

37. D. V. Thompson: "Liber de Coloribus Illuminatorum Siue Pictorum from Sloane Ms., No. 1754," *Speculum*, 1, (3), 1926, pp. 280-307.

graphs [I] (*viride de Gretia*) and [VI] (*viridi*). In paragraph [XV] explains how to prepare a green colour for books (*viridis color in libris*).

Within this period of history, the second of the treatises consulted is *De arte illuminandi.*[38] It is an anonymous text, probably written by an Italian monk and is edited in a clearer way than the previous ones. It gathers recipes for making natural and artificial colours, as well as processes for preparing supports and gilding, amongst other topics. It does not contain any information about verdigris.

Chronologically, the next text consulted is that of the Italian painter Cennino Cennini, *Il libro dell'arte* belonging to the end of the 14th century. It is considered to be the first modern treatise of painting and, in a very didactic way, describes different topics directly related to artistic practice, such as: the way to do a good drawing emphasing lights and shades; the compositions and nature of pigments; preparation of paintbrushes; mordants for gilding. Amongst other subjects, it also deals with the materials and procedures required for fresco and oil painting techniques. This author mentions that *verderame* (verdigris) is made by alchemy, using copper and vinegar.[39]

Documentation dating from the early Renaissance has also been consulted. In particular, the Giovanni Alcherio text, the anonymous *Strasburg Manuscript*, the also anonymous *Secreti per colori* or *Bolognese Manuscript,* the Manuscript of J. Le Begue and the *Trattato d'Architettura* of A. Filarette.

Giovanni Alcherio's work consists of two collections of recipes: *De coloribus diversis modis tractatur* and *De diversis coloribus*.[40] The author first writes it in 1398 and then corrects it in 1411. The original was transcribed by Le Begue in his own manuscript (*Manuscript of J. Le Begue*-1431). It contains recipes and instructions for the miniature technique, painting on board, on paper, on wall and on canvas. Recipes 300 and 301 are related to verdigris ink, which is named as *viride corrosivum* and *viridem non corrosivum*, respectively.[41]

The *Strasburg Manuscript*[42] (14th-15th C.) is the first text on artistic materials written in German. It describes the way to make and prepare pigments, the synthesis and mixture of inks and the making of glues, varnishes, gilding

38. Anonymous: *De arte illuminandi,* 14th C., Franco Brunello, Neri Pozza editore, Vicenza, 1975.

39. C. Cennini: *Il libro dell'arte*, 14th C., El libro del arte. Commented and notes by Franco Brunello, introduction by Licisco Magagnato, translated from Italian by Fernando Olmeda Latorre. Ed. Akal. Madrid, 2002, p. 100.

40. G. Alcherio: "De coloribus diversis modis tractatur," "De diversis coloribus" (Manuscript of Alcherio), 15th C., in Merrifield, M, P, *Medieval and Renaissance Treatises on the Arts of Painting: Original Texts with English Translations.* Vol. 1, Dover Publications, Inc. 1999, New York, pp. 258-321

41. *Ibidem*, pp. 284-289

42. Anonymous: *The Strasburg Manuscript*, 14th-15th C., Borradaile, V & R, Alec Tiranti, London, 1966.

and silvers. This manuscript gives recipes and instructions on making inks and dyes from verdigris.[43]

Secreti per colori (15th C.) or *Bolognese Manuscript* is Codex 165 of the San Salvatore Convent Library (Bologna). It refers, like most of the texts consulted, to the making of pigments, but also to the elaboration of binders, varnishes, artificial gems, ceramics and leather tanning, among other techniques.[44] Many of its recipes are related to verdigris and its uses as pigment, ink and dye. Recipes referring to verdigris pigment are: 82, 84 and 85 which describe the making of *viridem ramum*, 83 (*viridem*), 86 (*verde*), 95 (*verde bono*), 101 (*verde-verderamo depurgato*), and 227 (*verde ramo*).[45] In recipe B109, it is described how to prepare *tentura verde* to be used as ink and dye, and recipes 371 and 377 are referring to the making of *viridem erem* and *verderamo*, both, to be used as dyes.[46]

The last text consulted corresponding to the first half of the 15th century has been that of Jehan Le Begue (1431),[47] which consists of two parts; the first one is the author's original and comprises a dictionary of words and synonyms, *Tabula de vocabulis sinonimis et equivocis colorum*, and 118 recipes that make up the part called *Experimenta de coloribus*; these recipes deal with the elaboration of inks, pigments, gilding etc. The second part of the text transcribes treatises of former periods (*Ms. of Sant Audemar, Ms. of Heraclius,* and *Ms. of Archerio*). Jean Le Begue provides recipes deal with verdigris pigment, these are: 8 (*viride rami*) and 43 (*viridem rarum*).[48] Another ones are referring to verdigris ink: 28 (*aquam viridem*), 29 (*azurrum from viridis eris*) and 331[49] (*couler verde pour escrire*).[50] In recipe 40, this author describes the making of a verdigris dye (*colore viride*).[51]

Antonio Averliono Filarette is the author of the *Trattato d'architettura* (c. 1465). In the twenty-four book, this author shortly describes the colours and how to make them. Filarette mentions the verdigris (*verde di rami*) and only tells that it is made from copper.[52]

43. *Ibidem*, pp. 20-21, 22-23, 34-37, 42-43.

44. Anonymous: "Secreti per Colori" (*Bolognese Manuscript*), 15th C., in M. P. Merrifield, *Medieval and Renaissance Treatises on the Arts of Painting: Original Texts with English Translations*. Vol. 2, Dover Publications, Inc. New York, 1999, pp. 325-600.

45. *Ibidem*, pp. 418-421, 422-423, 426-427, 502-503.

46. *Ibidem*, pp. 430-431, 590-593.

47. J. Le Begue: "Manuscripts of Jehan Le Begue" (1431), in M. P. Merrifield, *Medieval and Renaissance Treatises on the Arts of Painting: Original Texts with English Translations*. Vol. 2, Dover Publications, Inc., New York, 1999, pp. 1-321.

48. *Ibidem*, pp. 48-49, 66-67.

49. This is one of the recipes that Le Begue include at the end of Alcherius' Manuscript.

50. J. Le Begue: "Manuscripts of Jehan Le Begue," *op. cit.*, pp. 58-61, 31-311.

51. *Ibidem*, pp. 64-65.

52. A. Averlino (Filarete): *Trattato d'architettura* (c. 1465), trad. y comentarios P. Pedraza, Ephialte, Instituto de Estudios Iconograficos, Vitoria-Gasteiz, 1990, p. 371.

Besides all the types of historical sources mentioned, it is important to point out a genre of books known as *Books of Secrets*. In general, this type of texts does not correspond to a specific type of books or to a specific period. Under this denomination there are a wide variety of texts: from alchemy writings done before Christ, medieval recipe books, to some texts written in the 17th and 18th centuries. Although it is not a general norm, in some cases, the word *secrets* appears in the title itself. For example, we have *The Book of Secrets* written by Albertus Magnus (12th C.) and which has already been mentioned above; in the introduction of the edition consulted and on the subject of books of secrets, it is said that literature of secrets was a popular tradition that is demonstrated by its survival through the centuries. This was due to a great extent to the transcriptions made and printed in the 17th century. Another book that had a great dissemination in its time and of which many translations and editions were made,[53] is the book by the Reverend Alexis of Piemonte called *De secretis.*[54] Within this group of books of secrets, we have the collection known as *Secretos raros de artes y oficios*,[55] made up of several volumes numbered III to XII. They were all published in Madrid between 1806 and 1807 at Villalpando's printing house, except for volume number VII (Repullés printing house). All of them deal, amongst other topics, with the way to make different inks or how to dye paper different colours.

In the text *De Secretis* there are two recipes related to the making of verdigris inks. On of them is recommended to write green letters and the other one is deals with to illuminate, write and paint.[56]

Besides the above-mentioned texts, Spanish treatises have also been consulted that are strongly connected with the texts already referred to and which are of compulsory consultation for our research group, since they refer to the artistic materials used in the Iberian Peninsula.

Etimologías of San Isidoro de Sevilla (560-636) are of interest due to their encyclopedic nature and their description of the origin of the names of the substances know at that time, some of which were related to the art trade. This author describes how to make verdigris pigment (*aeruginem*).[57]

53. W. Eamon: *Science and the Secrets of Nature: Books of Secrets in Medieval and Early Modern Culture*. N. J. Princeton, Princeton University Press, 1994, pp. 250-256.

54. G. Ruscelli: *Secretos del reverendo Don Alexo Piamontes / traducidos de lengua italiana en castellano, añadidos y remendados en muchos lugares en esta última impresión.* Traducción del italiano al castellano por Vázquez, A y López, M, Antonio Vázquez Impresos de la Universidad, a costa de Manuel Lopez, Alcala de Henares, 1640.

55. *Secretos raros de artes y oficios*, tomos III, IV y V. Madrid, imprenta de Villalpando 1806. *Secretos raros de artes y oficios*, tomos VIII, IX, X, XI y XII. Madrid, imprenta de Villalpando 1807. *Secretos raros de artes y oficios,* tomo VII. Madrid, imprenta de Repullés, 1807.

56. G. Ruscelli: *Secretos del reverendo Don Alexo Piamontes ..., op. cit.*, pp. 245-246.

57. San Isidoro de Sevilla: *Etimologías*, (Latin, Span. vers. and notes by José Oroz Reta and Manuela Marcos Casquero), Vols. I and II, Madrid, Biblioteca de Autores Cristianos, 1993, pp. 306-307.

The 12th century anonymous Madrid treatise, the *Codex Matritensis*,[58] to be found at the Madrid National Library classified as Manuscript 19, contains recipes for dyes, metallurgic and gilding works, amongst other subjects.[59] In this manuscript there are two recipes referring to making of verdigris pigment, which is named as *chrisocolle* (rec. XXXXI) and *eruginem* (rec. LXVIII).[60]

In the 13th century, Alfonso X the Wise compiles the *Lapidario.* It is a text on astrology and alchemy discovered by King Alfonso X who has it translated in 1234. As in the introduction of the edition consulted, the original consisted of two groups of works; the first comprising four treatises and, as for the second, only the index is left, which is why we know of its existence. The *Lapidario* describes three hundred and sixty stones, explaining their properties, symbolic character and place of origin. It shows a recipe to obtain verdigris (*azingar*).[61]

One of the most interesting Spanish mediaeval texts is the Manuscript H490. It is a Castilian technical book of prescriptions, attributed to Juan de Celaya and dates from the second half of the 15th C. (1460-1480). It is writing in Spanish and Latin languages and is preserved in the Montepellier Medicine Faculty. It is a miscellaneous text which deals with different matters such as: medicine, metallurgy, alchemy, astrology, artistic materials, and several crafts related to leather and glass. It contains recipes about making of colours, pigments and inks to be used for illuminating and writing. The recipe 36 of this manuscript is related to verdigris, although this is named as *azurum* . The procedure described is similar to that in *Ms. Boloñes* (rec. 45 – *ad faciendum azurrum*). In both recipes, a silver plate is used instead of copper; but the formation of a corrosion blue on the plate must be due to the presence of copper in the silver sheet. Recipes 30 and 36 are related to the used of verdigris (cardenillo or verdet) in the dying process of gold and glass, respectively.[62]

One of the first treatises in the Spanish language on painting materials is the anonymous text *Reglas para pintar* (late 16th century), discovered by the researcher R. Bruquetas (1988) at the Santiago de Compostela University Library. It is bound together with two other works: *Medidas del Romano* (1541) by Diegro de Sagredo and *Libellus artificiosus omnibus ìctoribus,*

58. J. M. Burnam: "Codex Matritensis," 12th C., Paleographic edition of a facsimile in black and white. Series II, Vol. VIII. University Press, Cincinnati, Ohio, 1912, pp. 5-47.

59. S. Kroustallis: *Edición crítica y estudio de un tratado de tecnología artística medieval: el "Codex Matritensis" 19*, Doctoral Dissertation. Directed by E. Ruiz García. Faculty of Geography and History, Department of Sciences and Historiographical Techniques, Complutense University of Madrid, 2005.

60. J. M. Burnam: "Codex Matritensis," *op. cit.*, pp. 19, 24.

61. Alfonso X.: *Primer lapidario*, s. XIII, Manuscrito h.I.15 of the El Escorial Library. Transcription by María Brey Mariño. Madrid, facsimile by Edilán, S.A. 1982, p. 128.

62. R. Córdoba de la Llave: "Un recetario técnico castellano del siglo XV: el manuscrito H490 de la Facultad de Medicina de Montpellier," *En la España Medieval*, 28, 2005, pp. 7-48.

satatuariis (1539). The three form a group of rules with a didactic purpose and in the first one the correct use of painting materials is described: preparation of supports, pigments, such as verdigris (cardenillo) and binders.[63]

It is not until the 16th century that we find a text related to the artistic practice, *Comentarios de la pintura*, written around 1560 by Felipe de Guevara.[64] It is about the materials required for oil paintings and mentions the colours needed for fresco painting. It also deals with the technique of illumination. This author refers to the use of verdigris.[65]

Special mention must be made of *Painters' Ordinances*, a series of provisions regarding the municipality to which they refer and which are aimed at regulating the various sectors of local, social and economic life, also including work activities. They usually provide interesting information for art specialists as they include rules regarding the artist community.[66] Among other issues, refer to the practical use of pigments and, obviously, the verdigris is included We must especially highlight *Ordenanzas de Córdoba de 1493*.[67]

Recipes for verdigris inks found in texts of the high/late middle ages and early renaissance

Many of the texts mentioned in the previous section contain recipes related with the elaboration of verdigris, or express reference is made to its use as a pigment. In some of them the process of preparing verdigris' ink and dye are also described. For this paper we have chosen those corresponding to the High/Late Middle Ages and Early Renaissance, the 12th-15th centuries. Basing ourselves on the revision of these documentary sources, it can be seen that there are two ways of obtaining these ink and dye: one is from the verdigris pigment (previously synthesized) and the other is from the solution obtained as a result of dissolving copper (shavings, powder or sheets), or its alloys (brass or bronze), in vinegar or wine. Independently of the initial substance, other ingredients can be added whose nature varies according to the recipe consulted. In Table 1 are showed recipes related to verdigris's inks and dyes from High Mediaeval and Early Renaissance texts.

63. R. Bruquetas Galán: "Reglas para pintar. Un manuscrito anónimo de finales del siglo XVI," *PH Boletín del Instituto Andaluz del Patrimonio Histórico*, 24, 1998, pp. 33-44.

64. F. de Guevara: *Comentarios de la pintura*, s. XVI, Segunda edición reproducida de la edición príncipe. Pórtico y revisión por Rafael Benet. Barcelona, Selecciones Bibliófilas, 1948.

65. *Ibidem*, p. 179.

66. S. Santos Gómez, M. San Andrés Moya: "Aportaciones de antiguas ordenanzas al estudio de técnicas pictóricas," *Pátina*, 10 y 11, 2001, pp. 266-285.

67. R. Ramírez de Arellano: "Miscelánea, Ordenanzas de pintores," *Boletín de la Real Academia de Bellas Artes de San Fernando*, año IX, 33, 1915, pp. 29-46.

The recipe used for obtaining verdigris ink is very important, since the nature and dosage of its components determine its colorimetric properties[68] and also affects the behaviour of the ink with aging.[69]

Some recipes will be commented in paragraphs below. The selected texts are: *Mappae Clavícula* (12th C.), Manuscript of Heraclius (10th-13th), *The Secrets of Calligraphy* of Fadlu'llah Ansari Wal Faruqi (1102), *The Innsbruck Manuscript* (14th C.), The Pietro Di Sant Audemaro Manuscript (13th-14th C.), The Alcherio Manuscript (15th C.), *Liber de coloribus iluminatorum sive pictorum*, the *Bolognese Manuscript* and *Experimenta de Coloribus* of Jean Le Begue (15th C.).

Within the first group of recipes, that is to say, those starting off with the verdigris pigment, we have the following texts:

Mappae Clavicula (12th C.), its recipe 92-B saying:[70]

> "A green writing. Put into vinager, efflorescence of copper, verdigris, and sulphur. Add gum and write."

This recipe specifically refers to the preparation for green writing. The basic ingredient is the verdigris pigment, although another copper compound called *efflorescence of copper* is named. It can be understood that this term refers to the products resulting from the corrosion formed on the surface of metallic copper, whose nature will vary in function of the environment. They could be copper chlorides[71] or carbonates,[72] green or bluish-green, respectively.

As for sulphur, it is necessary to point out that this is the only recipe in which this ingredient is mentioned. It is one of the few elements that is found in its native form in nature and that is known since Antiquity. Alchemists used it for making sulphurous waters, among other things, by means of which they carried out their works for faking gold.[73] Within the scope of pictorial materials, it was used since antiquity to obtain some pigments like vermilion and orpiment. Vermilion, chemically-speaking, is mercury (II) sulphide (HgS) of a

68. N. Sancho, S. Santos, J. M. de la Roja, M. San Andrés: "Colorimetrical study of verdigris inks. First results." *10th Congress of the International Colour Association AIC Colour 05*, Granada, 8-13 May, 2005, *Proceedings Book, Part 2*, pp. 1437-1440.

69. J. M. de la Roja, M. San Andrés, N. Sancho: "Tinta de verdigris: síntesis, envejecimiento y valoración de sus efectos sobre el papel." *II Congreso del GEIIC, Investigación en Conservación-Restauración*, Barcelona, 9-11 November, 2005, pp. 75-83, 497.

70. C. S. Smith, J.G. Hawthorne: "Mappae Clavicula, 12th C. A little ...," *op. cit.*, p. 40.

71. Chlorides are formed when copper (Cu) is in contact for a long period of time with the saline soil of desert regions, or in a damp environment full of marine aerosols (NaCl). D. Scott: *Copper and Bronze in Art. Corrosion, Colorants, Conservation*. Ed. The Getty Conservation Institute, Los Angeles. 2002, pp. 122-144.

72. The most important carbonates, both for their frequency and corrosion products, as well as for being some of the most important historic pigments in all painting schools, are azurite and malachite. D. Scott: *Copper and Bronze in Art ..., op cit.*, pp. 100-121.

73. H. M. Leiceste: *Panorama histórico de la química.* Spanish version by Federico García Portillo, edit. Alhambra, S. A. 1967, p. 46; E.R. Caley: "The Leyden Papyrus X," 3rd century AD. *Journal of Chemical Education*, 3, (10), 1926, pp. 1149-1166, recipe 89.

red colour and orpiment is an arsenic sulphide (III) (As_2S_3), whose colours range from yellow to greenish-yellow. Both sulphurs exists in nature, but can also be made.[74] However, in the case of the recipe in *Mappae Clavicula*, it is not clear which was the effect of this ingredient.

Lastly, the term *gum* refers to gum arabic, which is produced by several species of acacia. It is a polysaccharide gum produced by the barks of the acacia Senegal or other species of African acacia in the form of a highly viscous liquid. It is soluble in water and traditionally has been used as a binder in pigments and inks.[75]

Recipe XV of *Liber de coloribus iluminatorum siue pictorum* (14th C.) specifies:[76]

> [...] *in libris uero non teres, sed in uino bono albo et clarísimo siue aceto temperare permittes, et sic digito tantum fricabis, et statim totum uinum uirideum erit.* [...] [... for books you should not grind it, but let it soak in good wine, white and very clear, or in vinegar; and then rub it a little with your finger, and immediately all the wine will become green].

In this case, the verdigris pigment is soaked in *vinegar* or *white wine. Vinegar* is a sour liquid produced by the acid fermentation of wine. It contains acetic acid (CH_3COOH) (4-5 %) that, as it has already been indicated, is the reagent used to form the verdigris pigment on the copper sheet. In turn, verdigris is soluble in an acid environment, and that is why when it comes into contact with vinegar a solution is obtained which is verdigris ink. As for the other possibility indicated in this recipe, related with the mixture of the pigment with *white wine*, it is necessary to point out that in this case the dissolution would equally take place.

The recipe that appears in the *Innsbruck Manuscript* (c. 1330) says the following:[77]

> *Swer grün varb welle machen, der nem grünspat vnd siede daz in harn vnd alaun misch dar vnder vnd gumi ein tail; vnd veb damit* [...] [To make a green dye, take verdigris and boil it in urine and mix alum thereto and a portion of gum Arabic, and dye therewith].

Urine is mentioned very frequently in recipe books, being one of the ingredients used, for example, to prepare lacquers[78] and pigments, such as verdi-

74. N. Eastaugh, V. Walsh, T. Chaplin, R Siddall: *Pigment compendium. A dictionary of historical pigments.* Elsevier Butterworh-Heinemann. 2004, pp. 285, 386 y 387.

75. R. J. Gettens, G. L. Stout: *Painting Materials. A Short Encyclopedia.* Ed. Dover Publications, Inc., New York, 1966, pp. 27-28.

76. D. V. Thompson: "Liber de Coloribus Illuminatorum Siue Pictorum ...," *op. cit.*, 1926.

77. Anonymous: "Innsbruck Manuscript," *op. cit.*

78. M. I. Báez, M. San Andrés: "Las lacas rojas II: Historia de su empleo y preparación," *Pátina*, 10 y 11, 2001, pp.172-186.

gris. In principle, urine is of an acid nature, due to its content in uric acid ($C_3H_4N_4O_3$); however, in time it becomes basic. This is because urea (H_2NCONH_2), which is another of the components of urine, decomposes and produces ammonia (NH_3). Also, uric acid, when heated, is turned into urea and ammonia. Therefore, in Antiquity its use was based on its alkalinising nature and, also, on its capacity to act as a mordant fluidifier.[79] *Alum* is a double sulphate of hydrated potassium and aluminium ($Al_2 (SO_4)_3{\cdot}K_2SO_4{\cdot}24H_2O$). Traditionally, it has been given various names, many of them related with the geographical location of the mine from which the mineral (alunite[80]) was extracted; for example, Rock alum, Yemen alum, Roman alum, etc. It is a substance used frequently since Antiquity as a mordant in dyeing processes. In this case, its function in the recipe would be, in fact, that of a mordant, so that the verdigris ink would adhere more efficiently to the support. Finally, to make the ink more consistent, *gum arabic* is added.

Recipe 300 of the Alcherio Manuscript (15th C.) states:[81]

> [...] *Accipe viride aeris et modicum de faece vini sicca, quae dicitur in latino tartarus et in gallico gravella, et subtilia et tere super lapidem durum et planum insimul quae dicta sunt cum aceto* [...] [Take verdigris and a little of the dried lees of wine, wich in Latin is called tartarus, and in French gravelled, and pulverize it and grind both the ingredients together upon a hard and smooth stone with vinegar].

Dry wine lees refer to the white, crystalline deposits, soluble in water, which are formed in the vats where grapes are fermented. These deposits were formerly known as *cream of tartar* or simply *tartar*, also being called *argol*. Chemically-speaking, it is hydroxy acid,[82] specifically 2,3-dihydroxybutanedioic ($C_4H_6O_6$), which is found in nature in the form of potassium acid salt. Traditionally, tartar has been used as a mordant in dyeing. Therefore, in the case of the Alcherio recipe, it would be used to improve the adhering of the ink to the support. Vinegar would also be used to dissolve the ingredients (verdigris and tartar) and therefore to obtain the ink.

Recipe B-109 of *Bolognese Manuscript* (15th C.) says:[83]

> [...] *A fare tentura verde da scrivare. – Recipe 1b. doi de verderamo abrusciato et fanne polvere subtili et polla a distilare a lambico er serva laqua e de bona da scrivare et da tegnare filo* [...] [To make a green tincture for writing. – Take

79. C. Cennini: *Il libro dell'arte, op cit.*, notes p. 191.

80. Alunite is a mineral that, apart from alum, contains iron salts. It is dehydrated when heated at 500°C, dissolved in hot water and the solution obtained is left to cool.

81. G. Alcherio.: "De coloribus diversis modis ...," *op cit*, p. 284.

82. Hydroxy acids are very abundant in nature; for example: lactic acid in sour milk; malic acid in apples; tartaric acid in grapes; citric acid in citrus fruits.

83. Anonymous: "Il libro dei colori. *Bolognese Manuscript*," *op cit*, p. 430.

> 2 lbs of calcined verdigris, reduce it to fine powder, and distil it in an alembic, and keep the water that comes over and it is good for writing and dyeing thread].

When considering the possibility of reproducing this recipe some omissions can be observed in its redaction. In order to appreciate them, in the first place it is necessary to remember that there are two types of verdigris; the one which chemically corresponds to a hydrated copper acetate or hydroxyacetate (II), and the variety *viride salsum*, which is a copper chloride or hydroxychloride (II). If the verdigris used to start off with were a copper acetate (II), the calcination process could not be carried out, since this compound decomposes in acetic acid and cupric oxide of a black colour at relatively low temperatures (approximately 240°C). If it were a copper chloride it could be heated, but this compound is not very soluble in water, although it is in an acid environment. On the other hand, as indicated in this recipe, calcinated verdigris is subjected to a *still distillation* process. To be able to carry out this process, the verdigris would have to be previously dissolved. Since it cannot be copper acetate, due to its decomposition when heat is applied (both in calcination and distillation), the recipe probably refers to the variety known as *viride salsum*, although nothing is specified in this respect. On the other hand, nor is there any indication that the pigment has to be dissolved previously or of the solvent that should be used, which, in the case of *viride salsum* must be vinegar, since its acid nature favours the dissolution of the copper chlorides (II).

As to the second group of recipes, they include all those in which the solution obtained after treating the copper or its alloys with vinegar, is used to start off with. In this group, the copper or alloy used, usually are in the form of shavings or powder, although they may also be sheets or copper vessels.

Recipe XI of the Heraclius text (10th-13th C.) says:[84]

> *De viridi colore ad scribendum. Si quaeris viridi scriptura colore notari, acri commisum melli miscebis acetum; hinc valde calido vas ipsum sontege fimo. Sic et bissenis hoc extrahe solibus actis* [Of a green colour for writing. If you wish to embellish your writing with a green colour, mix vinegar together with strong honey, and them cover up the vase itself in very hot dung; and so take it out after twelve days shall have elapsed].

In this case, as a previous stage to obtaining the ink, the verdigris pigment is formed that is immediately dissolved in the vinegar itself. As for the *honey*, it is a highly viscous sweet-flavoured solution, composed of fructose, glucose and water. It easily absorbs the humidity of the atmosphere, which is why it has traditionally been used as a moistening agent and for plastifying. Honey was also used by the classics to slow down the high drying speed of tempera

84. Heraclius: "De coloribus et artibus," *op. cit*, p. 194.

paints.[85] As regards dung, its use is mentioned in recipes connected with pigments, such as verdigris itself and lead white. It was used as a heat-producing source.[86]

The text of *Fadlu'llah Ansari Wal Faruqi* (1102), *The Secrets of Calligraphy*, says:[87]

> "... Take one rati of naushagar (sal ammoniac) and half rati of copper scraps, put them in a pot and pur grape Vinegar drop by drop into the vessel and with the help of a stick, whose top shuld be flat, grind the admixture in the pot till it becomes zangar (verdigris)."

The term *sal ammoniac* refers to *ammonium chloride* (NH_4Cl); it was also formerly known as ammonium salt, *almojatre* and Armenian salt.[88] This salt was known by Persian alchemists, who were acquainted with two varieties; one coming from the mineral kingdom and the other obtained by distillation of animal products such as fur, although in this last case it must have been ammonium carbonate and not ammonium chloride.[89] The mineral product is a fine, crystalline white powder that does not melt when heated, but vaporises, being dissociated, and is used as a mordant. This salt has a corrosive effect on copper and was also used for making purpurin and as a mordant for gold.[90] Therefore, sal ammoniac and vinegar would speed up the corrosion process of the metal and the salt, in turn, would help the ink adhere better to the support.

Recipe 153 of the Pietro Di Sant Audemar text (13th-14th C.) says:[91]

> *De aqua vel viridi colore ad scribendum. – Si vero literas scribere volueris, pone viridem pulverem aeris in vino vel aceto, ut dictum est, et sic digito tantum fricabis, et statim totum vinum vel acetum virideum erit* [...] *Et si vis quod mirae pulcritudinis fiat, adde aliquantulum de croco et cum quieverit ...* [Of a green water, or colour, for writing. But if you wish to write letters, put the green powder of brass in wine or vinegar as aforesaid, and then stir it round a little with your finger only, and immediately the whole of the wine or vinegar will be green [...] But if you wish it to be very beautiful, add a little saffron ...]

85. C. Cennini: *Il libro dell'arte ..., op. cit.*, Note on p. 223.

86. It was also used as a source for producing carbon dioxide, which was produced during its fermentation and favoured the formation of pigments such as lead white [dihydroxy lead (II) bicarbonate].

87. Y. K. Bukhari: "Pigments," *op. cit.*

88. A. Alonso Barba: *Arte de los metales en qve se enseña el verdadero beneficio de los de oro, y plata por açogue, el modo de fvndirlos todos, y como se han de refinar y apartar unos de otros.* 17th C., Librerias París-Valencia S.L. 1998. (Facsimile of 1770). According to this author, the common people call it "Armoniac," because they believed this salt came from Armenia.

89. H. M. Leicester: Panorama histórico de la ..., *op. cit*, p. 78.

90. M. P. Merrifield: *Medieval and Renaissance Treatises on the Arts of Painting: Original Texts with English Translations.* Vol. 1 and 2, New York, Dover Publications, Inc., 1999, p. xxviii.

91. Pietro di Sant Audemar.: "Liber magistri petri de ...," *Ibid.*, p. 122.

For this recipe, one would start off with *green powder of brass*. Nowadays, brass refers to an alloy of copper and zinc of a pale yellow or reddish colour, depending on the copper content; however, this term could be used in the past to refer to other copper alloys. Anyway, the presence of this element is responsible for the formation of green-coloured corrosion products. Since these recipes state that the brass powder is green, the material must have been rusted and, therefore, already contained the copper (II) salts of this colour.[92] As for the dissolution of this green powder, according to the recipe, the process is done by putting it in contact with *wine* or *vinegar* and simply by shaking it, a green-coloured solution is rapidly obtained. As it has already been indicated, the verdigris that corresponds to the composition of copper (II) acetate is soluble in water, whereas if it is a copper (II) chloride it is not very soluble or insoluble, requiring an acid environment (like the one given by vinegar or wine) in order to carry out this process. Lastly, saffron is a colouring that has been greatly used throughout history. It produces solutions of a yellow-orange colour; therefore its addition would change the tonality of the ink.[93]

The last of the recipes commented is that of Jean Le Begue (1431), recipe 28 saying:[94]

> *Ad faciendum aquam viridem ad scribendum. – Accipe bonum acetum oncias ii., salis armoniaci ii., salis communis oncias ii., limature eris oncias ii., pone omnia in ampula vitrea per vj. dies, et fiet aqua viridis, quam cola et reserve* [To make a green ink for writing. – Take a good vinegar oz. ij., sal ammoniac oz. ij., common salt oz. ij., brass filings oz. ij., put them all together in a glass flask for six days, and it will make a green ink, which you must strain and keep for use].

The ink is obtained directly by treating brass filings with a mixture of *sal ammoniac* and *common salt* (i.e. sodium chloride-NaCl) and *vinegar*. This mixture has to react with the brass filings; the final result will be a mixture of copper (II) chlorides and acetates, which, as they form, will be dissolved in the vinegar. Since this chemical reaction is not immediate, the recipe indicates that it is necessary to wait *six days* to obtain the ink. Both salts act as a corrosive agent of the copper of the brass and, also, as a mordant.

Conclusions

From our overview of the medieval texts which describe ways of making verdigris ink, we have reached a series of conclusions.

92. In fact, one is starting off with the verdigris pigment.

93. N. Sancho, *et al.*, 2005, *op. cit.*.

94. G. Alcherio: "De coloribus diversis modis ...," *op. cit.*, p. 58.

There are two ways of preparing this ink. One of them is starting from the verdigris pigment (previously synthesized), and the other one is starting off with the solution obtained as a result of the dissolution of copper or its alloys, as for example brass or bronze in vinegar. Independently of the substance used, other ingredients can be added whose nature varies according to the recipe consulted.

Both types of recipes can appear in the same treatise. This is the case of the Jean Le Begue text Experimenta de Coloribus (1431) (recipes 28, 29 and 331).

It often occurs that the authors of the different texts copy each other's recipes. They mention the same ingredients and a similar process. This is the case of recipe 157 of the Pietro Di Sant Audemar Manuscript (13th-14th C.) and recipe XI of the Manuscript of Heraclius (10th-13th C.) and recipe 331 of Jean Le Begue text.

In the recipes consulted the exact dosage of each one of the ingredients of the recipes is not usually indicated; for example, Alcherio (15th C.), in recipe 300 mentions a little wine lees and the Innsbruck Manuscript (14th C.) talks of a large amount of gum arabic. As for the time that the ingredients should be left to make the ink in a satisfactory way, this fact is not usually specified in the texts, but, as an exception, we can mention Alcherio's recipe 28, which mentions six days for obtaining the ink and recipe 11 of Manuscript of Heraclius, which states twelve days.

Table 1.
Recipes to prepare verdigris's inks and dyes which appear in High Medieval and Early Renaissance texts.

Author	Text	Date	Name*	Recipe*	
				Ink	Dye
Heraclius (attributed to)	*De coloribus et artibus romanorum*	10th -13th C.	*Viridi colore*	XI a+b+c	
Anonymous	*Mappae Clavicula*	12th C.	Green Writing	92-B V+b+d+S+G	
Fadlu´llahAnsari Wal Faruqi	*Asrarul Khat (The Secrets of Calligraphy)*	1102	*Zangar*	(+) a+b+e	
Pietro Di Saint Audemaro	*Liber magister Petri de Sancto Audemaro de coloribus faciendis*	13th-14th C.	*Aqua viridi colore*	153 f+b/g+h	
			Viridi eris	157 a/f+b+c	
				160 a/f+b+i	
Anonymous	*De coloribus, naturalia exscripta et collecta (Ms. Amplonius Quarto 189)*	13th-14th C.	*Viridi*	19 j+b+c	
Anonymous	*The Innsbruck Manuscript*	c. 1330	*Grünspat*		(+) V+k+i+G
Anonymous	*Tractatus de coloribus (The Munich Manuscript)*	14th C.	*Viride grecum*		6 V+b+i
			Viridis	32 a+g	

Author	Text	Date	Name*	Recipe*	
				Ink	Dye
Anonymous	*Liber de coloribus illuminatorum sive pictorum (Ms Sloane 1754)*	14th C.	*Viride de Gretia*	I V+g+h	
			Viridi	VI a+b+c	
			Viridis color in books	XV V+b/g	
Giovanni Alcherio	*De diversis coloribus*	1398-1411	*Viride corrosivum*	300 V+b+ñ	
			Viridem non corrosivum	301 V+o+G	
Anonymus	*The Strasburg Manuscript*	14th-15th C.	*Grün*	(+) V+b+h+G	
			Spangrün	(+) V+b+h+m+ñ+G	
			Spangrün		(+) V+b
			Grün	(+) V+b+c+h+m+ñ+G	
Anonymous	*Secreti per Colori (The Bolognese Manuscript)*	15th C.	*Tentura Verde*	B109 V+W	B109 V+W
			Viridem erem		371 V+b
			Verderamo		377 V+l+m
Jean Le Begue	*Experimenta de Coloribus (Ms of Jean Le Begue)*	1431	*Aquam viridem*	28 b+p+e+f	
			Viridis eris	29 V+e+ñ	
			Couler verde pour escrire	331 a+b+c	
			Colore viridi		40 b+f+h+q
Juan de Celaya (attributed to)	Manuscript H490	1460-1480	*Cardenillo or verdet*		30 V+b+e+k+p+ r
			Cardenillo		34 V+e+i

* Some names correspond to the English translation of the original term; (+) The recipes are not numbered; Components: a (copper), b (vinegar), c (honey), d (efflorescence of copper), e (sal ammoniac), f (brass), g (wine), h (saffron), i (alum), j (bronze), k (urine), l (goat milk), m (yolk), ñ (tartar), o (gladiolus juice), p (common salt), q (Roman vitriol), r (saltpeter)
V: verdigris pigment; G: gum; W: water

Acknowledges

This work is part of the Project HUM2005-04618/ARTE, supported by the Ministry of Education and Science. The authors are grateful to Drs. Sonia Santos and José Manuel de la Roja, who are part of our research group.

Bibliography

ALBERTUS MAGNUS: "De Rebus metallicis et mineralibus," 13th C. *The Book of Secrets of Albertus Magnus*, edited by Michael R. Best and Frank H. Brightman, Claredon Press, Oxford, 1973.

ALCHERIO, G.: "De coloribus diversis modis tractatur," "De diversis coloribus" (Manuscript of Alcherio), 15th C., in Merrifield, M, P, *Medieval and Renaissance Treatises on the Arts of Painting: Original Texts with English Translations*. Vol. 1, New York, Dover Publications, Inc. 1999, pp. 46-111/258-321.

ALFONSO X.: *Primer lapidario*, s. XIII, Manuscrito h.I.15 of the El Escorial library. Transcription by María Brey Mariño. Madrid, facsimile by Edilán, S.A. 1982.

ALONSO BARBA, A.: *Arte de los metales en qve se enseña el verdadero beneficio de los de oro, y plata por açogue, el modo de fvndirlos todos, y como se han de refinar y apartar unos de otros*. 17th C., Librerias París-Valencia S.L. 1998. (Facsimile of 1770).

ANONYMOUS.: "Secreti per colori" (*Bolognese Manuscript*), 15th C., in M. P. Merrifield, *Medieval and Renaissance Treatises on the Arts of Painting: Original Texts with English Translations*. Vol. 2, New York, Dover Publications, Inc. 1999, pp. 325-600.

ANONYMOUS: "The Innsbruck Manuscript," 1330. On http://costume.dm.net/dyes/innsbruck/ [Consulted 30th March 2005].

ANONYMOUS: *De arte illuminandi*, 14th C., Franco Brunello, Neri Pozza editore, Vicenza, 1975.

ANONYMOUS: *The Strasburg Manuscript*, 14th-15th C., Borradaile, V & R, London, Alec Tiranti, 1966.

ANONYMUS:: "Composittiones ad tingenda musiva, pelles et alia, ad deaurandum ferrum, ad mineralia, ad chisographiam, ad glutina, quedam conficienda, aliquae artium documenta, ante anno nonagentos scripta," 8th C., (*Lucca Manuscript*), Codex Lucensis 490 of the Lucca Chapter Library, translated and notes in Swedish by H. Hedfors, University of Uppsala, 1932.

AVERLINO, A. (Filarete): Trattato dárchitettura (c. 1465), trad. Y comentarios P. Pedraza, , Ephialte, Instituto de Estucios Iconográficos, Vitoria-Gasteiz, 1990.

BÁEZ AGLIO, M.I.; SAN ANDRÉS MOYA, M.: "La práctica de la pintura a través de las antiguas fuentes documentales," *PH, Boletín del Instituto Andaluz del Patrimonio Histórico*, 34, 2001, pp. 64-77.

BÁEZ, M.I.; SAN ANDRÉS, M.: "Las lacas rojas II: Historia de su empleo y preparación," *Pátina*, 10 y 11, 2001, pp. 172-186.

BAILEY, K.C.: The *Elder Pliny's Chapters on Chemical Subjects*, Parts I, II, London, Edward Arnold & Co, 1929.

BLONDHEIM, S.: "An old Portuguese work on manuscript illumination," *The Jewish Quarterly Review*, XIX, 1928, pp. 97-135.

BORDINI, S.: *Materia e imagen. Fuentes bibliográficas de las técnicas de la pintura* (original title *Materia e imagine. Fonti sulle tecniche della pittura.* Ed. De Luca), Barcelona, Ed. del Serbal, 1995.

BRUQUETAS GALÁN, R.: "Reglas para pintar. Un manuscrito anónimo de finales del siglo XVI," *PH Boletín del Instituto Andaluz del Patrimonio Histórico*, 24, 1998, pp. 33-44.

BUKHARI, Y. K.: "Pigments," *Marg, Bombay*, 16, 1963, pp. ii-iii.

BURBNAM , J. M.: "Codex Matritensis," 12th C., Paleographic edition of a facsimile in black and white. Series II, Vol. VIII. University Press, Cincinnati, Ohio, 1912, pp. 5-47.

CALEY, E. R.: "The Leyden Papyrus X," 3rd century AD. *Journal of Chemical Education*, 3, (10), 1926, pp. 1149-1166.

CENNINI, C.: *Il libro dell'arte*, 14th C., El libro del arte. Commented and notes by Franco Brunello, introduction by Licisco Magagnato, translated from Italian by Fernando Olmeda Latorre. Ediciones Akal. Madrid, 2002.

CLARKE, M.: *The Art of All Colours. Mediaeval Recipe Books for Painters and Illuminators*, Archetype Publications, 2001.

CÓRDOBA DE LA LLAVE, R.C.: "Un recetario técnico castellano del siglo XV: el manuscrito H490 de la Facultad de Medicina de Montpellier," *En la España Medieval*, 28, 2005, pp. 7-48.

COVARRUBIAS, S. de: *Tesoro de la lengua castellana o española*, (ed. by Felipe C. R. Maldonado, rev. by Manuel Camarero), Madrid, Editorial Castalia, 2nd ed. 1995.

DE LA ROJA, J. M.; SAN ANDRÉS, M.; SANCHO, N.: "Tinta de verdigris: síntesis, envejecimiento y valoración de sus efectos sobre el papel." *II Congreso del GEIIC, Investigación en Conservación-Restauración*, Barcelona, 9-11 November, 2005, pp. 75-83, 497.

DE LA ROJA, J. M.; SAN ANDRÉS, M.; SANCHO, N.; SANTOS, S.: "Variations in the colorimetric characteristics of verdigris films depending on the process used to produce the pigment and the type of binding agent used in applying it," *Color Research and Application*, **32**, (5), 2007, pp. 414-423.

DE LA ROJA, J. M.; SAN ANDRÉS, M.; SANCHO, N.; SANTOS, S.; BAONZA V. G.: "Varieties of verdigris containing copper(II) hydroxychlorides and other compounds: preliminary results," Lacona IX. Lasers in the conservation of Artworks (London, 7-11 September, 2011), London, 2012, (in press)

DIOSCORIDES, P.: *Acerca de la materia medicinal y de los venenos mortíferos*, (original title *De materia medica*, 1st C., trans. into Spanish from Greek and com. by Andrés de Laguna, Salamanca, 1566, current edition with pres. by Agustín Albarracín), Madrid, Ediciones de Arte y Bibliofilia, 1983.

EAMON, W.: *Science and the Secrets of Nature: Books of Secrets in Medieval and Early Modern Culture*. N. J. Princeton, Princeton University Press, 1994.

EASTAUGH. N, WALSH. V, CHAPLIN. T, SIDDALL. R.: *Pigment compendium. A dictionary of historical pigments.* Elsevier Butterworh-Heinemann. 2004.

GETTENS, R.J.; STOUT, G.L.: *Painting Materials. A Short Encyclopedia.* Ed. Dover Publications, Inc., New York, 1966.

GIMÉNEZ PRADES, M.; SAN ANDRÉS MOYA, M.; DE LA ROJA, J. M.: *"El color y su significado en los documentos cartográficos del Cuerpo de Ingenieros Militares del s. XVIII,"* Ge-Conservación, 0, 2009, pp. 142-16.

GUAL CAMARENA, M.: *Vocabulario del comercio medieval*, Barcelona, El Albir, 1976.

GUEVARA, F. DE: *Comentarios de la pintura*, s. XVI, Segunda edición reproducida de la edición príncipe. Pórtico y revisión por Rafael Benet. Barcelona, Selecciones Bibliófilas, 1948.

HAWTHORNE, J.G.; SMITH, C.S.: *"On Divers Arts" of Theophilus. The Foremost Medieval Treatise on Painting, Glassmaking and Metalwork (11th-12th C.)* (original title *Schedula Diversarium Artium*), translated into English and notes by J. G. Hawthorne and C. S. Smith, Dover Publications, New York, 1979.

HERACLIUS: "De coloribus et artibus romanorum" (Manuscript of Heraclius), 10th-12th C., in M. P. Merrifield, *Medieval and Renaissance Treatises on the Arts of Painting: Original Texts with English Translations*. Vol. 1. Dover Publications, Inc. New York, 1999, pp. 166-257.

KROUSTALIS, S.: *Edición crítica y estudio de un tratado de tecnología artística medieval: el "Codex Matritensis" 19*, Doctoral Dissertation. Directed by E. Ruiz García. Facultad de Geografía e Historia, Departamento de Ciencias y Técnicas Historiográficas, Universidad Complutense de Madrid, 2005.

LE BEGUE, J.: "Manuscripts of Jehan Le Begue" (1431), in M. P. Merrifield, *Medieval and Renaissance Treatises on the Arts of Painting: Original Texts with English Translations*. Vol. 2, New York, Dover Publications, Inc. 1999, pp. 1-321.

LE BRUN, P.: "Recueil des essaies des merveilles de la peinture" (*Brussels Manuscript*, 1635) in M.P. Merrifield *Medieval and Renaissance Treatises on the Arts of Painting: Original Texts with English Translations*. Vol. 2. Dover Publications, Inc. New York, 1999, pp. 759-841.

LEICESTER, H. M.: *Panorama histórico de la química.* Spanish version by Federico García Portillo, edit. Alhambra, 1967.

LOMBARDI, C.; MARICONDA, M.; SALVATORI, S.; MINIERI, F. R.; SANTAMARIA, U.; VALORI-OIAZZA R.: "Experimental study on the production and use of artificial blue and green pigments (copper acetates and carbonates) from Antiquity to the sixteenth century," *Art et Chimie. La Couleur*, CNRS editions, Paris, 2000, pp. 31-37.

MERRIFIELD, M. P.: *Medieval and Renaissance Treatises on the Arts of Painting: Original Texts with English Translations.* Vol. 1 and 2. Dover Publications, Inc. New York, 1999.

MUÑOZ VIÑAS, S.: "Original written sources for the history of mediaeval painting techniques and materials: A list of published texts," *Studies in Conservation*, 43, 1998, pp.: 114-124.

RAMÍREZ DE ARELLANO, R.: "Miscelánea, Ordenanzas de pintores," *Boletín de la Real Academia de Bellas Artes de San Fernando*, año IX, 33, 1915, pp. 29-46.

RUSCELLI , R. (1640): *Secretos del reverendo Don Alexo Piamontes / traducidos de lengua italiana en castellano, añadidos y remendados en muchos lugares en esta última impresión*. Traducción del italiano al castellano por Vázquez, A y López, M, Antonio Vázquez Impresos de la Universidad, a costa de Manuel Lopez, Alcala de Henares, 1640.

SAN ANDRÉS, M.; DE LA ROJA, J. M.; SANCHO, N.; BAONZA, V. G.: "Verdigris, a Pigment with Different Hues. Relation between Chemical Composition and Colour," *Lasmac 2009 & IMRC09. Proceedings of the 2nd Latin-American Symposium on Physical and Chemical Methods in Archaeology, Art and Cultural Heritage Conservation*, Ed. Universidad Nacional Autónoma de México (UNAM), Universidad Autónoma de Campeche (UAC), Instituto Nacional de Antropología e Historia (INAH), Mexico, 2010, pp. 4-9.

SAN ANDRÉS, M.; DE LA ROJA, J.M.; SANTOS, S.; SANCHO, N.: "Patrones de identificación del verdigrís. Elaboración a partir de la reproducción de recetas antiguas," in *Fatto d'Archimia. Historia e Identificación de Pigmentos Artificiales*, Ed. Secretaría General Técnica. Subdirección General de Publicaciones, Información y Documentación (Ministerio de Cultura), Madrid, 2011, pp. 237-260.

SAN ANDRÉS, M.; J. M. DE LA ROJA, V. G. BAONZA, N. SANCHO: "Verdigris pigment: a mixture of compounds. Input from Raman spectroscopy," *Journal of Raman Spectroscopy*, 41, (11), 2010, pp. 1178-1186.

SAN ANDRÉS, M.; SANCHO, N.; SANTOS, S.; DE LA ROJA, J. M.: "Verdigrís. Terminología y Recetas de Preparación," in *Fatto d'Archimia. Historia e Identificación de Pigmentos Artificiales*, Ed. Secretaría General Técnica. Subdirección General de Publicaciones, Información y Documentación (Ministerio de Cultura), Madrid, 2011, pp. 199-236.

SAN ISIDORO DE SEVILLA: *Etimologías*, (Latin, Span. vers. and notes by José Oroz Reta and Manuela Marcos Casquero), Vols. I and II, Madrid, Biblioteca de Autores Cristianos, 1993.

SANCHO CUBINO, N.: *Revisión y reproducción de los antiguos métodos de obtención de los pigmentos de cobre.* Research project presented for the Advanced Studies Diploma

[Diploma de Estudios Avanzados-DEA]. Directed by M. San Andrés Moya. Facultad de Bellas Artes, Departamento de Pintura (Pintura-Restauración), Universidad Complutense de Madrid, Universidad Complutense de Madrid, 2004.

SANCHO CUBINO, N.; SANTOS GÓMEZ, S.; DE LA ROJA, J. M.; SAN ANDRÉS MOYA, M: "Variación cromática del verdigris en función de su composición química," *Optica Pura y Aplicada*, 37, 2004, pp. 119-123.

SANCHO, N.; SANTOS, S.; DE LA ROJA, J.M.; SAN ANDRÉS M.: "Colorimetrical study of verdigris inks. First results." *10th Congress of the International Colour Association AIC Colour 05*, Granada, 8-13 May, 2005, *Proceedings Book, Part 2,* pp. 1437-1440.

SANT AUDEMAR P. DI: "Liber magistri Petri de Sancto Audemaro de coloribus faciendis" (Manuscript of Pietro di Sant Audemar), 13th-14th C., in M. P. Merrifield, *Medieval and Renaissance Treatises on the Arts of Painting: Original Texts with English Translations.* Vol. 1. Dover Publications, Inc. New York, 1999, pp. 112-165.

SANTOS GÓMEZ, S.: SAN ANDRÉS MOYA, M., BALDONEDO RODRÍGUEZ, J. L., RODRÍQUEZ MUÑOZ, A: "Recetas de preparación del verdigris. Resultados preliminares de la obtención de la variedad conocida como *viride salsum*," *Pátina*, 12, 2003, pp. 41-52.

SANTOS GÓMEZ, S.; SAN ANDRÉS MOYA M.: "Aportaciones de antiguas ordenanzas al estudio de técnicas pictóricas," *Pátina*, 10 y 11, 2001, pp. 266-285.

SANTOS GÓMEZ, S.; SAN ANDRÉS MOYA, M; BALDONEDO RODRÍGUEZ, J. L.; RODRÍQUEZ MUÑOZ, A; DE LA ROJA DE LA ROJA, J. M.; GARCÍA BAONZA, V.: "Procesos de obtención del verdigris. Revisión y reproducción de antiguas recetas. Primeros resultados," *I Congreso del GEIIC. Conservación del Patrimonio. Evolución y Nuevas Perspectivas*, Valencia, 2002, pp. 383-388.

SCOTT, D.: *Copper and Bronze in Art. Corrosion, Colorants, Conservation.* Ed. The Getty Conservation Institute, Los Angeles, 2002.

SCOTT, D.; TANIGUCHI, Y.; KOSETO E.: "The verisimilitude of verdigris: a review of the copper carboxylates," *Reviews in Conservation,* 2, 2001, pp. 73-91.

Secretos raros de artes y oficios, tomo VII. Madrid, imprenta de Repullés, 1807.

Secretos raros de artes y oficios, tomos III, IV y V. Madrid, imprenta de Villalpando 1806.

Secretos raros de artes y oficios, tomos VIII, IX, X, XI y XII. Madrid, imprenta de Villalpando 1807.

SMITH, C. S.; HAWTHORNE, J. G.: "Mappae Clavicula (12th C.). A little key to the world of medieval techniques." *The American Philosophical Society,* New Series, 64, (4), 1974, pp. 1-122.

SORAPAN DE RIEROS, I.: *Medicina Española contenida en Proverbios vulgares de nuestra Lengua*, Granada, Martín Fernández Zambrano, 1st edition, 1616.

THEOPHILUS: "Schedula diversarium artium," 11th C., The various arts, translated from Latin with introduction and notes by C. R. Dodwell. Thomas Nelson & Sons LTD. C.R. Dodwell, 1961.

THOMPSON D. V.: "De coloribus, naturalia excripta et collecta, from Erfurt, Stadtbücherei, Ms. Amplonius Quarto 189 (XIII-XIV Century)," *Technical Studies in the Field of Fine Arts*, III, 1935, pp. 133-145.

THOMPSON, D. V.: "Liber de Coloribus Illuminatorum Siue Pictorum from Sloane Ms., No. 1754" *Speculum*, 1, (3), 1926, pp. 280-307.

THOMPSON, D. V.: "The *De Clarea* of the so-called *Anonymus Bernensis*," *Technical Studies in the Field of the Fine Arts*, I, 1932, pp. 8-19, 69-81.

THOMPSON, D. V.: "More medieval color-making: Tractatus de coloribus from Munich, Staatsbibliothek, Ms. Latin 444," *Isis*, 24, 1935-1936, pp. 382-396.

THOMSON D. V.: "More medieval color-making: Tractatus de coloribus from Munich, Staatsbibliothek, Ms. Latin 444," *Isis*, 24, 1935-1936, pp. 382-396.

GOLD AND ITS MANIPULATION ACROSS MEDIEVAL TREATISES

Eva López Zamora/Consuelo Dalmau Moliner

Introduction

Gold has been the most widely used metal throughout time. Its malleability allows it to adapt to numerous applications (objects of daily use, ceramics, jewels, textiles, adornment of different materials, works of art, etc.) and its longevity, shortage and difficulty of extraction have contributed to its high cost. On the other hand, its colour and sheen have been associated with light and divinity in numerous cultures.

Transformed into gold sheets, it has been used as artistic material for covering different surfaces. Gilding wood using gold leaf was already practised by the Egyptians, whose funeral sarcophagi have been kept in mint condition to the present day. Similarly, Byzantine art resorted to the same metal for wall decorations (paintings and mosaics) and on panel paintings. Though it reached its major brilliance in the medieval stage – especially in Gothic art –, where it was handled with great mastery, it formed an important part of the background of polychrome panels, demonstrating its great versatility.

The aim of this study centres on the analysis of medieval treatises, selecting references and recipes devoted to the manipulation of gold and its specific application on wood. The study also deals with the variety of materials and procedures known and used in this technique during that period.

These early medieval treatises are generally recipe books containing collections of recipes made mostly by monks in the monasteries, places where intellectual and artistic activity was concentrated, serving as the basis of the books of secrets. Therefore, they are written in simple and didactic form, making its use easily understandable for craftsmen who were starting out in the trade. From the 13th century onwards, late medieval texts became more specialised (more structured texts), focusing on the work carried out in craftsmen's workshops, where they were written.

Nowadays, they have become indispensable reference works, contributing extremely valuable information for a stylistic and technical understanding of works of art, since they gather artistic materials and their technologies together in chronological order. Nevertheless, the nature and particular features of these documents mean that they should be interpreted with caution.

Gold: characteristics and extraction process

The characteristics and methods of extraction of this precious metal, as well as the qualities attributed to it, are mentioned in numerous texts. Isidoro of Seville's *Etymologies* (6th century), based on the search for the origin of names, describes in detail the philological derivations and uses of gold, as well as a few other special qualities of the metal, in particular its sheen, how it reflects light, the various reddish tones it can have and the names it receives depending on these tones. The work also describes the different states in which gold might be found (as coins, crafted – in goblets or figures, for example –, as ingots and also as gold sheets).[1]

Its density, unlike the lightness of silver, is already mentioned in the 9th century in the *Mappae Clavicula*[2] and by Eraclius[3] in the low Middle Ages, whereas the monk Theophilus wrote about the origin of different types of gold used by goldsmiths and medieval painters[4] (Arabic gold, Spanish gold, sand gold or Havilah gold).[5]

For Alfonso X in *El Lapidario*, the best gold was found in Spain, in the shape of small nuggets mixed with the sand in rivers. Curiously, he attributes the sun's virtues to gold, regarding it as the most noble metal and holder of certain medicinal properties.[6] Its weight, hardness and sheen are equally remarkable, especially when it is cleaned and burnished.

1. Isidoro, Archbishop of Seville, *Etimologías* (7th century) (bilingual edition, Spanish version and notes by J. Oroz Challenges and M. A. Marcos Casquero, intr. by M. C. Díaz and Díaz), vol. II, books XI-XX, 2nd ed., Biblioteca de Autores Cristianos, Católica, Madrid, 1993, p. 301.

2. Anonymous, "Mappae Clavicula" (12th century). In C. S. Smith, J. G. Hawthorne, "Mappae Clavicula. A Little Key to the World of Medieval Technique." *The American Philosophical Society, Year Book, New Series*, 64, part 4, Philadelphia, 1974, p. 56

3. As Merrifield states, the authorship of this treatise is assumed to Eraclius, name of Greek origin. It is possible that the author was of the same origin or was a native of Italy, of the Duchy of Benevento's Lombardy, which was under the Greek and Saracen influence in the Middle Ages, nevertheless, nothing is known about his biography. It is probable that the text would have several authors from different periods as the date when it was written is unknown. Eraclius, "De coloribus et artibus Romanorum" (centuries X-XIII) ("Eraclius's Manuscript"). In Merrifield, M. P., *Medieval and Renaissance Treatises on the Art of Painting,* (1849); in edition Dover, New York, 1999, pp. 171 y 227.

4. Theophilus, "Schedula diversarium artium" (centuries XI-XII). In J. G. Hawthorne; C. S. Smith, *Theophilus. On Divers Arts. The Foremost Medieval Treatise on Painting, Glassmaking and Metalwork*, (1st ed. Chicago University, 1963), 2nd edition, Dover, New York, 1979, p. 118.

5. City that according to Genesis 2, 10-13, was bathed by the river Trod. *Santa Biblia*, Sociedades Bíblicas Editions, Madrid, 1966, p. 2.

6. "El Lapidario de Alfonso X" (13th century). In: M. Brey Mariño; J. L. Amorós Portolés, *El primer Lapidario de Alfonso X El Sabio. El códice y su texto* (facsimile edition of the Ms. H. I. 15 of the Library of El Escorial), 2 vols., Edilan, Madrid, 1982, pp. 21-22.

From the 13th century and, especially in the 15th century, during which artistic debate took on greater importance, these texts focus mainly on the description of the technique of gilding and its decoration, leaving aside references to the metal as a material together with the location and treatment of the raw material.

Nevertheless, the work of compiling ancient texts of artistic recipes continued, but specialized more on materials. This seems to be the reason why the *Bolognese Manuscript*, written in the 15th century, still contains a recipe describing how gold can be obtained from lapis lazuli.[7]

Another exception to this new situation of ancient manuscripts is *Manuscript H490*, which consists of three sections that include recipes 10 to 30 dealing with the mining industry and metallurgy, especially of gold and silver. It describes three mineral forms where gold is located and the numerous places on the Peninsula where it can be found, in rivers and as ore in metalliferous veins.[8]

The shortage and high price of gold meant that alloys were becoming a recurring theme in these treatises. They enabled imitation gold to be obtained, using in its place other less costly materials that were more widely available to craftsmen. The *Mappae Clavicula* for example, is one of the recipe books with the greatest amount of information on the components and mixes of gold with other predominantly chemical materials (although they are sometimes more like alchemy recipes and practically impossible to reproduce), as a kind of reminder for artists. It contains a wealth of combinations for obtaining metals resembling gold and meticulously describes different metallurgical operations, placing special emphasis on the reduction in the price of the raw material.[9] To a lesser extent it deals with processes of artistic utilization and constitutes a summary of detailed descriptions and samples of a technology prior to the medieval period. It also proposes a test to verify if the metal is in its pure state or whether it is an alloy of other metals.[10] The same treatise sets out two methods of purification for removing silver and other impurities.[11]

Alcherius also describes the method of purifying gold by means of a process involving similar ingredients and treatments.[12] The Spanish text *Ms. H490*

7. Anonymous, "Secreti per Colori" (15th century) ("Bolognese Manuscript"). In M. P. Merrifield, *Medieval and Renaissance Treatises on the Art of Painting,* (1849); in edition Dover, New York, 1999, p. 351.

8. R. Córdoba de la Llave, "Un recetario técnico castellano del siglo XV: el manuscrito H490 de la facultad de Medicina de Montpellier," *En la España Medieval*, 28, 2005, pp. 41-43.

9. Anonymous, "Mappae Clavicula," *rev. cit.*, p. 19.

10. It consists of smelting liquid alum, *Canopia's balsam* and the gold to be analysed: the oxidation produced indicates the presence of other mixed metals.

11. These impurities are eliminated as sulphurs by the emission of sulphurous gases, due to the application of heat. Anonymous, "Mappae Clavicula," *rev. cit.*, pp. 32 and 58.

12. Gold, previously ground with rock salt is mixed with sulphur dissolved in water exposed to the sun. The addition of salt (alum, iron sulphate , caustic soda ashes or common salt) when smelting the metal and alloys, served to make it more fluid and dissolve impurities. J. Alcherius, "De Coloribus Diversis Modis Tractatur in Sequentibus. De diversis coloribus" (14th century) ("Alcherius's Manuscript"). In M. P. Merrifield, *Medieval and Renaissance Treatises on the Art of Painting* (1849); in edition Dover, New York, 1999, p. 305.

gathers equally diverse processes for separating gold from the mineral that contains it, reflecting the methods commonly used in the medieval era (grinding the mineral with oils, adding flux and smelting; cementation; the method of making an amalgam of gold with mercury; the sulphur method).[13] The monk Theophilus also explains how to recover gold from copper or silver gilded objects by applying high temperatures.[14]

Obtaining and applying gold sheets

The technology of artistic gilding applied on various surfaces – especially on wood panel paintings – was developed mainly in the Middle Ages, the same as with decorating gilded surfaces using a wide range of methods such as niello, enamelling and distempering or the decorations engraved on them. Treatises from that period contain interesting information about pure gold leaf and its varieties, as well as on the diverse hues of this material or its imitations.

Gold was reduced to fine sheets by hammering, to adapt it to the surfaces to be gilded. This operation was called beating, and was the responsibility of the goldbeaters. The process consisted of several phases: refining, laminating and making the gold leaf.[15] It is a process that has been stood the test of time; the only change is that beating is no longer carried out manually, but using steel weights, and the purity of gold leaf now varies from 20 to 24 carats, instead of the previous 23 carats.[16]

The first references to the technique of manufacturing metal sheets and its application on different surfaces are recorded in the *Codex Lucensis*, also known as *Lucca's Manuscript*[17] and especially in the *Mappae Clavicula*. According to the author, the process starts with an ingot formed by the alloy of one ounce of gold and one ounce of silver that gave it a whitish tone, which was beaten to obtain a thin leaf that was then cut into square pieces.[18]

13. R. Córdoba de la Llave, "Un recetario técnico castellano del siglo XV: el manuscrito H490 de la facultad de Medicina de Montpellier," *En la España Medieval*, 28, 2005, pp. 41 and 47.

14. Theophilus, "Schedula diversarium ...," *op. cit.*, pp. 71 and 147.

15. Gold sheets are the result of reducing the metal ingot to thin sheet by laminating, the initial thinning phase, used by goldsmiths to coat other metals. "Leaves" are much thinner sheets of only a few microns of thickness, obtained by subjecting them to soldering and moulding, used not only by goldsmiths but also by painters and gilders on paintings and sculptures. This difference in the use of gold reduced down to different thicknesses can be seen in the treatises: the oldest texts mention mainly the use of sheets, whereas in the most modern reference is generally made to leaves.

16. The first visual reference of the practice of hammering gold to obtain leaves appears in the wall paintings of Re'hem's Egyptian tombs in Deir the-Gabrâwi (tomb no. 72), of 2300 B.C., and in Mereruka, Saqqara, of 2323-2291 B.C.

17. Anonymous, *Compositiones ad tigenda musiva* (8th century) ("Lucca's Manuscript," *"Codex Lucensis 490 of the Lucca Capitular Library"*) (translation and comments by H. Hedfors, *Almqvist and Wiksells Boktryckeri*, Uppsala, 1932, pp. 21-23.

18. Anonymous, "Mappae Clavicula," *rev. cit.*, p. 65.

The process of laminating and beating gold was widespread in the Middle Ages; it may have been the reason why Theophilus did not include it in his text.[19] Nevertheless, it is possible that this was the first occasion on which mention is made of the preparation of a gold leaf set, made using Byzantine papyrus, similar to those sold today. Two centuries later, Cennini[20] commented on the qualities of gold leaf, including his own preferences on where to acquire it and the most suitable types of leaves.

False gold leaf consists of any beaten metal that is not pure gold (it may contain small proportions of the noble metal or it may be of a similar colour), such as tin, silver, *oropel*[21] (brass leaf) or part gold (a thin sheet of gold beaten together with another thicker sheet of silver and then welded), from which gold colour leaves can be obtained. This likeness is achieved not only by similar colours of the alloy metals in these false leaves, but also by subsequently applying ink, paint or coloured varnish to give the appearance of gold.

Theophilus provides the only recipe for obtaining part gold,[22] whereas various other authors use sheets of tin, which they cover with a range of ingredients to give the desired tone, such as Eraclius[23] and the *Mappae Clavicula.*[24]

Audemar does not specify the type of metal that his sheets are made of and which he coated with several mixtures to imitate gold, justifying himself by citing the high price of the precious metal.[25] Le Begué, on the other hand, used sheets similar to traditional brass tinsel, obtained by fusing together brass, zinc oxide and tin.[26]

Gold powder. Inks and metal powders

For specific areas, artists were applying layers of colour that imitated gold and so avoided increasing the price of works because of the high price of this

19. Theophilus, "Schedula diversarium ...," *op. cit.*, pp. 29-30.

20. C. Cennini, *El Libro del* Arte (s. XIV), Akal, Madrid, 1988, pp. 174-175.

21. False gold leaf par excellence is *tinsel*, which has been called many names throughout history, such as *bronze leaf, metallic leaf* or *Dutch leaf.* It is an alloy consisting of copper, tin or zinc, with similar appearance to thin gold foil, but with the advantage of being easy to manage due to its higher density and weight. It has been traditionally used to decorate works in interiors not susceptible to damp, as otherwise it oxidizes and blackens, besides being unstable in air.

22. Theophilus, "Schedula diversarium ...," *op. cit.*, p. 156.

23. Eraclius, "De coloribus ...," *op. cit.*, pp. 221, 241.

24. Thin sheets are bathed in vinegar and alum and covered with a mixture of saffron and parchment glue, or saffron and orpiment, a mixture to which regurgitated grasses or linseed oil and glue can be added. Anonymous, "Mappae Clavicula," *rev. cit.*, pp. 36 and 44.

25. P. S. Audemar, "Liber Magistri Petri de Sancto Audemaro de coloribus faciendis" (13th century) ("Audemar's Manuscript"). In M. P. Merrifield, *Medieval and Renaissance Treatises on the Art of Painting* (1849); in edition Dover, New York, 1999, pp.161-163.

26. J. Le Begue, "Experimenta de Coloribus" (1431) ("Le Begué Manuscript"). In M. P. Merrifield, *Medieval and Renaissance Treatises on the Art of Painting,* (1849); in edition Dover, New York, 1999, p. 81.

precious metal. These inks were made from gold powder (authentic or fake), as well as vegetable and mineral pigments plus other ingredients and additives.

Eraclius[27] obtained inks for gilding different objects by mixing mercury and gold. Theophilus also explained how they are made and described their use for decorating books, mixed with fish glue.[28]

These inks could also be made using red lead, sand, gold and alum filings, as well as vinegar and orpiment (possibly to give a deeper yellow tone) dissolved in water-glue. The mixture obtained had the consistency of wax, so it was used not only on wood panels but also to seal envelopes, as described in the *Mappae Clavicula*. The anonymous author of this treatise prepared other inks incorporating other metals (copper, lead, tin) in addition to gold.[29]

The use of similar inks that contain gold continued up to the 15th century, when the *Bolognese Manuscript*[30] still records several simple formulae in which ground gold is mixed with vegetable juice or salt and honey. Nevertheless, Le Begue[31] only mentioned one recipe for yellow dye that included pure gold as an ingredient, alongside others using imitation gold.

In these manuscripts, there are many formulae of inks that simulate the colour of gold, but do not contain gold in their composition. Again, as in the case of gold inks, the *Mappae Clavicula*[32] is the recipe book containing the highest number of metal powders, though in this case production is more complex. Because they consist of a larger number of ingredients (plant juices – blackberry, fig or grenade –, brass powder, cadmium, *misy* – ferric sulphate, *elidrium* – arsenic sulphide resins, egg white, glue, orpiment, animal gall, saffron, sulphur, alum or sandarac). Other texts that propose compositions with similar ingredients to imitate gold (egg white, alum or sulphur, mercury, tin, ammonium salt), are *De arte illuminandi*,[33] the Le Begué text,[34] the *Bolognese Manuscript*[35] and the *Ms H490*.[36] These inks would have consequently lowered the price of pieces on sheets of copper, brass, tin, etc., both in goldsmithing work and panel paintings.

27. Eraclius, "De coloribus ...," *op. cit.*, pp. 221, 241.

28. Theophilus, "Schedula diversarium ...," *op. cit.*, pp. 28 and 34.

29. Anonymous, "Mappae Clavicula," *rev. cit.*, pp. 33 and 34-35.

30. Anonymous, "Secreti per Colori," *op. cit.*, pp. 465-467.

31. J. Le Begue, "Experimenta de ...," *op. cit.*, p. 65.

32. Anonymous, "Mappae Clavicula," *rev. cit.*, pp. 33 and 35-37.

33. Anonymous, *De arte illuminandi. E altri trattati sulla tecnica della miniatura medievale* (the second half of the 14th century) (Franco Brunello's translation and comments), Neri Pozza Editore, Vicenza, 1975, pp. 55-59.

34. J. Le Begue, "Experimenta de ...," *op. cit.*, pp. 57, 65.

35. Anonymous, "Secreti per Colori," *op. cit.*, pp. 458, 460-461, 465, 467, 473-475.

36. R. Córdoba de la Llave, *Un recetario ...*, *op. cit.*, pp. 45-46.

With regard to specific technologies for applying inks on metals, Eraclius[37] notes the need to use a knife to scrape the metallic surfaces that are to be coated with ink, a fact also mentioned by Alcherius,[38] together with the recommendation of applying ox gall in successive layers to improve adhesion.

The *Mappae Clavicula* proposes a simple method based on the application of mercury to give gilded colour to silver and copper,[39] while Le Begué describes two inks that are applied warm on metals and consist of sulphur, orpiment, and alum with urine.[40]

The process of gilding on wood

Wood provides one of the most frequent and adapted bases for artistic gilding. Proof of this are Byzantine painted icons and later Gothic panel paintings. The wide variety of recipes that refer to golden wood in which gold complements the polychromy of panel paintings, show how the methodology followed was perfected over the course of centuries.

The versatility of gold produces a range of different appearances in the end result, ranging from shiny to matte,[41] depending on the various adhesives and treatments applied. Shiny effect gilding is the one most frequently described in the texts, however, matte gilding is not clearly described until the 16th century, in spite of the fact that earlier texts contain numerous recipes using mordant for this technique.

Texts from this period give detailed descriptions of steps and precautions that should be taken during the phase of preparing the surface for gilding, the importance of these operations for the end result of the process was well known.

One of the first premises highlighted in relation to preparation is the one concerning weather conditions; guidelines on this topic are set out by several commentators: Audemar,[42] Alcherius[43] or Cennini.[44]

37. Eraclius, "De coloribus ...," *op. cit.*, p. 199.

38. J. Alcherius, "De Coloribus ...," *op. cit.*, p. 309.

39. Anonymous, "Mappae Clavicula," *rev. cit.*, p. 40.

40. J. Le Begue, "Experimenta de ...," *op. cit.*, p. 79.

41. Traditionally they have been grouped on the basis of the adhesive, that is to say, water-based (water gilding) or oily (mordant gilding). But some authors base the classification of the technique on the final appearance of the surface (shiny or matte). They therefore propose the term of burnished gold (shiny) as there are recipes for gilding in which the adhesive is water-based but do not include burnishing, and matte gilding; this classification consequently avoids any ambiguity.

42. P. S. Audemar, "Liber Magistri ...," *op. cit.*, pp. 155-157.

43. J. Alcherius, "De Coloribus ...," *op. cit.*, pp. 263 and 265.

44. C. Cennini, *El Libro ...*, *op. cit.*, pp. 153-156.

Early medieval treatises do not follow the methodology set out in several stages of preparation followed later and still used today (application of animal glue, various strata of thick and thin gypsum, and bole). According to the *Mappae Clavicula*[45] only a simple layer of saffron and egg is applied, whereas other versions are described in the Finnish text *Líkneskjusmíð*[46] and the *Bolognese Manuscript.*[47] In the case of coloured preparations, the strata of gypsum and priming are unified; this coloured priming may consist of bole or other materials such as tile powder.

At the end of the surface preparation phase, reliefs and paste work were applied, which comprised part of the gold decoration and imitated the work of the goldsmith. Cennini describes the production process minutely and sets out the necessary guidelines for subsequent dry and wet polishing with pumice powder[48] to prepare the surface to receive the priming on which the gold was laid.

The characteristics of gold and its final appearance are equally determined by the type of adhesive used; "lean," as in the case of water-based adhesives, or "fat," that includes drying oils. In this way, shiny gilding obtained by polishing gold with the burnisher, is done using water-based adhesives (*a la templa* or *al guazzo*) and generally begins with coloured priming. The meticulous description of the application process for the *templa* and gold sheets can be found in Cennini's text.[49] Both the latter and Theophilus before him emphasised the importance of making one's own workshop tools, including instructions on how to make the burnishers.[50]

Matte gilding with water-based mordant differs from shiny gilding in that the polishing stage is eliminated. This method is described by *Alcherius's Manuscript*,[51] which replaced egg white water-based (which provides the malleability required for burnishing) with glue, and the author wrongly states that this makes the surface more flexible.

Oil matte gilding, also without sheen, very resistant and destined for specific areas, is dealt with by Le Begué, who proposed two mordants resistant to the inclemency of the weather, formed by a mixture of pigments or loads, linseed oil, glaze and glue.[52]

45. Anonymous, "Mappae Clavicula," *rev. cit.*, p. 60.

46. U. Plahter, "Líkneskjusmíð, 14th Century Instructions for Painting from Iceland," *Norwegian Medieval Altar Frontals and Related Materials* (Acta ad Archaeologiam et Artium Historiam Pertinentia, vol. XI) (papers for the Conference in Oslo, 16th to 19th December 1989), Institutum Romanum Norvegiae, Giorgio Bretschneider Ed., Roma, 1995, p. 171.

47. Anonymous, "Secreti per Colori," *op. cit.*, pp. 469, 471, 473 and 477.

48. C. Cennini, *El Libro ...*, *op. cit.*, pp. 155, 158-160.

49. *Ibid*, pp. 169-174.

50. Theophilus, "Schedula diversarium ...," *op. cit.*, p. 114.

51. J. Alcherius, "De Coloribus ...," *op. cit.*, pp. 267-269.

52. J. Le Begue, "Experimenta de ...," *op. cit.*, p. 95.

With regard to gilding with false gold, Theophilus described the process of obtaining false leaf using tin covered with golden lacquer, before applying it to the surface. He also shows he was familiar with polishing with translucent paint, consisting of first applying tin leaf, which once laid was covered with some type of mixture to imitate gold.[53] The author of the *Líkneskjusmíð*[54] refers to silver leaves applied on wood, with a gilded ink to imitate gold leaf gilding.

The absence of sheen in false gilding is replaced by applying glazes. One of them is *auripetrum* as recommended by Eraclius,[55] which fulfils the double function of protecting the gilding and masking any possible areas where the plaster is visible. Once the gold is sealed and smoothed, varnishing can be also done with a layer of alum, water, mercury, ochre, saffron, glue and animal gall, as described in the *Mappae Clavicula.*[56]

Unlike silver, gold does not blacken since it does not undergo oxidation, but despite this, it is advisable to keep it clean of superficial dirt and the darkening produced by the passage of time to preserve its sheen and delicate appearance. Theophilus[57] advised soap diluted in water to clean dirt such as traces of powder or wax, and he eliminated darkening by rubbing the metal with a damp cloth and sieved coal powder, washing away any excess with water. To recover sheen he recommended rubbing it with a cloth and chalk.

Conclusions

The study of ancient documentary sources reveals that the process of gilding and its principal raw material, gold, has had a presence and major importance from the dawn of civilization. Its symbolic character, connecting it to different divinities in various cultures, contributed to the development of a methodology for manufacturing by-products of this metal that were more manageable and less costly, with the purpose of applying them to religious or profane objects.

Its artistic use reached its heyday in the Middle Ages, at first as a largely unrefined technique and for applications mostly related to metallurgy or to the art of goldsmithing. Later, gilding was refined to make it suitable for application in painting and sculpture, and this is reflected in the treatises, which clearly trace the gradual development of the manufacture of increasingly thin metallic leaves, while at the same time the range and origin of other materials involved was increased to imitate or achieve its aesthetic effects.

53. Theophilus, "Schedula diversarium ...," *op. cit.*, pp. 33-34.
54. U. Plahter, "Líkneskjusmíð ...," *rev. cit.*, p. 171.
55. Eraclius, "De coloribu s...," *op. cit.*, p. 225 and 241.
56. Anonymous, "Mappae Clavicula," *rev. cit.*, pp. 34 and 64-65.
57. Theophilus, "Schedula diversarium ...," *op. cit.*, pp. 157-158.

Artistic treatises were generally produced for teaching and were based on assumptions inherited from the past. In many cases, they are literal copies of others, or contain obvious disparities from transcriptions or subsequent mistranslations, deliberate or involuntary omissions of recipes, all of which may have caused confusion or error. This means that they need to be read and reviewed critically.

These texts show the survival of the tradition, in spite of the fact that over the course of the centuries it is apparent that the techniques and recipes described have evolved for the sake of achieving a more enduring and beautiful end result.

All this demonstrates the relevancy of the study of the methodology followed in ancient text sources, in order to gain an understanding of the particular features of the development of different artistic techniques. This knowledge can have an impact on the improvement of interventions for the conservation and restoration of gold surfaces, while at the same time facilitating dating and the stylistic, iconographic and iconological understanding of the variety of artistic manifestations, that contain gold as one of their artistic elements, in any of the forms described.

Finally, it can be verified once again that treatises based on the knowledge of the craft mostly contain widely practised recipes that have remained valid over time, as can be seen in texts from subsequent periods. The coincidence between the practices described in these texts and work produced using the methodology described is very much in evidence today when they are studied using modern methods of analysis.

Acknowledgements

Thanks are due to the Complutense University of Madrid for awarding a pre-doctoral scholarship to Eva López Zamora, and to the Department of Painting (Painting-Restoration) of the Fine Arts Faculty for their collaboration.

Bibliography Consulted

ALCHERIUS, J., "De Coloribus Diversis Modis Tractatur in Sequentibus. De diversis coloribus" (14th century) ("Alcherius's Manuscript"). In Merrifield, M. P., *Medieval and Renaissance Treatises on the Art of Painting* (1849); in edition Dover, New York, 1999, pp. 258-321.

ANONYMOUS, *De arte illuminandi. E altri trattati sulla tecnica della miniatura medievale* (the second half of the 14th century) (Franco Brunello's translation and comments), Neri Pozza Editore, Vicenza, 1975.

ANONYMOUS, *Compositiones ad tigenda musiva* (8th century) ("Lucca's Manuscript," *"Codex Lucensis 490 of the Lucca Capitular Library"*) (translation and comments by H. Hedfors, H), Almqvist and Wiksells Boktryckeri, Uppsala, 1932.

ANONYMOUS, "Mappae Clavicula" (12th century). In Smith, C. S; Hawthorne, J. G., "Mappae Clavicula. A Little Key to the World of Medieval Technique." *The American Philosophical Society, Year Book, New Series*, 64, part 4, Philadelphia, 1974, pp. 3-122.

ANONYMOUS, "Secreti per Colori" (15th century) ("Bolognese Manuscript"). In Merrifield, M. P., *Medieval and Renaissance Treatises on the Art of Painting* (1849); in edition Dover, New York, 1999, pp. 325-600.

AUDEMAR, P. S., "Liber Magistri Petri de Sancto Audemaro de coloribus faciendis" (13th century) ("Audemar's Manuscript"). In Merrifield, M. P., *Medieval and Renaissance Treatises on the Art of Painting* (1849); in edition Dover, New York, 1999, pp. 112-165.

CENNINI, C., *El Libro del* Arte (14th century), Akal, Madrid, 1988.

CÓRDOBA DE LA LLAVE, R., "Un recetario técnico castellano del siglo XV: el manuscrito H490 de la Facultad de Medicina de Montpellier," *En la España Medieval*, 28, Complutense University, Madrid, 2005.

"El Lapidario de Alfonso X" (13th century). In: Brey Mariño, M; Amorós Portolés, J. L., *El primer Lapidario de Alfonso X El Sabio. El códice y su texto* (facsimile edition of the Ms. H. I. 15 of the Library of El Escorial), 2 vols., Edilan, Madrid, 1982.

ERACLIUS, "De coloribus et artibus Romanorum" (centuries X-XIII) ("Eraclius's Manuscript"). In Merrifield, M. P., *Medieval and Renaissance Treatises on the Art of Painting* (1849); in edition Dover, New York, 1999, pp. 166-257.

ISIDORO, Archbishop of Seville, *Etimologías* (7th century) (bilingual edition, Spanish version and notes by J. Oroz Challenges and M. A. Marcos Casquero, intr. by M. C. Díaz and Díaz), vol. II, books XI-XX, 2nd ed., Biblioteca de Autores Cristianos, Católica, Madrid, 1993.

LE BEGUE, J., "Experimenta de Coloribus" (1431) ("Le Begué Manuscript"). In Merrifield, M. P., *Medieval and Renaissance Treatises on the Art of Painting* (1849); in edition Dover, New York, 1999, pp. 1-321.

MERRIFIELD, M. P., *Medieval and Renaissance Treatises on the Art of Painting* (1849); in edition Dover, New York, 1999.

PLAHTER, U., "Líkneskjusmí ð, 14th Century Instructions for painting from Iceland," Norwegian Medieval Altar Frontals and Related Materials," (Record ad Archaeologiam et Artium Historiam Pertinentia, vol. the XIth) (papers for the Conference in Oslo, 16th to 19th December 1989), Institutum Romanum Norvegiae, Giorgio Bretschneider Ed., Rome, 1995, pp. 157-171.

Santa Biblia, Ediciones Sociedades Bíblicas, Madrid, 1966.

THEOPHILUS, "Schedula diversarium artium" (centuries XI-XII). In Hawthorne, J. G.; Smith, C. S., *Theophilus.On Divers Arts. The Foremost Medieval Treatise on Painting, Glassmaking and Metalwork* (1st ed. Chicago University, 1963), 2nd edition, Dover, New York, 1979.

THEOPHILUS' *SPANISH GOLD* AS EVIDENCE OF SPAIN'S ROLE IN THE DISSEMINATION OF TECHNIQUES

Spike Bucklow

Medieval Spain was a sustained and celebrated centre of learning. It had a major role in providing the means whereby the heritage of Greece – maintained and developed by the Islamic tradition – entered Christendom. The Christian tradition then enthusiastically adopted and adapted Greek science and philosophy. Spanish Jews, Christians and Muslims made original spiritual, philosophical and scientific contributions.[1] Medieval Spain also provided an ambience in which enormous quantities of translations were undertaken. For example, in Toledo, Gerard of Cremona and William of Moerbeke translated hundreds of texts between them.[2]

It follows that medieval Spain probably had a role to play in the spread of craft treatises. The dissemination and influence of spiritual, philosophical and scientific texts originating in, and passing through, Spain has been widely studied. This conference is a welcome contribution to the study of craft treatises, which have hitherto been relatively neglected – a neglect that is symptomatic of the low status enjoyed by the crafts in modern Europe.

The classical academic tools of study have been honed on texts by authorities such as Aristotle and Ptolemy. Some of these tools are also appropriate for approaching craft treatises. However, differences in form and function between texts that relate to the practical crafts and those that relate to more theoretical subjects mean that the same tools are not always appropriate or have not been applied.

Some classical tools of study such as philology and linguistics can be used to trace Spain's influence in craft treatises (See, for example, the paper by Luis Afonso and Antonio Cruz on Ibn Hayyim's *O Livro de Como se Fazem as Cores*, in this volume). Another example that shows Spain's influence in a

1. M. R. Menocal, *Ornament of the World*, Little Brown and Co, Boston, 2002.

2. E. Grant (ed.), *A source book in medieval science*, Harvard University Press, Cambridge, MA, 1974, pp. 35-41.

widely disseminated text is the *Mappae Clavicula*. Some copies of this manuscript have recipes that contain ingredients referred to by Arabic names, and transliterated with a 'z'. Thus, lead is rendered 'arrazgaz' (modern transliteration, al-raṣāṣ), tin is 'alcazar' (al-qaṣdīr) and sulphur, 'alquibriz' (al-kibrīt).[3] This suggests that the recipes, or the ingredients, may have had an Arabic origin. It also suggests the participation of a Spanish scribe at some stage in the manuscript's transmission.

However, pursuing such evidence is at present hampered by the fact that the vast majority of craft treatises are not readily available to scholars. The classical tools of study can be applied to individual craft treatises, but cannot – at present – be applied with equal success or confidence to generating stemma of craft treatises. On the other hand, craft treatises have the advantage of being associated with the products of craft activities. The study of surviving works of art – material culture – can therefore compliment the study of treatises and manuscripts.

One durable product of a craft tradition that illustrates the influence of Spain on the dissemination of techniques across Europe is Spanish leather. Spanish leather has a long history of use for mural hangings, screens, upholstery, altar frontals, ecclesiastical vestments, footwear, gloves, pouches and caskets.[4] In the mid-eighteenth century, English craftsmen were manufacturing Spanish leather as a luxury decorating and furnishing material.

Before being produced in England, Spanish leather was imported in some scale. In the mid-seventeenth century, one Spanish leather workshop in Amsterdam (Le Maire) consumed 16,000 French calf-skins per year and 20,000 leaves of silver every six weeks in its production. Earlier, in the sixteenth century, Spain was also a major exporter of leather and produced quantities for its own domestic market. For example, Cordoba craftsmen delivered 10,490 pieces in a single commission for Philip III's palace in Valladolid.[5] In Spain it was not known as Spanish leather – it was guadamecies. The name comes from Ghadames, a North African town near the present Libyan/Tunisian/Algerian borders and a centre of production from about 600AD.

The pattern of manufacture of Spanish leather shows the import of an Islamic craft into Spain and its consequent export across Europe. The distribution of this craft technique has parallels with what was implied by the *Mappae Clavicula* with its transliterated ingredients and eventual Europe-wide circulation. Given that the study of craft treatises can be complemented by the study

3. *Mappae Clavicula*, 195-203, in C. S. Smith and J. G. Hawthorne, "The Mappae Clavicula," *Transactions of the American Philosophical Society*, 64, 4, 1974, p. 57-8.

4. J. W. Waterer, *Spanish Leather, a history of its use from 800 to 1800*, Faber and Faber, London, 1971.

5. E. Koldeweij, "How Spanish is Spanish leather?" *IIC Madrid Congress*, IIC, London, 1992, pp. 84-8.

of crafted products, I would like to consider paintings in the light of two recipes written by a practising northwest German craftsman in the 1120's.[6]

Theophilus', *On Divers Arts*, contains two recipes that are described as Spanish. They are Spanish green[7] and Spanish gold.[8] Theophilus also refers to Spanish brass in the context of casting silver and stamping designs.[9] This is a material that Theophilus uses but does not define – we must assume that it is either brass from Spain, or brass made to an undefined Spanish recipe. However, Theophilus describes both Spanish green and Spanish gold and they beg the question – what is Spanish about them?

Unfortunately, with these two products, the Spanish leather parallel is unhelpful. Production of Spanish leather was fashion-led and mobile – its distribution can be traced chronologically and geographically. However, Spanish green's distribution cannot be traced because it is more or less ubiquitous. Both the recipe and the product were found all over Europe– its popularity only declined in the eighteenth century. Spanish gold's distribution cannot be traced because the product is actually non-existent.[10]

Since, in these cases, a stemma of manuscripts and a material culture approach are both problematic, we will look at Spain's influence by examining the ideas implied in the recipes – examining conceptual relationships between recipes and within recipes.

Spanish green gives no indication of any explicitly Spanish ingredients or practices. Theophilus presents it as a minor variant on the preceding recipe (which is also for a green) and it is followed by another variation on the same recipe (this time for a white pigment). Both these variants, Spanish green and ceruse,[11] became very widespread artists' pigments. But Theophilus treats them both as conceptual dependents of salt green.[12] Salt green is much less commonly encountered than Spanish green or ceruse and is rarely reported in the conservation literature. Green pigments containing both copper and chlo-

6. L. White, "Theophilus redivivus," *Technology and Culture*, 5, 1964, pp. 224-33.

7. Theophilus, *On Divers Arts*, I, 36, tr. J. G. Hawthorne and C. S. Smith, Dover, New York, 1979, p. 41.

8. Theophilus, *On Divers Arts*, III, 48, tr. J. G. Hawthorne and C. S. Smith, Dover, New York, 1979, p. 119-20.

9. Theophilus, *On Divers Arts*, III, 30 & 75, tr. J. G. Hawthorne and C. S. Smith, Dover, New York, 1979, pp. 106 & 154.

10. For the textual tradition, see: C. Opsomer and R. Halleux, "L'alchimie de Théophile," *Comprendre et maitriser la nature au Moyen Age. Mélanges d'Histoire des Sciences en l'honneur de Guy Beaujouan*, Genève 1994, pp. 437-459. My thanks to D. Oltrogge and M. Clarke for bringing this work to my attention.

11. Theophilus, *On Divers Arts*, I, 37, tr. J. G. Hawthorne and C. S. Smith, Dover, New York, 1979, p. 41-2.

12. Theophilus, *On Divers Arts*, I, 35, tr. J. G. Hawthorne and C. S. Smith, Dover, New York, 1979, p. 41.

rine – such as were found in Zuccaro's *Calumny of Apelles* (HKI 2077) – are usually identified as naturally occurring atacamite (itself rare) rather than synthetic salt green. So rare is salt green that many modern reference sources conflate it with Spanish green.[13] We treat salt green as a variant on Spanish green – inverting the relationship Theophilus outlined. For Theophilus, it seems that the green with greatest practical importance was the least theoretically important of the two greens.

Theophilus' Spanish green survived in the modern Dutch (spaans groen) and German (grünspan) but became known in English as verdigris and in French as vert-de-gris or Greek green. In Spain it has a variety of names including cardenillo, a word related to cardenal, the word for bruise.[14] The *Mappae Clavicula* (transmitted, at least in part, through Spanish hands) calls the pigment Greek green. The reason for calling it Greek is obvious. The pigment is one of the few unambiguously described by Theophrastus and other ancient Greeks.[15] The reason for calling it Spanish is not quite so obvious.

However, Spain was a major conduit whereby knowledge of the Greek world passed into medieval Europe. So could it be that some craftsmen recognised the country through which knowledge of the pigment entered medieval Europe? And might others have recognised the country from which knowledge of the pigment supposedly originated? We can only speculate. But there is a technical detail about Theophilus' two greens that reinforces the conceptual link between Greece as (the theoretical) originator and Spain as (the practical) disseminator of the recipe.

Spanish green is simply a salt-free version of salt green and the presence or absence of salt raises questions about alchemy.[16] European alchemy was one of the medieval sciences that derive directly from Greek science and Spain was a recognised centre of alchemical learning. The identification of a pigment with Spain or Greece could therefore imply an alchemical connection. Its alchemical origins were explicitly acknowledged some centuries later.[17]

It is possible that Spanish green and ceruse were presented as minor variants on salt green because they both only have two ingredients – either copper or lead and either vinegar or urine (in oak). Salt green, the conceptually superior pigment, on the other hand, has three ingredients – copper and vinegar plus salt (in honey). For Theophilus, a well-educated Benedictine,[18] the pigment whose

13. N. Eastaugh, V. Walsh, T. Chaplin and R. Siddall, *The Pigment Compendium*, Elsevier, London, 2004.

14. J. Casares, *Diccionario ideológico de la lengua Española*, Gustavo Gili, Barcelona, 1959.

15. J. Hill, *Theophrastus, History of Stones*, Davis, London, 1774, p. 225.

16. Albertus Magnus (attrib.), *Libellus de Alchemia*, cited in G. Roberts, *The Mirror of Alchemy*, British Library, London, 1994, p. 111.

17. C. Cennini, *The Craftsman's Handbook*, LVI, tr. D. V. Thompson, Dover, New York, 1960, p. 33.

18. J. Van Engen, "Theophilus Presbyter and Rupert of Deutz," *Viator*, 11, 1980, pp. 147- 64.

composition reflects the Holy Trinity would be more important than the two related pigments that do not reflect the Trinity.[19] There are also three ingredients in the other recipe that Theophilus calls Spanish.

Like the first recipe, Spanish green, Spanish gold also gives no indication of any Spanish ingredients or practices. Spanish gold has been dismissed as "nonsense"[20] that has "attracted far more attention than it merits" and has been interpreted as a poorly understood account of making brass or cementation.[21] (This interpretation is very questionable since Theophilus elsewhere demonstrates a mastery of both brass[22] and cementation[23]). However, like the presence of salt in Theophilus' more important green, the transmutation of a base metal into gold again suggests an alchemical connection.

Analysis of the recipe in alchemical terms suggests that Spanish gold is a primarily conceptual statement. It might be nonsense from a practical point of view, but from a theoretical point of view it is important. And, just as the theoretically important salt green spawned the practically important Spanish green, so Spanish gold also spawned a practical pigment – Mosaic gold.

Mosaic gold is not mentioned (or known?) by Theophilus but it is a very practical recipe and the product looks like shell gold. The recipe may again have some alchemical connection, since it is the 'gold of Moses' and Moses was considered to be one of the early alchemists.[24] Mosaic gold is not the only pigment that is conceptually dependent upon the theoretical content of Theophilus' Spanish gold recipe – its theoretical influence can be discerned in a family of synthetic yellow pigments (lead-tin yellow, Naples yellow, etc).[25]

The conceptual relationship between recipes can be summarised by noting that, like Spanish gold, Mosaic gold is the conversion of a base metal (tin) into gold using three ingredients. Cennini does not claim that Mosaic gold is a metal – he calls it a "colour" "similar to gold."[26] But he hedges his bets by placing the recipe amongst those for gilding, not those for pigments. Cennini identifies the three ingredients that convert tin into gold as sulphur, mercury and salt. So the recipe has an obvious debt to Jabir's Sulphur-Mercury theory

19. Albertus Magnus, *De Caelo et Mundi, ab initio,* cited in V. F. Hopper, *Medieval Number Symbolism*, Columbia University Press, New York, 1938, p. 94.

20. L. White, *Medieval Religion and Technology*, University of California Press, 1978, p. 100.

21. J. G. Hawthorne and C. S. Smith in Theophilus, *On Divers Arts*, III, 48, tr. J. G. Hawthorne and C. S. Smith, Dover, New York, 1979, p. 119, n.1.

22. Theophilus, *On Divers Arts*, III, 66, tr. J. G. Hawthorne and C. S. Smith, Dover, New York, 1979, pp. 143-4.

23. Theophilus, *On Divers Arts*, III, 33 & 34, tr. J. G. Hawthorne and C. S. Smith, Dover, New York, 1979, pp. 108-10.

24. C. Crisciani, "The conception of alchemy as expressed in the *Pretiosa Margarita Novella* of Petrus Bonus of Ferrara," *Ambix*, 20, 3, 1973, p. 171.

25. S. Bucklow, *The Alchemy of Paint*, Marion Boyars, London, 2009, pp. 75-108.

26. C. Cennini, *The Craftsman's Handbook*, T., M., CLIX, tr. D. V. Thompson, Dover, New York, 1960, p. 101-2.

and to the "trinity of natural phenomena." The same structure is evident in Theophilus' earlier recipe for Spanish gold.

In Spanish gold, copper (rather than tin) is converted into gold with a tripartite mixture of vinegar, the dried blood of a red haired man and the ashes of a basilisk. Vinegar is unremarkable – we have seen it used with copper in Spanish green – but the dried blood is a bit suspicious and the basilisk ash makes it evident that this is not a practical recipe. (Theophilus describes how to breed basilisks in his recipe – they hatched from hens' eggs incubated by toads and fed on soil in containers that keep the craftsman safe from their fatal glance, hiss and breath.)

The unlikely ingredients and impractical details (note that the egg is incubated by a cold-blooded creature) identify the recipe as a philosophical speculation. In the thirteenth century, basilisk ash was identified as a product with apotropaic properties[27] and as an alchemical elixir.[28] It has recently been interpreted as Mercury – Jabir's watery principle in his theory of metals.[29] Such an interpretation would imply that the other suspicious ingredient might correspond to Sulphur, Jabir's other principle. And, in the relevant paradigms (Aristotelian elements, Galenic humours and the doctrine of signatures) the dried blood of a red-haired man is indeed appropriate to represent the fiery principle.

Spanish gold is a medieval version of a modern thought experiment. Theophilus does not intend that we risk a hideous death by breeding basilisks any more than Einstein intends us to risk riding a beam of light whilst looking at our face in a mirror.

The analysis of surviving works of art suggests that salt green was not widely made in practice. The ingredients of Spanish gold suggest that it could not be made in practice. The occurrence of salt green with Spanish green and the occurrence of Spanish gold with perfectly functional metalworking recipes in the same treatise suggest that craft manuals are multivalent texts. Some recipes are to be taken literally whilst other recipes have other values.

Hawthorne and Smith found it "astounding that as practical a man as Theophilus should include [such a] fantastic chapter"[30] as Spanish gold. However, there is no evidence – codicological or otherwise – to suggest that the recipe can be dismissed. It seems from the evidence that tendencies to assign individual recipes, or even entire treatises, to exclusive categories such as "practical"

27. Albertus Magnus, *On Animals*, 25, 2, 13, tr. K. F. Kitchell and I. M. Resnick, John Hopkins University Press, Baltimore, 1999, vol.2, p. 1720.

28. John of Trevisa's translation (1398) of Bartholomaeus Anglicus, (c.1240) *De Propreitatibus Rerum*, 18, xvi, Clarendon Press, Oxford, 1975, vol. 2, p. 1154.

29. A. Wallert, "Alchemy and Medieval Art Technology," in *Alchemy Revisited*, ed. Z. R. W. M. von Martels, Leiden, 1990, p. 161.

30. J. G. Hawthorne and C. S. Smith in Theophilus, *On Divers Arts*, III, 48, tr. J. G. Hawthorne and C. S. Smith, Dover, New York, 1979, p. 119, n.1.

or "fantastic," etc. should be resisted. Craft treatises contain recipes that lie somewhere on a spectrum from the practical to the fantastic, but the practical and the fantastic were evidently not uncomfortable bed-fellows for their authors, even if they are for some later translators and editors. (See the paper by Manfred Lautenschlager, in this volume, for an example of the co-existence of 'serious' and 'frivolous' recipes within a single manuscript).[31]

This brings us to the question of craft treatises' intended audiences. Hawthorne and Smith's footnote implies a dichotomy – practical is not fantastic, and visa versa. It also suggests an assumed ability to work with one's hands or an ability to work with one's head – but not with both. Such an assumption underlies the historic neglect of the crafts by academia, as noted above. But the assumption does not sit easily with the multivalent character of other medieval literary forms and it can be tested by the examination of crafted products and the scientific analysis of material culture.

Comparing Spanish gold to a relativistic thought experiment does not imply that its meaning was necessarily beyond the comprehension of the average painter. Scientific analysis of paintings suggests that artists' materials could be used in a philosophically sophisticated manner. It was noted that Mosaic gold looks very similar to shell gold (powdered gold dispersed in a transparent medium). Where Mosaic gold occurs in conjunction with shell gold or burnished gold, it does not appear to be used as a substitute for gold. Rather, it is used as a distinct material that extends the range of golden effects without intending to replace real gold. For example, Arcangelo di Cola da Camerino:

> "Deliberately restricted the use of burnished gold leaf to unearthly matters like the sky and halos, and used Mosaic gold for the tangible matters ... [such as] the dress of Christ and the side of the boat."[32]

Whilst they may look similar, the artists' use of real gold and Mosaic gold is differentiated. Arcangelo's use of these materials reflects a statement by the third century Alexandrian scholar, Origen.

> "Let us take it then that true gold denotes things incorporeal, unseen and spiritual; but that the likeness of gold, in which is not the Truth itself but only the Truth's shadow, denotes things bodily and visible."[33]

31. One suspects that had Theophilus wished, he might have annotated the salt green and Spanish gold recipes in red – indicating that they elaborated upon principles. Using the same convention, the ceruse and Spanish green recipes might be annotated in green – indicating that they were trivial operations.

32. L. Speleers, "An early example of the use of Mosaic gold," *Kunsttechnologie*, 13, 1, 1999, p. 52.

33. Origen, *Cant.* 2, p.159 lines 3-4 cited in D. Janes, *God and Gold in Late Antiquity*, Cambridge University Press, Cambridge, 1998, p. 79.

Arcangelo di Cola da Camerino's differentiation of visually similar materials is not unique. The unknown illuminator of the fourteenth century *Macclesfield Psalter* (Fitzwilliam Museum, MS 1-2005) used Mosaic gold to depict St John the Baptist's fur. In this manuscript, which is richly gilded, the pigment is restricted to the representative of the Old Testament.

These examples serve to indicate that at least some practising painters had philosophical insights into their materials. In practice, Mosaic gold was not used as a substitute for real gold, it was treated in a manner appropriate to its philosophical nature. And its philosophical nature is evident in its recipes – the transmutation of a base metal with a tripartite mixture, a conceptual dependent of the fantastic Spanish gold.

Spanish green, on the other hand, was a more prosaic pigment. Conceptually, it is merely a rust, like lead white and iron red. Rusts were seen as the consequence of the diseased or corrupt state of base metals, the rustlessness of gold being evidence of its health and noble incorruptibility. Petrus Bonus drew parallels between metals and the diseases suffered by people with humoural imbalances.

> "... corrupt metals suffer from four different kinds of leprosy ... Iron is affected with leprosy from corruption of the bile, copper from corruption of the blood, tin from corruption of the phlegm, and lead from simple melancholic corruption."[34]

The connection between human health and the behaviour of metals would have aided the assimilation of technical knowledge about pigments. Artists' understanding of their materials was influenced by their understanding of their health. And we can see that knowledge deployed in paintings. For example, it is traditional for painters to represent those who torment Christ as imbalanced. Sometimes this is achieved through the general ugliness of the tormentors. Sometimes they are shown with specific problems such as noses swollen by drink or eroded by syphilis. But in a Seville altarpiece of circa 1500, *The road to Calvary* (HKI 610), Juan and Diego Sanchez showed imbalances in the Galenic humours (Fig. 1).

Christ's most prominent persecutors are characterised by humoural imbalances. One is choleric, with an imbalance of fire/bile and is shown as an aggressive person with a 'fiery' temper. One is sanguine, with an imbalance of air/blood, shown as a vacant-looking 'air-head'. One is phlegmatic, dominated by water/phlegm, shown dissipated and 'washed-out'. The last is melancholic, dominated by earth/black bile, dark and 'weighed-down'. Christ's persecutors

34. Petrus Bonus, *The New Pearl of Great Price*, ed. J. Lacinius, tr. A. E. Waite, Vincent Stuart, London, 1963, p. 271.

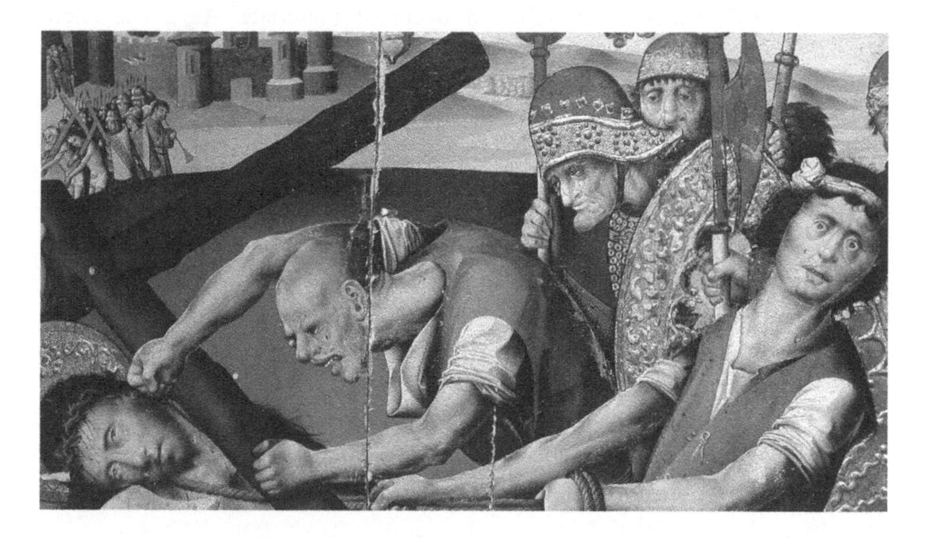

Fig. 1. Detail (during treatment) of – J. & D. Sanchez, *The Road to Calvary*, Fitzwilliam Museum, Cambridge.

are evidence that Juan and Diego were aware of medical theories. But is there evidence of any connection between the humours and the metals?

The bodies of the phlegmatic and melancholic persecutors are hidden behind shields. However, the choleric and sanguine tormentors are quite central to the image. The choleric individual is dressed in a red tunic. Red is the colour of the rust that Petrus Bonus associated with the metal afflicted by corruption of the bile – iron. The sanguine individual is shown in a green tunic. Green is the colour of the rust associated with the metal afflicted by corruption of the blood – copper. And the green tunic is not only the same colour as copper's rust, it is actually painted with copper's rust.

(Of course it just may be coincidence that the person afflicted by a corruption of the blood is shown clad in a garment depicted with the metal afflicted by a corruption of the blood. But the Spanish painters would have called the pigment cardenillo, a name that reinforces connections with blood. Blood is a red fluid, copper is a red metal[35] and the Spanish name of copper's rust was related to bruising, a type of corruption of blood.)

These details take us away from the subject of Spain's role in disseminating craft treatises. But sufficient has been said to suggest that Arcangelo di Cola da Camerino, Juan and Diego Sanchez, and the unknown illuminator of the

35. Theophilus, *On Divers Arts*, III, 47, tr. J. G. Hawthorne and C. S. Smith, Dover, New York, 1979, p. 119.

Macclesfield Psalter were informed users of materials. As a Benedictine author as well as a practicing craftsman, Theophilus was probably better informed than most. The presence in his treatise of recipes for Spanish green and Spanish gold suggests that he had some awareness of Aristotelian physics and Jabirian alchemy. The fact that he chose to call these recipes Spanish suggests that he may have been formally acknowledging Spain's role in disseminating those ideas across Europe.

Acknowledgements

I would like to thank Juan Acevedo for discussions and for the etymology of cardenillo.

THE *PRATO HAGGADAH*: AN INVESTIGATION INTO THE MATERIALS AND TECHNIQUES OF A HEBREW MANUSCRIPT FROM SPAIN IN RELATION TO MEDIEVAL TREATISES

Silvia A. Centeno/Nellie Stavisky

Introduction

A grant from The Dr. Bernard Heller Foundation made it possible for The Library of the Jewish Theological Seminary to undertake the conservation of one of its outstanding illuminated manuscripts, the *Prato Haggadah* (MS 9478). The haggadah is the text read by Jews on the first night of the holiday of Passover, commemorating the exodus from Egypt. To our knowledge, no non-destructive has previously been undertaken on any early Hebrew codex from Spain. Therefore, the disbinding of the haggadah in preparation for treatment presented an unparalleled opportunity to learn what materials and techniques had been used and relate them to artist's manuals of the period.

The only known medieval Jewish treatise, *Libro de como si facem as côres* was written in Loulé, Portugal, by Abraham b. Judah ibn Hayyim no later than 1462.[1] The manuscript, housed in the Biblioteca Palatina in Parma,[2] is written in Portuguese using Hebrew characters, and gives instructions for the preparation of various forms of gold, pigments, binders, glue and varnish. A translation into English appeared in 1932 in an article by D.S. Blondheim.[3] According to a colophon at the end of the ninth chapter, the manuscript was completed in 1262.[4] The Parma manuscript is written on paper with a fifteenth century watermark which Benjamin Richler dates to 1462 in his catalogue of *Hebrew Manuscripts in the Biblioteca Palatina*,[5] making it 200 years later than

1. *See* "Ibn Hayyim, Abraham ben Judah," in *Encyclopaedia Judaica*, Jerusalem: Keter, n.d.

2. Catalogue no. Parma 1959; De Rossi 945. Fols 1r-20r of the manuscript.

3. D.S. Blondheim, "An Old Portuguese Work on Manuscript Illumination," *The Jewish Quarterly Review*, Vol.XIX, 1928-1929, pp. 97-134.

4. Blondheim, *op.cit.*, p. 97.

5. Benjamin Richler, ed., *Hebrew Manuscripts in the Biblioteca Palatina*, Jerusalem, 2001, Catalogue No. 1561, pp. 472-3.

the colophon indicates, and about a century and a half later than the *Prato Haggadah*. Richler believes the date on the colophon may be a mistake. The research of Luis U. Alfonso and Antonio J. Cruz also suggests that this is a fifteenth-century collection of recipes, culled from earlier works, rather than a carefully composed treatise.[6] Some recipes, for instance the ones for vermilion and verdigris,[7] are very similar to those found in the *Mappae Clavicula.*[8] The author obviously had a familiarity with the practices of contemporary artists and with available treatises and recipes, which may, in turn, be copies of previous works.[9] The fact that Abraham b. Judah was writing in Judaeo-Portuguese must indicate that he was making this information available to an audience of Jewish artists and illuminators who were excluded from membership in the guilds.

Medieval painting treatises have a history going back to ancient times. Later works are generally based on earlier ones.[10] They added directions and assisted the medieval art practitioner in his search for appropriate ingredients found in the world of nature, helping to convert them into useful materials for his workshop.[11] Recipes written in diverse geographical areas,[12] at different times, are strikingly similar. The pigments, inks and other materials identified in the *Prato Haggadah* by different analytical techniques correspond to those frequently used by medieval artists.

Description

The *Prato Haggadah* is an unfinished illuminated manuscript, lacking a colophon (Fig. 1).

6. Faculdade de Letras, Universidade de Lisboa and Instituto Politecnico de Tomar, respectively. Paper presented at the International Symposium *Craft Treatises and Handbooks: the Dissemination of Technical Knowledge in the Middle Ages*. Cordoba, Spain, October 6-8, 2005.

7. Blondheim, *op. cit.*, pp. 126 and 124, respectively.

8. Cyril Stanley Smith and John G. Hawthorne, eds., *Mappae Clavicula: A Little Key to the World of Medieval Techniques* in Transactions of the American Philosophical Society, Philadelphia, July 1974. See Introduction.

9. Mark Clarke, *The Art of All Colours: Medieval Recipe Books for Painters and Illuminators*, London: Archetype Publications, 2001, p. 14ff. Clarke points out that "[e]arly medieval texts derived a certain amount of content from classical literature' and continued to be copied, with adaptations, pertinent to the changing needs of the period or the experience of the author, throughout the Middle Ages." Clarke also describes the Abraham ben Judah's manuscript (pp. 17-18).

10. Cyril Stanley Smith and John G. Hawthorne, *op.cit.* See Introduction.

11. *Ibid.*, pp. 20-22. The text says: "Before mixing the ingredients, weigh them, then make a good mixture, grind them well on marble and mix them well in amounts appropriate to your operation" (Chapter 192). And a bit later: "We list all those things made from flowers of land or sea, also from plants; we list in this way their qualities or their uses ..." (Chapter 192).

12. For some non-European treatises, see: Gulnar Bosch, John Carswell and Guy Petherbridge, *Islamic Binding and Bookmaking*, Chicago: University of Chicago, 1981; and Nancy Puriton and Mark Watters, "A Study of the Materials Used by Medieval Persian Painters," *Journal of the American Institute of Conservation*, 30:2 (1991), pp. 125-144.

Fig. 1: Fol.33r. "Hallalujah."

It consists of 85 octavo leaves written on fine quality calf parchment. The pages are arranged according to Gregory's rule, in which facing pages alternate as either both flesh or hair side, with the hair side on the outside of the quires. Most of the 11 quires comprise 8 leaves. Folios 1-53 are written in a square Sephardic script with a reed pen (*calamus*), the writing implement used in the Iberian peninsula, southern France, the Middle East, Asia Minor and North Africa. An added square Italo-Ashkenazic section, fols. 54-68, uses the more flexible quill, the pen used in Christian Europe[13] (Figs. 2a and 2b). There are no catchwords or vocalization. The illuminations do not necessarily follow a consecutive order, and sometimes are interspersed with pages of plain text. When the Library received the codex in 1964, the twentieth-century leather binding was causing damage to the manuscript.[14]

13. Malachi Bet-Arie, *Hebrew Codicology*, Paris, 1976, pp. 13-14; Ada Yardeni, *The Book of Hebrew Script*, Jerusalem: Carta, 1997, p. 90.

14. The original cover may have been made of parchment. We can also surmise that it may have been red-brown leather over wooden or paste boards, tooled and incised within a double outline in a continuous geometric strap-work pattern, typical of *mudejar* bindings of the period. It would be likely that the binding was executed by a Jewish artisan. See Henry Thomas, *Early Spanish Bookbindings: XI-XV Centuries*, London, Bibliographical Society, 1939, especially pp.xx-xxii. We should note that there are no Hebrew manuscripts in their original bindings that pre-date the 15th century. See Colette Sirat, *Hebrew Manuscripts in the Middle Ages*, Cambridge University Press, 2002, p. 115.

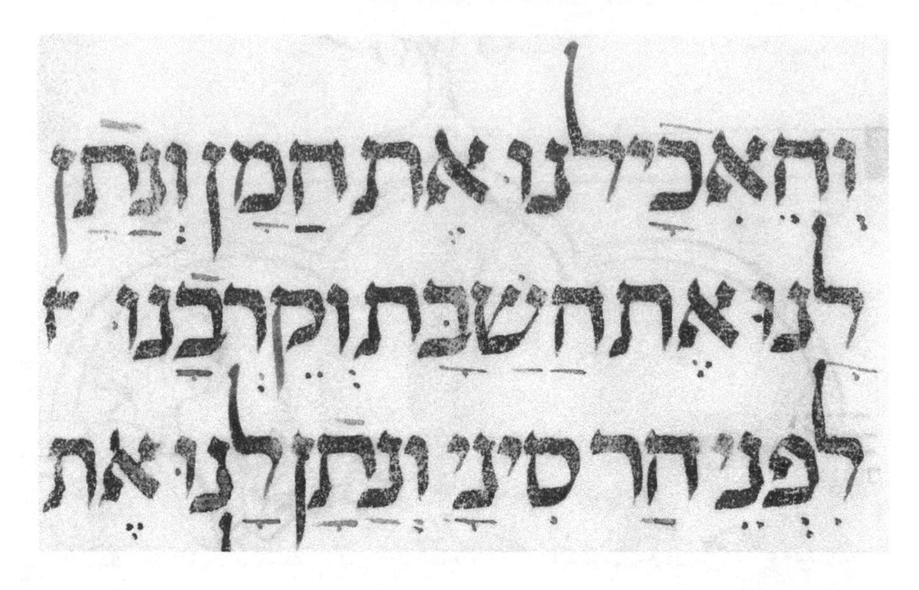
והאכילנו את המן ונתן
לנו את השבת וקרבנו
לפני הר סיני ונתן לנו את

Fig. 2a: Fol.27r. Sephardic script.

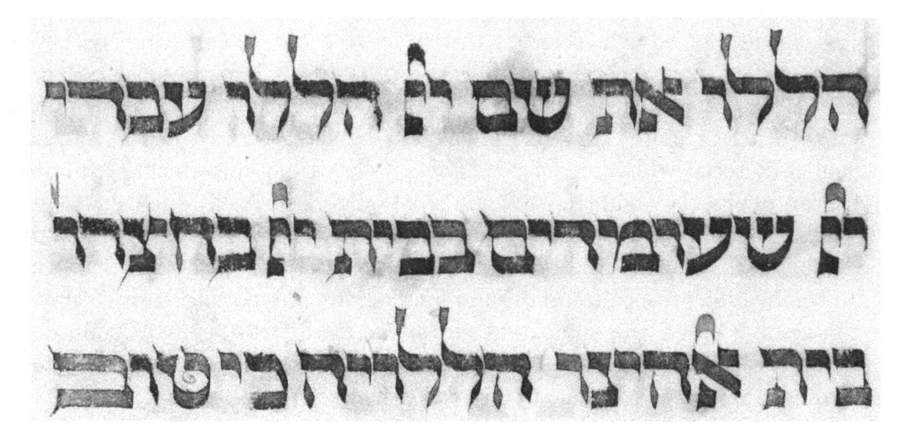
הללו את שם יי הללו עבדי
יי שעומדים בבית יי בחצרי
בית אלהינו הללויה כי טוב

Fig. 2b: Fol.59r. Square Ashkenazic script.

The haggadah is especially fascinating because it gives an insight into the process of making a manuscript in the Middle Ages: some pages are complete (Fig. 3a), some unfinished (Fig. 3b), others virtually complete, and yet others somewhere in between with preparatory drawings, some gesso and color. Many of the pages have illuminated initial word panels, comparable to illuminated initials in Christian or secular manuscripts.

Fig. 3a: Fol.40v. Completed page.

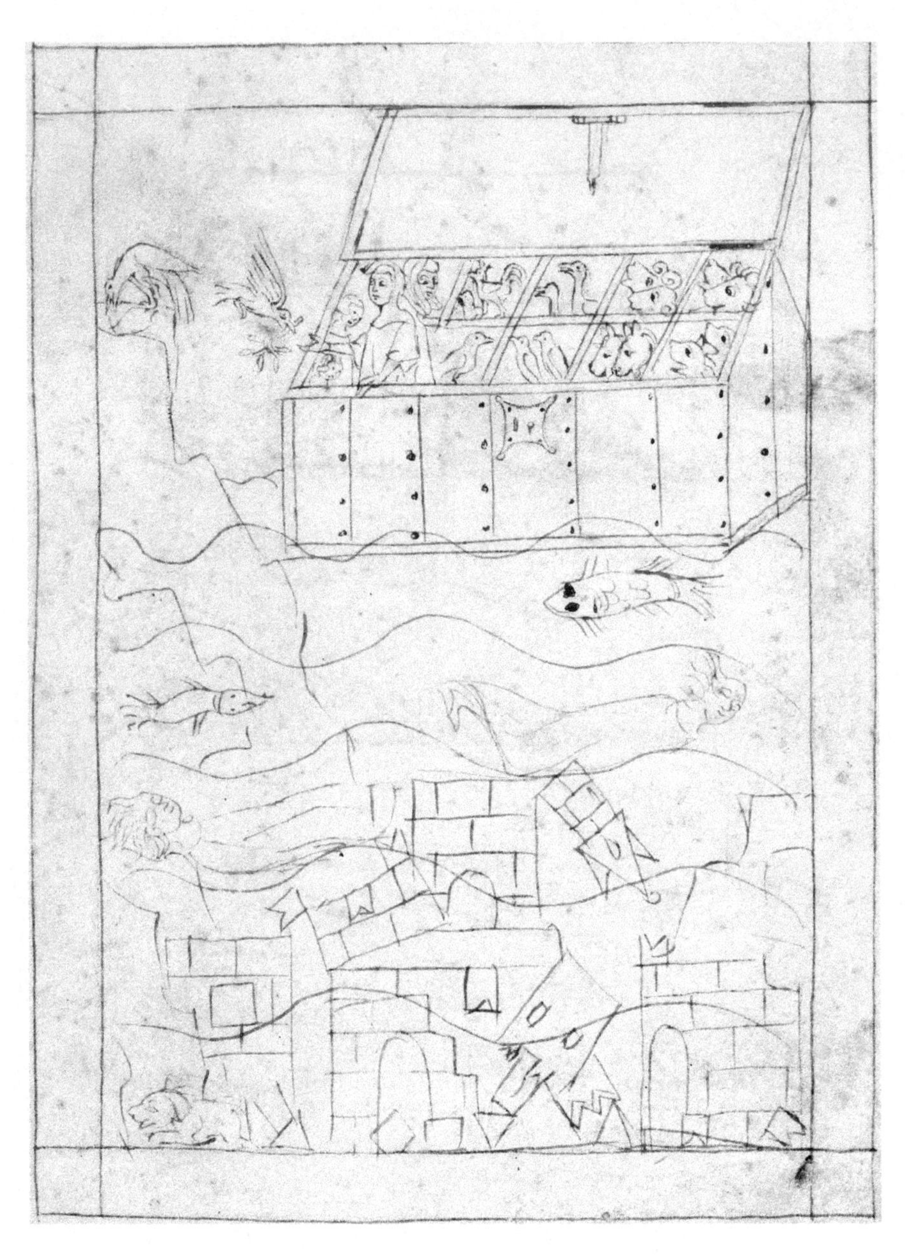

Fig. 3b: Fol.84v. Unfinished page.

Throughout, illustrations accompany the text, such as the illustrations of *matza* (Fig. 4) and *maror* (bitter herbs).[15] Margins are replete with fanciful drawings of hybrid creatures, imaginary birds, grotesques, drolleries and climbing vines. Preparatory drawings from *Genesis*, including scenes from the Noah story, appear at the end of the manuscript in unfinished folios.

Fig. 4: Fol.29r. Matza.

In the Christian tradition,[16] after the master planning and layout was determined, the text was written by the scribe. The master draftsman made the pre-

15. Fol.30r.

16. This is the general Christian tradition from approximately the early thirteenth century, after the trade took over book production from the monasteries. Christopher Clarkson, personal communication, September 2004.

paratory drawings and the word panels were sketched out. The quires were distributed to various specialists, some more skilled than others. These artisans laid down gesso to cushion the gold leaf, applied gold and silver leaf, burnished the gold and ultimately added pigments. First the backgrounds were painted, and then more highly skilled artists modeled the flora, fauna and other images, and completed the figures.

The Hebrew tradition was somewhat different because the scribe had a dominant role in laying out the manuscript.[17] It was he who was familiar with the text, which is considered the salient element in a Hebrew manuscript. The scribe also determined thc words for the word panels, especially since in medieval books these markers helped the reader locate important passages, or specific places in the text, much as chapter headings play this role in later books (Fig. 5).[18] The illumination's function was to adorn the text.

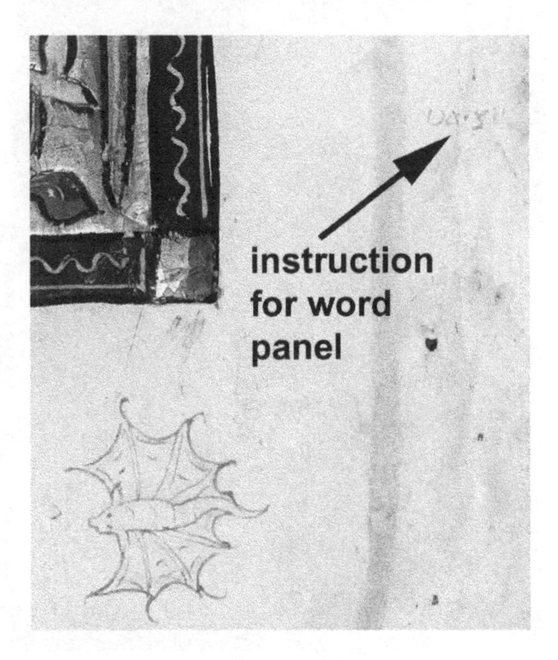

Fig. 5: Fol.16r. Detail: instruction for word panel.

Although all surviving Spanish haggadot are written on parchment, it should be noted that fine quality, long fibered, Arab style linen paper sized with rice starch continued to be produced in Spain in the late thirteenth century by Arabs and Jews, even when ownership of many mills was acquired by the

17. David Goldstein, *Hebrew Manuscript Painting*, London: The British Library, 1985, p. 7.

18. Christopher de Hamel, *The British Library Guide to Manuscript Illumination: History and Techniques*, University of Toronto Press, 2001, Chapter 1.

monasteries.[19] There are Spanish Hebrew manuscripts written on a combination of paper and parchment[20] and also many manuscript fragments and letters from the Cairo Geniza from the medieval period written on paper.

Inks

Medieval manuscripts in Europe were generally written in iron gall ink.[21] There are literally hundreds of recipes. Basically, all of them consist of mixing a tannin-containing material, such as a decoction of gall nuts or crushed tree bark, with an iron salt and a gum solution, generally gum arabic.[22, 23]

Carbon based inks, which could be erased, were most certainly used by Jews as early as biblical times.[24] The first Jewish mention of iron gall ink is in the *Mishnah*, redacted at the beginning of the 3rd century C.E., where a differentiation is made between "non-permanent" and "permanent" writing materials. An "acceptable" ink, permanent and a deep black,[25] could be obtained by adding vitriol, an ore containing iron sulfate and other minerals, to gall nut juice, that is an iron gall ink.[26, 27]

19. Oriol Valls I Subira, "Three Hundred Years of Paper from Spain from the Tenth to the Thirteenth Century," *The Paper Maker*, Vol. 34, No. 1, 1965, pp. 29-35.

20. JTS, MS 7242, *Aboda Zara* [1290]; and JTS, MS 8184, Avicenna [1468].

21. Iron gall ink may have been used for writing before the first millennium AD, however, the earliest known detailed recipe for ink combining an iron salt with the bark of the hawthorn wood, containing tannin is described by Theophilus, *On Divers Arts*, New York: Dover 1963, p. 42-43. Such inks may have been used on parchment as early as the second century C.E. Alfred Fairbank, *The Story of Handwriting*, Watson-Guptill, 1970, p. 90.

22. Mark Clarke, *op. cit.*, Mary P. Merrifield, *Medieval and Renaissance Treatises on the Arts of Painting*, New York: Dover, 1967, p. 60; Sigmund Lehner, *Ink Manufacture, including Writing, Copying, Lithographic, Marking, Stamping, and Laundry Inks*, Trans. from the German 5th ed. by Arthur Morris and Herbert Robson, London 1914 (first English ed. 1902).

23. Materials containing hydrolyzable tannins are the ones more suitable for ink production. These contain large amounts of gallic acid derivatives incorporated into a polymeric molecule, i.e. gallotannins. In fermentation or acid conditions some of this gallic acid is liberated by hydrolysis. Vincent Daniels, "The chemistry of Iron Gall ink," *Postprints of the Iron Gall Ink Meeting*, A. Jean. E. Brown ed., The University of Northumbria, Northumbria, 2001, pp. 31-36. According to C. Krekel, gallic acid reacts with ferrous sulfate to form a colorless ferrous gallate complex at first. In the presence of air, this complex is oxidized to ferric gallate. Free ferric ions, already present in the ferrous sulfate-containing mineral used, catalyze the formation of the dark colored ferric pyrogallate compound. Christoph Krekel, "The Chemistry of Historical Iron Gall Inks," *International Journal of Forensic Document Examiners*, Jan/Dec 1999, Vol. 5, pp. 54-58. The gum solution functions as a suspension agent for the insoluble pigment particles; it modifies the viscosity, and therefore the flow of the ink, and it binds the ink to the substrate. V. Daniels, *op.cit.*, pp. 31-36.

24. See "Writing" in *Encyclopaedia Judaica*, *op.cit.* See also *Numbers*, 5:23 where the ink used by the priest is erased in water, which indicates that it was not an iron-gall ink since it washed off from the parchment easily.

25. Babylonian Talmud, Tractate *Gittin, 19a.* The Schottenstein Edition, Mesorah Publications, 1993. See also "ink" in *The Jewish Encyclopedia.*

26. A common source of iron sulfate was 'iron vitriol' that, depending on the location where the ore was mined, may contain other sulfates such as copper, manganese, zinc and aluminum sulfates. Johan G. Neevel, "(Im)possibilities of the Phytate Treatment of Ink Corrosion," *Contributions to Conservation*, Jaap A. Mosk and Norman H. Tennent eds., London: James & James, 2002, pp. 74-86.

27. Marcus Jastrow, *A Dictionary of the Targumin, the Talmud Babli and Yerushalmi, and the Midrashic Literature*, New York: Pardes, n.d., Vol. II, p. 1382.

Moses Maimonides the great medieval codifier of Jewish law who lived in Spain and later in Egypt, discusses the preparation of ink in his *Mishnah Torah*. He repeats that an "acceptable" ink could be obtained by adding vitriol. Maimonides recommends using a carbonaceous black obtained by condensing the vapors of oils, with tree sap, a drop of honey, and then adding it to gallnut juice or vitriol before writing.[28]

Iron gall ink does not assume its full intensity of color, brown-black to purple-black hue, until after the aqueous mixture is applied to the substrate.[29] This would have been seen as a shortcoming by the scribes. The recipe in the *Mishnah Torah* addressed this limitation by adding a carbon-based black pigment that would confer the ink with a dark color at the time the text was written.

A number of medieval recipes are interspersed in various Hebrew manuscripts, usually in collections of medicinal or miscellaneous works, often on blank pages or at the end of a section.[30] These generally have detailed instructions, and call for gall nuts, gum arabic and vitriol, the primary ingredients of iron gall inks. Others add white wine, rain water, or crushed salt to reduce viscosity.[31, 32, 33, 34] XRF (X-ray fluorescence) elemental analyses were carried out on the inks present in representative folios with Sefardic script (ink I) and

28. Moses Maimonides, *Mishne Torah: Hilchot Tefillin UMezuzah VSefer Torah*, translation with commentaries and notes by Rabbi Eliyahu Touger, New York/Jerusalem: Moznaim Publishing Corporation, pp. 14-16.

29. Johan G. Neevel, "(Im)possibilities of the Phytate Treatment of Ink Corrosion," pp. 74-86 in *Contributions to Conservation*, J.A. Mosk and H. Tennant ets., London, James & James, 2002, p. 75.

30. Benjamin Richler, the Institute of Hebrew Microfilmed Manuscripts (IMHM) at the Jewish and National University Library (JNUL), has kindly directed us to a number of microfilms of these manuscripts.

31. Vatican Apostolica, ebr. 509, fol. 54a. We are indebted to Benjamin Richler for directing us to this recipe, and to Dr. Jay Rovner for helping with the translation.

32. "Take 4 liters of clear and white wine and put into it 4 ounces of crushed gall nuts. Put on the fire until it boils well. And put in 2¼ ounces of crushed gum arabic and boil this all together until it has all been reduced by half, stirring constantly while the mixture is on the fire. When it has been reduced by half, remove it from the fire and wait until it cools and is lukewarm. Afterwards, strain it through a thin cloth or a fine strainer. Afterwards, put in 1¼ ounces of crushed vitriol and mix well, and leave to stand for half a day. Afterwards, strain it again and put the ink in a glass vessel and it will be good, God willing. And if you boil it again, add some of the glass, and it will be more golden. And in the summer, because of the heat, you need only add 2 ounces of gum, and it is customary to always put in 1½ ounces of vitriol." Colette Sirat, *Hebrew Manuscripts of the Middle Ages,* Nicholas de Lange ed., Cambridge University Press, p. 112; Fol. 51v. Parma 2486, De Rossi 1420, Richler 1553. Also cited by Monique Zerdoun Bat-Yehouda, *Les Encres Noir au Moyen Âge*, Paris, 1983, pp. 277-8, where the author dates it as fifteenth century.

33. Colette Sirat, *op. cit.*, p. 112.

34. "... take 1 litre of *gall-nut* and 16 liters of *rain-water* in winter; if in summer 20 litres of rain-water. Crush the gall-nuts a little and leave them to soak for a while in the water. Then boil them in the same water; in winter, reduce by two-thirds, in summer by a half. Filter, then take 2 ounces of *gum Arabic* and 3 ounces of *vitriol*, pound together well, and pour into the water. Heat gently without boiling. Stir well. Filter again through a good filter. This will give a good ink. God willing. If the ink be too thick, add a little *crushed salt*. If it be too runny to be used, add a little gum and pounded vitriol and the ink will be good. God willing," Sirat, *op. cit.*, p. 111. See also, Bat Yehouda, *op.cit.*, p. 276.

Italo-Ashkenazic script (ink II), respectively. The corresponding results and experimental conditions are summarized in the Appendix (Table 1).

The presence of iron gall ink can be inferred from the presence of Fe (iron) and S (sulfur) characteristic peaks in the XRF spectra, as well as from the peaks due to various other elements present in lower percentages such as Cu (copper), Mn (manganese), and Zn (zinc), their relative amounts depending on where the vitriol ore was mined.[35]

On five folios in the Spanish text (fols. 3r, 30v, 50v, 51r and 51v) Fe and S were detected in the ink I areas, indicating the use of iron gall ink. In fol. 49v, the presence of Zn and Cu, in addition to Fe and S, could be due to the use of an ink I batch prepared with a different ore or to a contamination of the ink-making utensils.[36]

The elemental compositions of the ink II areas analyzed were less consistent than the corresponding ink I areas described above. Fe and S are present in the ink used in folio 54r, pointing to the use of an iron gall ink. While Fe was detected in all the ink II areas examined, the peaks for S on fols. 55r and 68r have approximately the same intensity as the corresponding signals for the "bare" parchment areas, so it is not possible to determine whether sulfur S is present as a constituent of the ink used. Therefore, for these folios, the results from XRF analyses do not allow a tentative identification of the source materials. A thorough characterization of the organic components in the inks would be necessary to explain these results, though no non-invasive techniques are currently available for this purpose.

No visual difference between inks I and II was observed under UV illumination, although they were clearly fabricated by different hands.

Preparatory drawings

The preparatory drawings are of remarkably high quality, often displaying great detail. Some were later delineated by sharper lines executed in dilute ink, occasionally changing the initial sketch slightly to achieve a crisper design.

35. We are grateful to Ad Stijnman from the Instituut Collectie Nederland (ICN), Amsterdam, for this suggestion.

36. The areas probed with XRF for the present study were approximately 1cm in diameter, big enough to probe both the ink areas and a portion of their surrounding 'bare' parchment areas, therefore it is unlikely that this is due to the migration of sulfur to the areas surrounding the inscription, as observed by other authors using PIXE mapping. See Céline Remazeilles, Véronique Quillet, Thomas Callingaro, Jean Claude Dran, Laurent Pichon, Joseph Salomon, "Pixe elemental mapping of original manuscripts with and external microbeam. Application to manuscripts damaged by iron-gall ink corrosion," in *Nuclear Instruments and Methods in Physics Research* B 181 (2001) 681-887.

Unfinished folios gave us the possibility of analyzing the materials used for the preparatory drawings by XRF. In some faint initial sketches examined[37] the characteristic peaks of lead were observed, suggesting the use of lead point (Fig. 6).

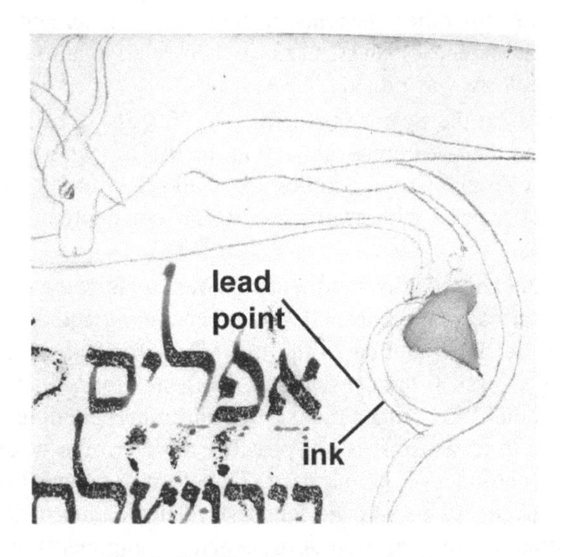

Fig. 6: Fol.53r. Preparatory drawing.

Lead point, or plummet, was a lead alloy fabricated into an instrument with a point, often used for lines or preparatory drawings in illuminated manuscripts. Theophilus mentions lead point as a drawing instrument for delineating the design before carving ivory.[38]

The sharper preparatory drawings examined[39] were found to contain Fe, S, and Zn, confirming the use of an ink.

Illuminations

Gesso

From the early thirteenth century, gold leaf was usually applied over a raised gesso layer. A variety of methods are described in early sources for preparing the gesso.[40] Gesso, in a liquid form, was painted in several layers onto

37. Fols. 27v, 53r, 45v.

38. Theophilus, *op. cit.*, p. 187-188.

39. Fol. 45v.

40. De Hamel, *Manuscript Illumination, op cit.*, p. 71; Mary P. Merrifield, *op. cit.*, p. 152, 154, and 238; *The Strasburg Manuscript. A Medieval Painters' Handbook*, Translated from the old German by Viola and Rosamund Borradaile, London: Alec Tiranti, 1966, p. 27; Theophilus, *op. cit.*, p. 29-31.

the areas to be gilded, building up the cushion for the gold. Ibn Hayyim describes a recipe in which ochre, chalk, white lead, and egg white are used,[41] and recommends that the gold leaf be laid on without waiting for the gesso to dry, the egg white functioning as an adhesive.

The gesso in the *Prato Haggadah* is a light, salmon color (Fig. 7). When gesso areas[42] were examined by *in-situ* Raman spectroscopy, only gypsum (calcium sulfate) was detected. A microscopic fragment of gesso that had detached[43] was analyzed by SEM-EDS (scanning electron microscopy-energy dispersive X-ray spectrometry). Using this technique, a large amount of Br (bromine) was detected along with relatively large amounts of Ca (calcium) and S, small amounts of Fe, Si (silicon), K (potassium) and Al (aluminum), and trace amounts of P (phosphorus). These results suggest that the gesso is composed mainly of a mixture of gypsum and a bromine salt. In red spots present in this sample, larger amounts of Fe, Si and Al were observed, suggesting the presence of iron ochre. The mixture of gypsum and iron ochre is consistent with recipes recommended in medieval treatises. The source of Br still remains unexplained.[44]

Fig. 7: Fol.53r. Gesso.

41. "Chapter XXV: If one wishes to work with gold or other colors and [to know] how they ought to be mixed, he should do it in this way, which is well tested. In the first place, when you wish to apply gold, take ochre and white lead in equal parts and a little chalk. And you mix these things and grind them thoroughly and very well in white of egg which is very thin and very light. And if [the resulting mixture] is very stiff, pour over it water and grind it very well. And if it is very light pour upon it white of egg. Test this on a piece of parchment. On the place where you wish to apply the gold, first apply this preparation with a brush, and before it dries place the gold upon it. Then burnish it very well with the tooth of a swine or horse," Blondheim, *op. cit.*, p. 129.

42. Fols. 27v, 50v, 51r, 51v and 53r.

43. Fol. 51r.

44. Some doubts have been expressed as to the findings of bromine in the sample examined, since this is an unusual finding. However, bromine is definitely in the sample of gesso examined. There is the possibility for confusion between bromine and aluminum with EDS (energy dispersive X-ray spectrometry) analysis due to peak overlaps for these two elements. However, there are also higher energy peaks for bromine alone which are not shared with aluminum or any other element. These higher peaks are clearly present in the X-ray spectra obtained from the sample (Personal communication, Mark T. Wypyski, Research Scientist, Department of Scientific Research, The Metropolitan Museum of Art. December 29, 2005).

In some of the spots analyzed by EDS a higher proportion of Al to Si than expected for iron ochre was observed and this could be due to the presence of small amounts of a "lake" pigment, that was historically made by precipitating organic pigments or dyes upon an inert, inorganic substrate such as alum or gypsum.[45] The identity of organic pigments or dyes cannot be obtained by SEM-EDS or XRF elemental analysis.

Gold and silver

Gold leaf is abundantly present throughout the manuscript; XRF analysis also showed the presence of Ag (silver), for example in the wicked son's helmet (Fig. 8).[46] Where the tarnished Ag has fallen away, both a preparatory drawing and some pinkish pigmentation are visible, the latter possibly the residue of gesso which fell away together with the loss of parts of the damaged Ag helmet. Instructions for burnishing gold leaf are mentioned by Theophilus and Ibn Hayyim.[47, 48]

Fig. 8: Fol.5v. Silver leaf.

45. Rutherford J. Gettens and George L. Stout, *Painting Materials.* New York: Dover, 1966, p. 124.

46. Fol. 5v.

47. Theophilus, *op. cit.*, p. 36-37.

48. Blondheim, *op.cit.*, p. 133.; Harold J. Abrahams, "A Thirteenth-Century Portuguese Work on Manuscript Illumination," in *Ambix: The Journal of the Society for the History of Alchemy and Chemistry*, Vol. XXVI, No. 2, July 1979, pp. 93-99. Abrahams made a translation of selected passages of Abraham b. Judah ibn Hayyim's manuscript and explained them in terms of modern chemistry.

Pigments

The author of the Portuguese medieval Jewish treatise sums up:

> "Know that there are ten principal colors: azure, orpiment, and vermilion, green, carmine, cufi [?], sunflower, saffron, red lead, white lead and brazil-wood."[49]

Azure is a general term for blue colors that may correspond, for example, to azurite or ultramarine.[50] In addition to the colors and pigments included in this list, gold, silver, and the pigments verdigris, ultramarine, indigo, "black" and lac are mentioned in the treatise.

As many paint colors representative of the manuscript's palette as possible were chosen for Raman and/or XRF examination (Fig. 9a and 9b). Minimal handling was set up as a priority when choosing a folio for analysis. The results are summarized in the Appendix (Table 2).

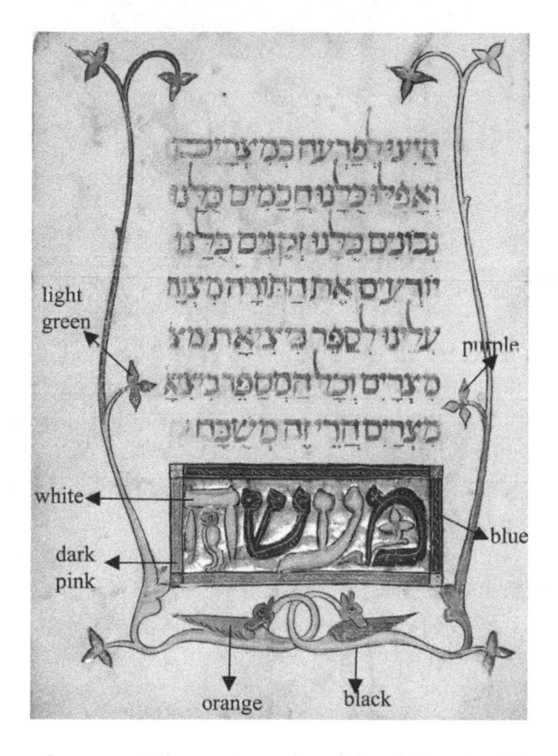

Fig. 9a: Location of representative spots analyzed for different paint colors in folio 3r.

49. Blondheim, *op.cit.*, p. 130.

50. S. M. Alexander. Glossary of Technical Terms, in Merrifield, *op. cit.*, p. (*xiv*); Nicholas Eastaugh, Valentine Walsh, Tracey Chaplin and Ruth Siddall, *The Pigment Compendium: A Dictionary of Historical Pigments*. Elsevier Butterworth-Heinemann, 2004, pp. 32-33.

Fig. 9b: Location of representative spots analyzed for different paint colors in folio 5r.

Azurite was detected in the blue areas (Fig. 10). This pigment was finely ground for its use in the word panels,[51] though coarser particles with a similar hue, possibly applied at a later date, were observed in the unfinished page displaying two figures holding a *matzah.*[52]

Lead white was identified by Raman spectroscopy in decorative details on top of other colors, especially in word panels (Fig. 10),[53] and it was also found to be present in mixtures with other pigments and/or possibly dyes, for example in the purple areas.[54]

Lead white, one of the most important medieval pigments, is sparkling and bright; it has remarkable opacity and density, giving it excellent covering power and obviating the need for repeated layers of paint.[55] Lead white-con-

51. E.g. fol. 3r.

52. Fol. 29r.

53. Fol. 3r.

54. Fols. 3r, 5r, 6v.

55. Daniel V. Thompson, *The Materials and Techniques of Medieval Painting*, New York: Dover Publications, 1956, p. 94.

taining paint can darken when exposed to polluted air, and is usually prone to cracking and flaking away from the substrate, a phenomenon that has been related to its interaction with the binding medium.[56] The lead white in the *Prato Haggadah* is, in most areas, fine and bright. It has not darkened, though when observed at high magnification fissures and areas of loss are apparent.

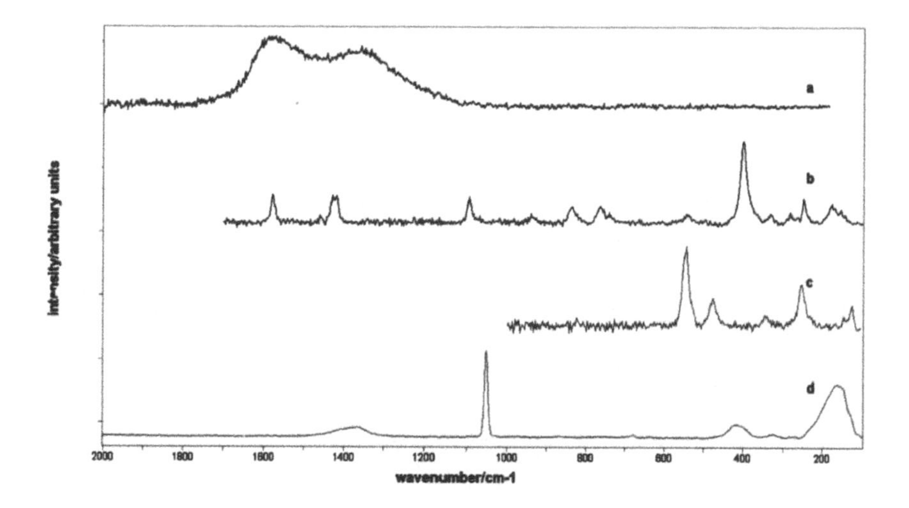

Fig. 10: Raman spectra recorded in paint passages in folio 3r: black, identified as a carbon-based black (a); blue, identified as azurite (b), orange, identified as a mixture of red lead and vermilion (c) and white, identified as lead white (c). All spectra recorded with λ_0: 514.5 nm.

Raman spectra could not be obtained for the coloring matter present in purple, light pink or dark pink areas (Fig. 11).[57] As mentioned above, lead white was identified in purple flower motifs, and also in the light pink areas,[58] while the white pigments gypsum and calcite (calcium carbonate), or gypsum with no calcite, were detected in dark pink areas.[59] These results suggest the presence of lake pigments that are below the detection limit of *in-situ* Raman spectroscopy.

56. Silvia A. Centeno, Marcelo I. Guzman, Akiko Yamazaki-Kleps and Carlos O. Della Vedova. "Characterization by FTIR of the Effect of Lead White on Some Properties of Proteinaceous Binding Media," *Journal of the American Institute for Conservation (JAIC)*, 43, 2004, pp. 139-150.

57. Fols. 3r, 5r, 6v, 36r.

58. Fol. 36r.

59. Fols. 3r, 5r, 36r.

In the Jewish Portuguese treatise, several recipes are described for making a "rose" color including brazilwood, carmine and lac.[60] Brazilwood[61] and lac, together with kermes were by far the most important red dyestuffs used in medieval times for making lakes.[62] Though several recipes existed in medieval times for making purple colors,[63] purple is not mentioned by Ibn Hayyim.

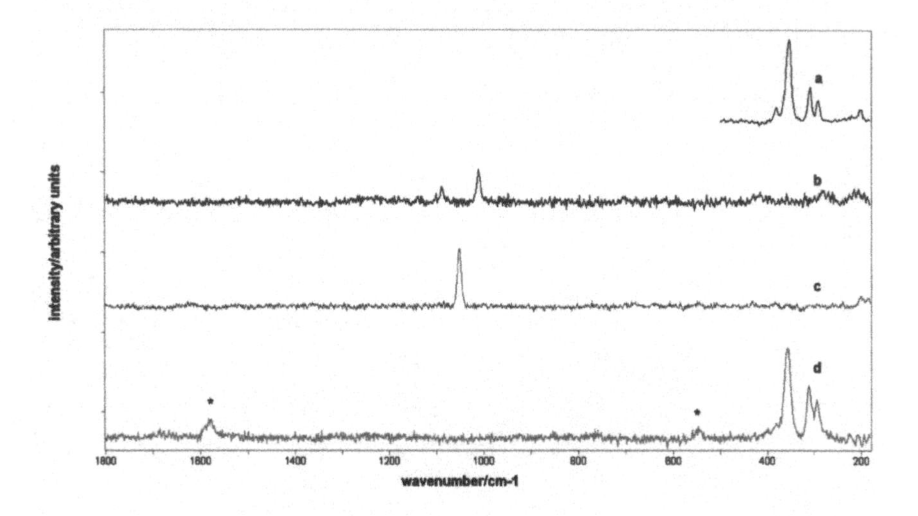

Fig. 11: Raman spectra recorded in paint passages in folios 5r (a-c) and folio 36r (d). Yellow pigment identified as orpiment, λ_0: 514.5nm (a); material detected in the darker pink areas identified as a mixture of calcite and gypsum, λ_0: 785nm (b); lead white detected in the purple areas, λ_0: 785nm (c); and pigments used to achieve the green color in folio 36r, identified as a mixture of orpiment and indigo, λ_0: 785nm (c). In spectrum c, the symbol * indicates peaks due to indigo.

The red-orange color used in border decorations[64] was found to be a mixture of vermilion and red lead. By the fourteenth century, vermilion was an important pigment in manuscript illumination. It could be artificially made by combining mercury and sulfur,[65] and it is also present in the mineral cinnabar.

60. Blondheim, *op. cit.*, p. 123 and p. 130.

61. According to Hofenk de Graaff, a Spaniard called Kimichi wrote about brazilwood in 1190, *braza* meaning "blaze," and she cites a number of recipes. Judith H. Hofenk de Graaff, *The Colourful Past: Origins, Chemistry and Identification of Natural Dyestuffs*, London: Archetype Publications, 2004, p. 142.

62. Thompson, *op. cit.*, pp. 108-9. See also Eastaugh *et. al.*, *op. cit.*

63. *Ibid.*, pp. 155-9.

64. Fol. 3r.

65. Thompson, *op. cit.*, pp. 78-9.

Red lead was often used to extend the more costly vermilion,[66] as observed in the *Prato Haggadah.*[67] Red lead alone was used in an illustration probably painted at a later date.[68] Although both vermilion and red lead may blacken, there is no evidence of this in the *Prato*, where the color remains exceptionally bright and vibrant.

The light green pigment used in floral border decoration[69] gave a detectable Raman signal though it could not be matched to any pigment reference sample available. XRF analysis revealed the presence of copper and this result suggests the use of verdigris, for which a large number of recipes were suggested in medieval treatises.[70]

The Hebrew-Portuguese manuscript has two recipes. One is very similar to one in the *Mappae Clavicula*, that recommends placing a copper foil and hot, strong vinegar in a pot, the mouth of which is smeared with honey and sealed with earth or manure. After thirty-one days the pot is opened and the verdigris is scraped out.[71] This is followed by another recipe which tempers the green with "a little saffron well ground."[72]

The green paint in the garment of one of the central figures on fol. 36r, probably painted at a later date, was found be a mixture of the organic blue pigment indigo and orpiment (Fig. 12). A mixture of these two pigments was also commonly used in medieval times to achieve green tones.

Orpiment was identified in several yellow areas.[73] This poisonous pigment was commonly used in book illumination before the fourteenth century, though

66. Thompson, *op cit.*, p. 100-101.

67. Fol. 3r.

68. Fol. 36r.

69. Fol. 3r.

70. "If you wish to make Byzantine green, take a new pot and put sheets of the purest copper in it; then fill the pot with very strong vinegar, cover it, and seal it. Put the pot in some warm place, or in the earth, and leave it there for six months. Then uncover the pot and put what you find in it on a wooden board and leave it to dry in the sun," *Mappae Clavicula, op. cit.*, p. 27; Merrifield, *op. cit.*, especially pp. 126-127, 418.

71. Blondheim, *op. cit.*, p.124.

72. "Chapter XII: Take a wide bowl and half fill it with putrid urine and take a basin of brass, very well washed from the bottom to the top, and place it above the bowl in such a way that the urine does not come within two fingers' distance of the bottom of the basin. Let the bottom of the basin be smeared with good honey, and the basin itself be half full of the same urine. And above the basin place upside down another bowl. Above the bowl place coverings for pack mules, and pour the urine from the basin into the bowl placed at the bottom. And go to the bottom of the basin and you will find the honey which you placed on it turned into verdigris. And scrape it with a spatula and keep it in paper. If you wish to make more green, smear the bottom of the basin with honey and do as you did before, and do so as often as you please. And for tempering this green, when you wish to work with it, grind it previously very thoroughly and place in it a little saffron well ground, and temper it with gum-water, for there is no devil who can take away its worth from this color," Blondheim, *op.cit.*, p. 124.

73. Fols. 5r, 5v.

medieval treatises from the twelfth and fifteenth centuries noted an incompatibility with lead and copper-containing pigments.[74]

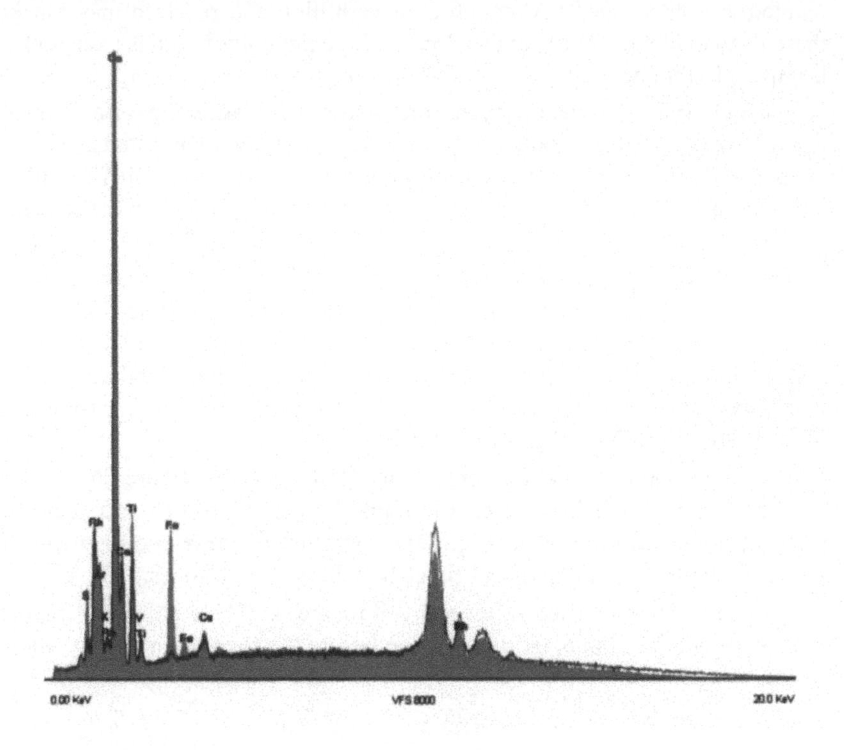

Fig. 12: XRF spectra of ink I (red) and 'bare' parchment (blue) in folio 3r showing the presence of S and Fe in the ink. The presence of characteristic peaks for Ti is due to the mattboard used to support the folio during the measurements.

In some recipes, Ibn Hayyim recommends mixing inorganic with organic pigments, such as red lead with carmine or brazilwood. It cannot be ruled out that dyes, like carmine or brazilwood, are present in some of the paint areas analyzed, though below the detection limit of *in-situ* Raman spectroscopy.

Conservation

The conservation, primarily involving removal of adhesives from the spine-folds and consolidation of the ink, pigments and gesso, has stabilized the man-

74. Thompson, *op. cit.*, p. 169; Eastaugh *et.al.*, *op. cit.*, pp. 285-6. "Now if you want to know which pigments conflict with each other, here it is. Orpiment is not compatible with folium, green, red lead," *Mappae Clavicula, op. cit.*, *xi*, pp. 27-8.

uscript and hopefully will forestall deterioration in the future. After re-binding, the manuscript will be stored in a controlled environment.

Conclusions

The disbanding and conservation of the manuscript presented an opportunity to observe the working methods of the scribe, the artist and illuminator, and the relationship between them.

The examination of selected inks, pigments, gesso, gold and silver leaf representative of those used in the entire manuscript proved to be consistent with the palette of other manuscripts from the period. The materials identified in the *Prato Haggadah* are similar to those described in medieval treatises, including that of the Judaeo-Portuguese treatise of Abraham b. Judah ibn Hayyim. The evidence presented confirms that Hebrew manuscript production was related, and similar, to production in the Christian environment, from which it borrowed both materials and techniques.

Appendix

Summary of the Raman and X-ray flourescence analyses

The pigments, inks and other materials used by the artists and scribes in the *Prato Haggadah* were analyzed in a non-invasive manner, that is without the removal of any sample material, by Raman spectroscopy and/or by X-ray fluorescence (XRF). The only exception was a microscopic fragment of gesso detached from fol.51r that was analyzed by SEM-EDS (scanning electron microscopy-energy dispersive X-ray spectrometry).[75]

Raman spectra were recorded with a Renishaw Raman System 1000, configured with a Leica DM LM microscope. The excitation sources used were a diode laser (785 nm) and an argon ion laser (514.5 nm). The laser beam was focused on different areas of the manuscript quires placed in the microscope stage using a x50 objective lens, allowing spatial resolution in the order of 3 microns. The power reaching the sample area was set between 0.2 and 10 mW using neutral density filters. Accumulation times were set between 10 to 200 seconds. Identification of the different materials was carried out by comparing the Raman spectra of the unknown to those of reference compounds and to previously published data.[76]

Qualitative X-ray fluorescence (XRF) analyses were performed *in-situ* with a Jordan Valley open-architecture 3600 system. The spot size analyzed was approximately 1cm in diameter, therefore it was not possible to focus the beam within the fine lines of the script or small pigment areas without the interference from the surrounding areas. Multiple spots, usually from four, for the pigment areas, to ten for the ink areas, were analyzed. In addition, several spots of "bare" parchment areas in each folio were analyzed under the same experimental conditions as the pigment or ink areas in order to use them as background spectra. Consistent experimental parameters were used in all the folios for both the ink or pigment and the "bare" parchment areas: direct unfiltered radiation from a Rh target, 30-40kV, 200 seconds for the preset time, medium throughput, and dead-times between 35 and 40 %. The elements reported as ink or pigment components in the tables below consistently gave peaks with intensities at least 30 % above the corresponding peaks for the "bare" parchment areas. Ti and Ca were detected in some of the additional supports used to back the quires during the XRF analyses.

75. The SEM-EDS analyses were carried out by Mark T. Wypyski, Research Scientist from the Department of Scientific Research at The Metropolitan Museum of Art.

76. Ian M. Bell, Robin J. H. Clark, Peter J. Gibbs, "Raman spectroscopic library of natural and synthetic pigments," *Spectrochimica Acta, Part A*. 1997; 53, pp. 2159-2179; Lucia Burgio, Robin J. H. Clark. "Library of FT-Raman spectra of pigments, minerals, pigment media and varnishes, and supplement to existing library or Raman spectra of pigments with visible excitation," *Spectrochimica Acta, Part A*. 2001; 57, p. 1491.

The Raman spectra of the ink areas presented strong fluorescent backgrounds, therefore only XRF analyses were carried out on representative folios with Sephardic script (ink I) and Italo-Ashkenazic script (ink II), respectively. These results are summarized in Table 1. A representative XRF spectrum of the ink present in folio 3r is shown in Figure 12.

A summary of the results of Raman spectroscopy and/or XRF obtained in different paint passages are presented in Table 2.

Raman spectra could not be obtained for the coloring matter present in the purple, light pink or dark pink areas such as the ones reported in Table 2. This suggests that organic pigments or dyes may have been used. The organic dyestuffs present in similar paint colors in other manuscripts have been reported to be difficult to identify by *in-situ* Raman spectroscopy due to the high molar absorptivities for bands in the visible range that make it possible to use them in very low concentrations.[77] Chromatographic techniques such as HPLC (high-performance liquid chromatography) are well established for the identification of natural organic pigments or dyes, though the technique requires the removal of sample material from the work of art. Other likely components of the lakes, such as alum, are also below the detection limit of *in-situ* Raman spectroscopy.

In the lighter initial sketches and the sharper preparatory drawings, Pb was detected by XRF revealing the use of lead point. The sharper preparatory drawings were found to contain Fe, S and Zn, confirming the use of an ink.

A discussion of these findings is presented in the main text of the article.

Folio	Script	Elemental composition
3r	Ink I	Fe, S
30v	Ink I	Fe, S
49v	Ink I	Fe, S, Zn, Cu
50v	Ink I	Fe, S
51r	Ink I	Fe, S
51v	Ink I	Fe, S
54r	Ink II	Fe, S, Zn, Cu
55r	Ink II	Fe, Cu, Zn
68r	Ink II	Fe, Cu

Table 1. Elemental composition of the inks used in different folios of the *Prato Haggadah* as determined by *in-situ* XRF.

77. Katherine L. Brown, Robin J.H. Clark, “The Lindisfarne Gospels and two other 8th Century Anglo-Saxon Insular manuscripts: pigment identification by Raman microscopy,” *Journal of Raman Spectroscopy*, 2004, 35, pp. 4-12.

Folio	Color	Results	Fig. #
3r	Black	Carbon-based black (Raman).	Fig. 10a
	Blue	Azurite (Raman).	Fig. 10b
	Red-orange	Red lead and vermilion (Raman).	Fig. 10c
	White	Lead white (Raman).	Fig. 10d
	Dark pink	Calcite, gypsum (Raman) and, possibly, an organic pigment or dye.	-----------
	Purple	Lead white (Raman) and, possibly, an organic pigment or dye.	-----------
	Light green	Cu, Pb (XRF) Cu-containing pigment.	-----------
5r	Yellow	Orpiment (Raman).	Fig. 11a
	Dark pink	Calcite, gypsum (Raman) and, possibly, an organic pigment or dye.	Fig. 11b
	Purple	Lead white (Raman) and, possibly, an organic pigment or dye.	Fig. 11c
5v	Yellow	As, S (orpiment, XRF).	-----------
6v	Purple	Lead white (Raman) and, possibly, an organic pigment or dye.	-----------
29r	Blue	Azurite (Raman).	-----------
36r	Blue	Azurite (Raman).	-----------
	Green	Indigo and orpiment (Raman).	Fig. 11d
	Orange-red	Red lead (Raman).	-----------
	Light pink	Lead white (Raman) and, possibly, an organic pigment or dye.	-----------
	Dark pink	Ca, S, and K (XRF). Gypsum and, possibly, an organic pigment or dye.	-----------

Table 2. Pigments identified in different folios of the *Prato Haggadah* as determined by *in-situ* Raman spectroscopy and/or XRF analyses.

Castilian Recipes for the Manufacture of Writing Inks (15th-16th Centuries)

Teresa Criado Vega

Ink is any substance that, by the means of an instrument such as a brush, adheres mechanically or chemically to writing, drawing, colouring, or printing materials.[1] Ink is a mixture of a colorant, a solvent, and a binder (vegetal gum, honey, egg white, oil or other semi-elaborated products), with the addition of an acid to ensure penetration and fixation onto the writing material.[2] All inks thus have several common components, such as the colorant (a natural or artificial pigment); the solvent, which allows ink to flow from the writing tool to the writing material (oil, water, alcohol); the binder, to provide the coloured particles with internal mechanical cohesion and adherence (vegetable gums, gelatines, glue, honey, sugar, egg white); and the mordant, to provide adherence in metallo-acid inks. Additionally, inks may also contain thickeners; humidifiers, used to control the drying process; antiseptics or anti-fermenters (vinegar) to prevent the growth of moulds and bacteria in the organic components; perfumes (musk); anti-freezers; glazes; and additional mordents (white wine).[3]

Black writing inks are divided into three major categories: carbonic, metallo-acid, and mixed. The carbonic type came first, but the most commonly used in the Middle Ages were metallo-acid black or ochre inks. The chemical process for the production of a black colour from the reaction of tannin and a metallic salt was already well known in Antiquity, and was profusely used in Pharaonic Egypt for leather tanning and fabric colouring.[4] The use of tannin

1. M. E. Rodríguez Díaz, "Técnicas de escrituras y del manuscrito," *Historia de la técnica y de la ciencia en la corona de Castilla*, (Coord. Luis García-Ballester), Valladolid, Junta de Castilla y León, 2002, volume II, p. 589.

2. A. Riesco Terrero, *Introducción a la paleografía y la diplomática general*, Madrid, 2000, pp. 262.

3. M. Romero Tallafigo, *Arte de leer escrituras antiguas: paleografía de lectura*, Huelva, 2003, pp. 43-44.

4. S. Kroustallis, "Escribir en el siglo XVI: recetas de la tinta negra española," *Torre de los Lujanes: Boletín de la Real Sociedad Económica Matritense de Amigos del País*, 48, 2002, pp. 99-112.

for writing purposes is mentioned by Philo of Byzantium in the 3rd century BC, Dioscorides in *De Materia Medica*;[5] and Pliny, who in his *Natural History* (1st century AD) alludes to a black liquid obtained by mixing iron salt, vegetal galls and pomegranate skin. The Leiden Papyrus (3rd century AD) and Isidore of Seville in his *Etymologies* (6th century AD) also mention the use of gills and vitriol to prepare the inks.[6]

The wider dissemination of recipes and formulae for the manufacture of ink, however, had to wait until the 12th century, possibly starting with the well-known 12th century treatise by Theophilus, *Schedula diversarum artium.*[7] Chapter 38, dedicated to ink (*encaustum*), points out that the infusion of tannin obtained from the soaking of hawthorn wood must be mixed with a metallic salt.[8] Other references can be found in Ramón Llull's *Llibre de contemplació*, in Joanot Martorell's *Tirant lo Blanch* (Chapter 289), in *De rebus metalicis, De cose minerali e metalliche, libri 5*, wrongly attributed to Albertus Magnus,[9] and in numerous codices, for example in "Miscellaneous 87," Vercelli's cathedral archive, in manuscript 490, Lucca's Biblioteca Capitolare, in "Vaticano Reginense Latino 124," the Vatican Library or in manuscripts 93 and 513, Biblioteca Antoniana.[10]

Everything seems to indicate that the substitution of carbon-based inks with metallo-acid recipes coincides with the transition from parchment to paper as the main writing material. Metallo-acid inks are better suited to paper, offering better fixation and permanence than on animal based materials, and also a clearer colour contrast. Additionally, the high alkalinity present in parchment can destabilise the ink and cause it to vanish.[11] In Stefanos Kroustallis's opinion Spain may have played a significant role in the substitution of one kind of ink for another. We know that in Islamic countries, both in the Near East and in northern Africa, both kinds were in use from the 10th century.[12] Their reci-

5. Mª M. Cárcel Ortí and J. Trenchs Odena, "La tinta y su composición. Cuatro recetas valencianas (siglos XV-XVIII)," *Revista de Archivos, Bibliotecas y Museos*, 72, 1989, pp. 417-418.

6. S. Kroustallis, "La Escritura y sus materiales: pigmentos, tintas e instrumentos.," in *El soporte de la lengua* (Coord. María del Carmen Hidalgo Brinquis), Nájera, Patronato Santa María de Nájera, 2008, p. 147.

7. J. G. Hawthorne and C. S. Smith, *Theophilus'on Divers Arts*, Dover Publ., New York, 1967.

8. M. Romero Tallafigo, *Arte de leer escrituras antiguas: paleografía de lectura*, Huelva, 2003, p. 45.

9. Mª M. Cárcel Ortí and J. Trenchs Odena, "La tinta y su composición. Cuatro recetas valencianas (siglos XV-XVIII)," *Revista de Archivos, Bibliotecas y Museos*, 72, 1989, pp. 419-420.

10. VV.AA., "Anciennes recettes d'encre," *Bibliothèque de l'École des Chartes*, 72, 1911, pp. 437-438.

11. S. Kroustallis, "Escribir en el siglo XVI: recetas de la tinta negra española," *Torre de los Lujanes: Boletín de la Real Sociedad Económica Matritense de Amigos del País*, 48, 2002, pp. 99-112.

12. M. Levey, *Medieval Arabic Bookmaking and its relation to early chemistry and pharmacology*, Transactions of the American Philosophical Society, New Series, vol. 52, part. 4, Philadelphia, 1962.

pes include the earliest mention of the systematic use of vegetal galls for the production of tannin. This seems to suggest that al-Andalus brought not only paper to the West, but also the most appropriate inks to use with it. The Spanish recipes are the first to mention the use of vegetal galls in the manufacture of metallo-acid inks;[13] additionally, the earliest recipes are often full of Arabic words and terminology.[14]

Most late medieval manuscripts are written in metallo-acid inks. The main ingredients are a tannin-rich vegetal extract and iron sulphate, resulting in an intense black colour.[15] In this kind of ink, the colour is caused by the chemical reaction of an acid in contact with a metal on the writing material. The acid oxidises the paper, colouring it, while "biting" it to the point of becoming fixed, so that it cannot be removed with water. Occasionally, soot is also used to improve the pre-fixing colouring.[16]

Galls are among the basic ingredients, although sometimes they were substituted with fruit, wood and bark (holm oak, chestnut), pomegranate skin, alum, salt, ammonia and berries. From the early 13th century, galls became the universal ingredient in Western Europe, and the other alternatives disappeared almost completely. This is because the galls' high concentration of tannic acids makes them ideal for mixing with larger amounts of iron and obtaining a more intense black colour. In Spain, the Islamic influence and the climate also favoured their continued use; the utilisation of this ingredient instead of others, such as honeysuckle, may indeed explain why the manuscripts from the Iberian Peninsula show a more intense black colour than others from central and northern Europe.[17]

Metals come second in the list of major ingredients, normally in the form of copper (blue vitriol) or iron sulphate (green vitriol). The latter was usually preferred, because though the black colour was less intense, the ink was more fluid and stable.

Another ingredient is the binder, which prevents the salts from forming lumps or decanting to the bottom of the container. The most commonly used were gum, honey, egg white, olive oil, and linseed oil, and of these the most commonly used was gum.[18] Although the sources regularly refer to it as gum

13. In his *Libre de contemplació en Deu* (cap. 291, 21) Ramón Llull offers the earliest mention of the use of ferro-tannic inks in the Iberian Península, saying that ink is made of galls, gum, vitriol and water (Raimundo Lulio, Beato, *Libre de contemplació en Deu*, Jordi Gayá Estelrich, Ed. Font, 1989).

14. S. Kroustallis, "Escribir en el siglo XVI: recetas de la tinta negra española," *Torre de los Lujanes: Boletín de la Real Sociedad Económica Matritense de Amigos del País*, 48, 2002, pp. 99-112.

15. S. Kroustallis, "La Escritura y sus materiales: pigmentos ...," *El soporte de la lengua*, p. 150.

16. M. Romero Tallafigo, *Arte de leer escrituras antiguas: paleografía de lectura*, Huelva, 2003, p. 45.

17. D. Carvalho, *Forty centuries of inks*, The Banks Law Publishing Co., New York, 1904.

18. S. Kroustallis, "La Escritura y sus materiales: pigmentos ...," *El soporte de la lengua*, p. 150.

Arabic, it cannot be taken for granted that it was in all cases obtained from acacia, because it can also be extracted from other resinous plants.[19]

The solvent, on the other hand, could be wine (red or white), water, beer,[20] vinegar or a mix of several of these. Wine solves the tannin, dries the ink (because of the alcohol) and preserves the gum Arabic; vinegar acts as an antiseptic, preventing the growth of mould in the container or in the paper itself.[21] Water was used in common inks, to write on paper and in the summer, because it evaporates more slowly than wine.[22]

Along with these ingredients, certain other additives aimed at changing some of the characteristics of the ink. Pomegranate skin, fig tree wood, carbon ink or other colorants (for example, indigo) could be added to improve the intensity of colour or even perfume (myrrh, incense).[23]

Spanish medieval manuscripts are rich in recipes for the production of black writing inks. The manuscripts recording these are preserved in the Spanish Biblioteca Nacional,[24] the Biblioteca del Palacio Real, Madrid,[25] the Real Academia de la Historia,[26] the University of Salamanca,[27] the archive in the Monasterio de Guadalupe (*Libro de los Oficios*); and the library of the Faculty of Medicine, University of Montpellier[28]. Additionally, we must also highlight the recipes located in the Archivo Histórico Provincial, Córdoba[29] and the Alejo Piamontés' *Libro de Secretos*. All these recipes are dated to between the 15th and the 16th centuries, and the ingredients and processes described are very similar.[30]

In general, Spanish recipes are considerably conservative and follow the Arabic tradition, as shown by the use of pomegranate skin or walnuts, which offered a high tannin content and resulted in a bright ink. This Arabic influence

19. In the Spanish recipes the terms "gum Arabic" and "gum" are interchangeable.

20. Beer is not used in any of the Spanish recipes.

21. M. Romero Tallafigo, *Arte de leer escrituras antiguas: paleografía de lectura*, Huelva, 2003, pp. 45.

22. S. Kroustallis, "Escribir en el siglo XVI: recetas de la tinta negra española," *Torre de los Lujanes: Boletín de la Real Sociedad Económica Matritense de Amigos del País*, 48, 2002, pp. 99-112.

23. S. Kroustallis, "La Escritura y sus materiales: pigmentos ...," *El soporte de la lengua*, p. 150.

24. BNE, mss. 9226 and 9028.

25. BPR, mss. II/657 and II/1393(6).

26. BRAH, Fondo Salazar y Castro, ms. K-50.

27. BUS, mss. 1562 and 354.

28. University of Montpellier, Faculty of Medicine Library, ms. H-490 (edit. R. Córdoba, "Un recetario técnico castellano del siglo XV: el manuscrito H-490 de la Facultad de Medicina de Montpellier," in *En la España Medieval*, 28, 2005, p. 32).

29. AHPC, Sección de Protocolos Notariales de la ciudad de Córdoba (quoted by R. Córdoba, "Un recetario técnico castellano del siglo XV ...," p. 32).

30. M. Zerdoun, *Les encres noires au moyen age*, París, 1983, pp. 247-290.

can also be detected in the terminology (for example, the use of the term "aziche" to refer to metallic salts) and in the practice of altering ingredients or their proportion depending on whether the writing material the ink was aimed for was paper or parchment.[31] According to Stefanos Kroustallis, however, the similarities are clearer regarding medieval than 16th century recipes, when a tendency to adopt Italian practices can be detected.[32]

Regarding ingredients, in Castilian recipes galls are clearly predominant in line with the Mediterranean tradition. In northern Europe, galls were replaced with the fruits, woods and barks obtained from different plant species (for example hawthorn).[33] Regarding the metallic salts, iron sulphate (so-called Roman or green vitriol) was generally preferred, although many recipes merely mention vitriol without giving further details. Among the binders, gum Arabic seemed to be predominant, followed by egg white. For solvents, wine, vinegar and water were used alone or in combination, while beer is not mentioned in any of the recipes. As to extra additives, these Castilian recipes mention sugar, poppy, ram bile, pomegranate skin, alum, indigo, myrrh, incense, perfumes and bleach. Indigo would improve colour intensity, while perfume would give the ink a pleasant aroma.

The amount of tannin, metal, binders and solvents required for ink manufacture are expressed in different units: solids are generally expressed in ounces and pounds, and liquids in *azumbres*, *cuartillos*, pounds and ounces. Other additives are given with such expressions as "a little," "a handful" or, simply, the number of units required.[34] In general terms, inks used a similar proportion of galls and vitriol, with a smaller amount of gum Arabic; these were the main ingredients, and other additives were not added in significant proportions.

The manufacturing process generally started with the introduction of the galls into a solvent (wine or water) to macerate for a few days, after which the remainder of the ingredients were incorporated to the resulting solution. Sometimes all ingredients were mixed at the first stage, to be left to macerate in the sun.[35] The extraction of tannin from the galls could be achieved by maceration in wine or water, but boiling offered a quicker method. In Castile, maceration

31. M. Levey, *op. cit.*, nota 4. The ink recipes recorded by Ibn Badis in northern Africa in the 11th century show these similarities clearly.

32. S. Kroustallis, "Escribir en el siglo XVI: recetas de la tinta negra española," *Torre de los Lujanes: Boletín de la Real Sociedad Económica Matritense de Amigos del País*, 48, 2002, pp. 99-112.

33. S. Kroustallis, "La Escritura y sus materiales: pigmentos ...," *El soporte de la lengua*, p. 150.

34. BNE, mss. 9226, fol. 106 and ms. 9028, fol. 1; BUS, ms. 354, fol. 225v; Libro de los Oficios, Monasterio de Guadalupe, fol. 203r-v (edit. M. C. Hidalgo, coord., *El Libro de los oficios del Monasterio de Nuestra Señora de Guadalupe*, Badajoz, 2007, vol. 1, p. 370).

35. BPR, ms. II/1393, fol. 1; BNE, ms. 9226, fol. 106.

was more common than boiling, in contrast with what is laid down in Northern European recipes. Maceration is more efficient in extracting tannins, resulting in a darker black colour, especially if the ink was left to oxidise before use. The maceration time was highly variable, and it was shorter if wine was used because alcohol favours the extraction of tannin. Often, the final stage of the process involved gentle boiling to enhance the mix and improve its shine.[36] Once the solution resulting from the maceration of the galls was ready the remaining ingredients were incorporated, at the same time or one by one, before exposing the mix to the sun[37] or to the gentle warmth of a charcoal fire (hot inks) or,[38] simply, left to rest.[39] The final ink could be strained or directly poured into the ink well, bladder or jar.

Regarding the treatment received by the galls, in most cases they were macerated for several days in advance[40] but in some cases all ingredients were put into the water or wine at the same time.[41] In other examples, not only the galls but other ingredients were left to macerate in wine, in as many containers as ingredients were macerated.[42] Some of the recipes warn that once the maceration of the galls was finished, they had to be removed from the solution before the remaining ingredients were added.[43] Others indicate that the mix of galls

36. S. Kroustallis, "Escribir en el siglo XVI: recetas de la tinta negra española," in *Torre de los Lujanes: Boletín de la Real Sociedad Económica Matritense de Amigos del País*, nº 48, 2002, pp. 99-112.

37. Cold inks, in BNE, ms. 9226, fol. 19, 71, 106, 112; *Libro de Secretos*, Alejo Piamontés, fol. 161 r and v; *Libro de Secretos*, Alejo Piamontés *(Libro cuarto de los secretos de diversos hombres ...)* fol. 41.

38. Hot inks, in BNE, ms. 9226, fol. 6; *Libro de Secretos*, Alejo Piamontés, fol. 161; *Libro de Secretos*, Alejo Piamontés (*Libro cuarto de los secretos de diversos hombres ...*), fol. 41; BRAH, ms. K-50; Libro de los Oficios del monasterio de Guadalupe, Recipe 1, fol. 202r-v (edit. M. C. Hidalgo, coord., *El Libro de los oficios del Monasterio de Nuestra Señora de Guadalupe*, Badajoz, 2007, vol. 1, p. 368); Universty of Montpellier, Medicine Faculty, ms. H-490 y Archivo Histórico Provincial de Córdoba (edit. R. Córdoba, "Un recetario técnico castellano del siglo XV," pp. 32 and 46-47).

39. BPR, ms. II/657, fol. 1; BNE, ms. 9028; Libro de los Oficios del monasterio de Guadalupe, Recipes 2, 3 , fol. 202r-v and 203r-v (edit. M. C. Hidalgo, coord., *El Libro de los oficios del Monasterio de Nuestra Señora de Guadalupe*, Badajoz, 2007, vol. 1, pp. 369-370).

40. BPR, ms. II/657, fol. 1; BNE, ms. 9226, fols. 1, 2, 5, 6 and 7; BUS, ms. 1562; *Libro de secretos*, Alejo Piamontés (*Libro cuarto de los secretos de diversos hombres ...*), p. 41; BRAH, ms. K-50; Universty of Montpellier, Medicine Faculty, ms. H-490 (edit. R. Córdoba, "Un recetario técnico castellano del siglo XV," pp. 46-47).

41. BPR, ms. II/1396; BNE, ms. 9226, fols. 3, 4 and 8.

42. BNE, ms. 9028; Libro de los Oficios del monasterio de Guadalupe, Recipes 1, 2 and 3, fol. 201r-v, 202rv and 203r-v (edit. M. C. Hidalgo, coord., *El Libro de los oficios del Monasterio de Nuestra Señora de Guadalupe*, Badajoz, 2007, vol. 1, p. 370); Archivo Histórico Provincial de Córdoba (edit. R. Córdoba, "Un recetario técnico castellano del siglo XV," p. 32).

43. BNE, ms. 9226, fols. 19, 71, 106 and 112; Biblioteca de la Real Academia de la Historia, ms. K-50, fol.146v; Archivo Histórico Provincial de Córdoba (edit. R. Córdoba, "Un recetario técnico castellano del siglo XV," p. 32).

and solvents should be fired in order to obtain a more concentrated substance.[44] One recipe even says that galls must be fried prior to maceration.[45]

Vitriol was commonly incorporated to the tannin-solvent solution.[46] Some recipes recommend wetting it before adding it to the mix, whereas others indicate that it must be incorporated to the solution and left to rest for some time before adding the remaining ingredients.[47] One case suggests exposing the vitriol to the sun, inside a cloth and hanging it from a cord, after which it must be introduced into a boiling pot and remain there until the mixture produces foam.[48] The gum, for its part, had to be pulverised, soaked and milled prior to incorporation.[49] As to the moment when it should be added to the solution, the recipes vary: after the gall maceration or at the end, along with the vitriol.[50]

The solvent used for the maceration of the galls was in most cases white,[51] and less frequently, red wine.[52] Only occasionally are more details on the characteristics of the wine given: young, mild, or weak.[53] Less frequently, the solvent used is water,[54] and on one occasion it is specified that this must be rain water.[55]

Among other additives, poppies are mentioned. The poppies must be left to soak in water all day, and the resulting water added to the maceration of the remaining ingredients.[56] Another recipe mentions indigo, which must be added at the end, after sitting for some time in wine.[57] Pomegranate skin appears in

44. Libro de los Oficios del monasterio de Guadalupe, Receta 1, fol. 201r-v (edit. M. C. Hidalgo, coord., *El Libro de los oficios del Monasterio de Nuestra Señora de Guadalupe*, Badajoz, 2007, vol. 1, p. 368).

45. *Libro de Secretos*, Alejo Piamontés, fol. 161r-v.

46. BPR, ms. II/657, fol. 24; BNE, ms 9226, fols. 19, 71 and 112; *Libro de Secretos*, Alejo Piamontés, fol. 161r-v.

47. BNE, ms. 9226, fol. 106; *Libro de secretos*, Alejo Piamontés, fol. 41; Archivo Histórico Provincial de Córdoba (edit. R. Córdoba, "Un recetario técnico castellano del siglo XV," p. 32).

48. BUS, ms. 354, fol. 225v.

49. BNE, mss. 9226, fol. 19 and 9028, fol. 1.

50. BNE, ms. 9226, fols. 71, 106 and 112; BPR, ms. II/657, fol. 24; BNE, ms. 9028, fol. 1; *Libro de Secretos*, Alejo Piamontés, fol. 161r-v; Archivo Histórico Provincial de Córdoba (edit. R. Córdoba, "Un recetario técnico castellano del siglo XV," p. 32).

51. BPR, ms. II/1393, fol. 1; BNE, mss. 9028, fol. 1 y 9226, fol. 35 y 71; *Libro de Secretos*, Alejo Piamontés, fol. 161r-v; BUS, ms. 1562, fol. 234r; BRAH, ms. K-50, fol. 146v; Libro de los Oficios del monasterio de Guadalupe, Recipes 1, 2 and 3, fol.s 202r-v y 203r-v (edit. M. C. Hidalgo, coord., *El Libro de los oficios del Monasterio de Nuestra Señora de Guadalupe*, Badajoz, 2007, vol. 1, pp. 368-370).

52. BPR, ms. II/657, fol. 24.

53. BNE, ms. 9226, fols. 71, 106 and 112.

54. BNE, ms. 9226, fol. 19; BUS, ms. 354, fol. 225v; Libro de los Oficios del monasterio de Guadalupe Receta 1, 2, fol. 201r-v y 202r-v (edit. M. C. Hidalgo, coord., *El Libro de los oficios del Monasterio de Nuestra Señora de Guadalupe*, Badajoz, 2007, vol. 1, pp. 368-369).

55. *Libro de secretos*, Alejo Piamontés, fol. 41.

56. BNE, ms. 9028, fol. 1.

57. BNE, ms. 9226, fol. 106.

two recipes: in the first it is added in the middle of the process, along with the vitriol,[58] and in the other it is incorporated at the end.[59]

Quite apart from the documents already mentioned, certain books published in Castile in the 16th century offer recipes for the producttion of – for the most part – metallographic inks. For example, the *Arte Subtilíssima* published by the calligrapher and scholar Juan de Yciar,[60] who invented the so-called Spanish "bastarda" calligraphy, still in use, after the inspiration drawn from Italian models – Henricis, Tagriente and Palatino. The work includes two recipes for the manufacture of ink to write on paper and on parchment.[61] The first is very similar to the aforementioned Castilian recipes: the ingredients are analogous (Roman vitriol, galls, gum Arabic, and finally pomegranate skin to enhance the colour), and the process is equally familiar: the galls must be left to macerate in the sun for two days after which the vitriol is added and left for another two days before the incorporation of the gum Arabic; the resulting concoction will be left to rest for a day before boiling with pomegranate skins. In the second recipe, "to write on parchment," water must be replaced with wine, but the other ingredients are the same (vitriol from Flanders, galls from Valencia, gum Arabic); vitriol, galls and gum must be left to soak separately in wine for a week, after which the solution where the galls have been soaking must be boiled for fifteen minutes over a gentle fire. Once this is done, the vitriol and gum Arabic solutions must be strained through a thick cloth and added while stirring the mix with a stick.

Juan de Yciar was followed by the Sevilian Francisco Lucas, one of the most important Spanish calligraphers, who in 1571 issued from the workshop of Francisco Guzmán in Toledo the *Instrucción muy provechosa para aprender a escrevir, con aviso particular de la traza y hechura de las letras redondilla y bastarda.*[62] Although Lucas' book does not include recipes for the production of ink, it advises the use of wine and gives some recommendations for the preservation of the ink inside the well. Another Spanish handbook on calligraphy, written by Ignacio Pérez, was *Arte de escrevir con cierta industria e*

58. BUS, ms. 354, fol. 225v.

59. BNE, ms. 9226, fol. 19.

60. Juan de Yciar, *Recopilacion subtilíssima, intitulada orthographia pratica por la cual se enseña a escriuir perfectamente, ansí por pratica como por geometría todas las suertes de letras que mas en nuestra España y fuera della se usan*, Zaragoza, Bartolomé de Nájera's printing press, 1548; new editions in 1550, 1553 and 1555 under the title *Arte subtilissima por la qual se enseña a escrevir perfectamente*: this is the earliest treatise published on the matter in Spain. He lived between 1515 or 1523, and 1590 (J. Durán, *Introducción al facsímil "Arte sutilísima, por la cual se enseña a escribir perfectamente hecho y experimentado y ahora de nuevo añadido por Juan de Icíar Vizcaíno.* Año 1553", Valladolid, Junta de Castilla y León – Ayuntamiento de Valladolid, 2002).

61. A. Mut Calafell, "Nuevas aportaciones sobre la tinta en Mallorca," *Mayurqa*, 22, 1989, pp. 849-864.

62. Francisco Lucas, *Arte de escrevir Francisco Lucas ...: dividida en quatro partes ...* Madrid, Alonso Gómez, 1577.

invención,[63] the seventh chapter of which includes three recipes for black ink: for the winter; for the summer; and for a formula which did not require the ingredients to be boiled. Besides these calligraphic treatises we also find ink recipes in works of a religious nature, for example by Andrés Flórez[64] who included a recipe for black ink[65] in his *Doctrina cristiana del Ermitaño y el niño*, published in Madrid in 1546 and reprinted in Valladolid in 1552.[66]

Recipes for the production of black inks are frequent in all regions during the middle ages. Antonio Mut Calafell has made a survey of the recipes published in the Crown of Aragon between the 16th and the 18th centuries. There is no major difference between these recipes and the Castilian ones, regarding either ingredients (galls, vitriol, gum, water, vinegar, and wine) or processes (cold and with the aid of fire).[67] In Valencia, 15th century recipes also mention galls, vitriol, gum, and alum.[68]

The comparison between black ink recipes from Spain and those from other countries shows strong similarities in this period, with the use of the same or highly comparable ingredients. The recipes compiled by Monique Zerdoun, dating to between the 13th and the 14th centuries, mention basic ingredients such as the galls – which would be pre-treated by boiling[69] and maceration[70] – solvents such as wine and beer,[71] and additives such as pomegranate skin, iron

63. Ignacio Pérez, *Arte de escrevir con cierta industria e invencion para hazer buena forma de letra, y aprenderlo con facilidad*, Madrid, Imprenta Real, 1599.

64. A. Esteve Serrano, *Estudios de teoria ortográfica del español*, Murcia, Universidad de Murcia, 1982, p. 26.

65. This same recipe is also included in the work by Bartolomé José Gallardo, *Ensayo de una Biblioteca española de libros raros y curiosos*, Madrid, Imprenta de M. Rivadeneyra, 1866, p. CLXXIV. The recipe says: "A warning: good ink is made with white wine, and common ink with water, preferably from a stagnant pool. To one cuartillo of water we must add one ounce of broken galls; this must be set to boil until one third of the water has evaporated, then strained and an ounce of vitriol and a quarter of an ounce of gum added. Keep the tepid mix well stirred. If a greater quantity is required, use always these proportions."

66. Andrés Flórez, *La Doctrina cristiana del Ermitaño y el niño*, Valladolid, Casa de Sebastián Martínez, 1552.

67. A. Mut Calafell, "Nuevas aportaciones sobre la tinta en Mallorca," *Mayurca: revista del Departament de Ciènces Històriques i Teoria de les Arts*, 22, 1989, pp. 849-864.

68. Mª M. Cárcel Ortí and J. Trenchs Odena, "La tinta y su composición. Cuatro recetas valencianas (siglos XV-XVIII)," *Revista de Archivos, Bibliotecas y Museos*, 82, 1979, p. 422.

69. British Library; ms. Harley 3915, f. 148v–149r; British Library, ms. Lansdovne 397, f. 6r; Biblioteca Vaticana, ms. Vat. Lat. 598, f. 61r; Pierre de Saint-Omer, *De coloribus faciendis*, Manuscrito de Jehan le Begue, nº 173 (M. Zerdoun, *Les encres noires au moyen age*, pp. 249-253).

70. Library, École de Chartres, 86, 1925, p. 484; Pierre de Saint-Omer, *De coloribus faciendis*; Manuscrito de Jehan le Begue, 303 and 47 (M. Zerdoun, *Les encres noires au moyen age*, pp. 255-259).

71. British Library, ms. Sloane 4, f. 3v.; British Library, ms. Arundel 507, f. 100v; British Library, ms. Sloane 416, ff. 87v, 140v and 141r; Bodleian Library, Oxford, Ms. Canon mise 128, f. 119r; Biblioteca Laurenziana, Flroence, ms. Ashmolean 349, f. 55r.; Biblioteca Nazinoale, Florence, Fondo Palatino, ms. 850, ff. 77v-78r; Biblioteca Riccardiana, Florence, ms. 1246, f. 79r; Biblioteca Riccardiana, Florence, ms. 1247, ff. 26v-27r and 49r (M. Zerdoun, *Les encres noires au moyen age*, pp. 260-275).

oxide, and rock alum. This author has also published several 15th century recipes which she refers to as "Italian," with the presence of vitriol, galls and gum as the main ingredients, and pomegranate skin, laurel leaves and crushed glass as additives.[72] For the 16th century, the situation remains pretty much the same, with the repetition of the same main ingredients and additives, among which bark and rose water are now also included. During this century there was a preference for macerating the galls, and boiling is only recorded in isolated instances. This is contrary to what the evidence tells us about the 13th and 14th centuries, but it coincides with the documented practices for the 15th century.[73]

This is once again shown in the Manuscript of Bologna, edited by Mary P. Merrifiel. The manuscript records three recipes, two of which only reflect the basic ingredients (tannin, binders, solvents and salts), whereas the third also includes pomegranate and ash bark, walnut roots and alum. The two former recipes also use fire throughout the process, similarly to the method set forth in the *Libro de los Oficios* (Monasterio de Guadalupe); the latter use sun exposure and fire, in line with most Castilian recipes. We must remember that pomegranate and alum are also frequently listed among the additives in Castilian recipes.[74] The Fondo Palatino in Florence's Biblioteca Nazionale preserves some ink recipes in which the same basic ingredients can be found: galls, water or wine, vitriol, gum Arabic; among the additives, rock alum and pomegranate skin, common in Castilian recipes, and incense and tallow candle soot mixed with water of roses. Equally, the ink is produced either by boiling or by exposure to the sun.[75]

In conclusion, Castilian ink recipes show strong similarity to others published in the rest of Europe or Northern Africa during the middle ages. The similarities are, at any rate, much more prominent than the differences, which may be explained by local peculiarities, such as the use of beer in northern Europe or the use of galls in the Mediterranean. In the modern period, ink continued to be manufactured using the same methods, and for this reason the recipes dated between the 16th and the 18th centuries reflect the same ingredients and the same manufacturing steps, making the production of black ink a rather conservative process over time.

72. Biblioteca Palatina, Parma, ms. Hebrew 402, 7, f. 4r (M. Zerdoun, *Les encres noires au moyen age*, pp. 276-282).

73. Les livres de Hierome Cardanus ... des Plantes, p. 187; Amphiareo of Ferrara, p. 200; Manuscript of Paduove: Ricette per far ogni sorte di colore – quoted in M. P. Merrifield, *Original Treatises dating from the Twelfth to the Eighteenth Centuries on the Arts of Painting*, New York, 1967 (reprint.), p. 677 (M. Zerdoun, *Les encres noires au moyen age*, pp. 290-292 y 302-303).

74. Recipes 303 and 303a, *De diversis coloribus in sequenti tractatus et primo modus prohemi* de Jehan Le Begue, and Recipe 374 *Segreti per colori,* Chapter VIII, Manuscript of Bologna (edit. M. P. Merrifield, *Original Treatises dating from the Twelfth to the Eighteenth Centuries on the Arts of Painting*, Nueva York, 1967 (reprint.), pp. 289-291 and 590).

75. Mss. Pal. 796 (fols. 49r-50r), Pal. 850 (f. 21v), Pal. 857 (f. 115r), Pal. 885 (f. 259v) and Pal 886 (f. 61r) (edit. G. Pomaro, *I Recetari del fondo Palatino de la Biblioteca Nazionale Centrale di Firenze*, Milan, 1991, pp. 109-115).

APPENDIX

1. BPR, ms. II/657, f. 24r.

In order to prepare ink without using fire, take thirty ounces of red wine, put in three ounces of galls from the Levante, well ground, and leave it to rest for three days, stirring six or seven times a day. After the three days you will add two ounces of Roman vitriol and one ounce of gum Arabic, well ground. After mixing well, you must leave it to sit for twenty four hours and you should find the mixture turned into good ink.

2. BPR, ms. II/1393, ff. 1r.

Ink. Take half an *azumbre* of white wine, three ounces of galls and one of *caparrosa* (iron sulphate), two of gum Arabic, and if you want also a bit of honey. Everything must be well ground and added to the wine, leaving the mix to rest for eight days, before straining it.

3. BNE, ms. 9028, f. 1.

Recipe to prepare ink. In order to make an *azumbre* of ink we need one and a half *azumbres* of white wine. It must be divided into three pots, of half an *azumbre* each. To the first, we must add four ounces of galls, finely chopped; to the second, four ounces of *caparrosa*; and to the third, poppies. These mixes must be left to concoct for ten days, stirring three or four times with a branch from a fig tree. After ten days the galls and *caparrosa* mixes must be put together, and that of the poppies must be cleaned, that is, removing the poppies and adding two ounces of gum Arabic; once the gum has dissolved all mixes should be put together with some ram bile. After letting this sit for three or four days we can write with it. Use the same proportions for whatever amount, if God is not adverse.

4. BNE, ms. 9226, f. 35r.

Recipe for black ink for paper or parchment. Leave six ounces of chopped galls to macerate for four or five days in one and a half *azumbres* of water or wine. After this, take them out, and put five ounces of ground *caparrosa* into the same wine and three ounces of gum Arabic which must be in lumps, and only ground after they have sat in the water for a while. Finally, put everything together and stir well, before leaving it in the sun for two or three extra days.

5. BSU, ms. 354, f. 255v.

Recipe for ink. Take one *azumbre* of water and some hematite and bring the water to boil until a quarter of the water has evaporated, and the water is the colour of red wine. Add four ounces of lead-coloured galls and stir with a branch from a fig tree or a cane – not of resin wood, such as pine – for 20 days. Afterwards, take 4 ounces of good *caparrosa*, sun-dried and ground, inside a cloth, and put it hanging from a thread and without touching the bottom of the container. After eight days, put the mix to the fire and leave it until some foam appears on the surface. Remove it from the fire and leave it to rest indoors. Some pomegranate skins will also be added at this point. Note that no gum is necessary, unless the ink results so thin that it goes through the paper, in which case add two ounces of ground gum, stirring well at the same time to ensure that it mixes well.

6. BRAH, ms. K-50, f. 146v.

Recipe for fine ink. Take four ounces of well ground galls and soak in white wine inside a glazed pot; [...] leave it to rest for six days, stirring twice a day. After this, the mix should be well strained and brought to a very gentle fire, removing it once a white foam forms on top, adding two ounces of gum Arabic stirring slowly, and finally adding two ounces of finely ground *caparrosa*, and this should do it, and the ink be good.

A SUIT OF ARMOUR PRODUCED BY FIVE WORKSHOPS: WALLACE COLLECTION A20

Alan Williams/David Edge

The medieval knight, like the Viking warrior, and the Roman legionary before him was protected by a defence of "mail," interlinked metal rings which covered the body, and he could afford, his arms and legs as well. It was a form of armour that was flexible, easy to repair and recycle, although labour intensive to make, but while effective against edged weapons, it was less so against missile weapons, because the links retained rather than deflected the points. So mail was first reinforced with, and then eventually replaced by, plates.

The homogeneous suit of plate armour developed during the late 14th century, and remained the most popular personal defence for those who could afford it, for 250 years.[1]

The production of relatively large plates of iron or steel placed considerable demands upon the metallurgical capacities of Europe, but first North Italy, and later South Germany, rose to these demands, and the most capable workshops produced armour consistently made of steel plates whose properties were controlled by judicious heat-treatment.[2] Of course, a good deal of armour was produced which was not of the highest quality, and makers' marks might have been thought to provide the customer with some assurance.

The 15th century suit of armour which is now in the Wallace Collection, London (cat.no. A.20) was put together by a 19th century dealer for a collector who wanted a "Gothic" armour.[3] The different parts are of very different origins:

1. The sallet (open-faced helmet) was thought to be Italian (second half of the 15th century) and has a maker's mark of a crowned orb. This may have

1. C. Blair, "European armour."

2. A. Williams, "The Knight and the Blast Furnace" (Leiden, 2003).

3. J. G. Mann, "Wallace Collection Catalogues: European arms & armour" (London, 1962), p. 7-9.

been the mark of one Martin Rondell, born in Milan, who became a citizen of Bruges in 1464.[4] It also occurs on one of the helms in St.George's Chapel, Windsor.[5]

2. The breastplate was made in two parts. They were thought to be German (second half of the 15th century). The backplate is not original.

3. The arm-defences are German in form. Only the upper right vambrace (arm-defence) is in fact original. The gauntlets are South German of the late 15th century. The right gauntlet bears the clover-leaf mark of the Treytz family of Innsbruck. The left gauntlet has traces of a similar mark.

4. The leg-defences belong to the early 16th century. The greaves did not originally fully enclose the leg, as did some German tournament armours of the 16th century, but have been altered at a later date. The poleyns (knee-defences) are also German in form.

5. The sabatons (foot-defences) have the marks IHS and VRBAN. This was probably the mark of an Italian armourer of the 15th century.

Wallace Collection A6 is a mail shirt that is often neglected as it is hidden inside the composite suit of plate armour, Wallace Collection A20. There is no evidence that these two armours ever belonged together.[6]

When this composite armour was dismantled for an exhibition, the opportunity to examine parts of this armour by metallography (microscopic examination) was taken, and a very wide disparity in the metal of the different components was found.

Seven of the plates were examined in cross-section by placing them on an inverted microscope:

1. The sallet.
2. The upper breastplate.
3. The backplate (19th century).
4. The left gauntlet.
5. The left poleyn.
6. The left greave.
7. the left sabaton

4. A. V. B. Norman, "Wallace Collection Catalogues: European arms & armour-supplement" (London, 1986), p. 1-2.

5. Research in progress.

6. Research in progress.

Hardness of metals

The hardness of metals can be measured on the Vickers Pyramid Hardness (VPH) scale whose units are $kg.mm^{-2}$. The hardness of pure copper after annealing is around 50 VPH, but after hammering cold it can be raised to 115 VPH. (Pure iron is only around 90 VPH). Alloying copper with 10 % tin to form "tin-bronze" can raise this to 230 VPH.[7] The hardness of iron increases as it absorbs more carbon, and becomes steel. But steel does not become superior to bronze until it is quenched, when it can become very much harder.

What can be seen microscopically ?

Ferrite crystals are pure iron, and appear as irregular white areas. Only the grain boundaries are actually visible. Hardness = 80-120 VPH.

Pearlite is a mixture of iron and iron carbide containing up to 0.8 % C. It has a grey, lamellar appearance, and a hardness up to 250-300 VPH. It may be formed when red-hot steels are cooled in air.

The form of solid iron stable at high temperatures (austenite) can dissolve a good deal of carbon. On equilibrium cooling, carbon is rejected as iron carbide and the low temperature form of iron (ferrite) is formed. A steel containing 0.8 % C transforms completely to pearlite at a constant temperature. Most steels of below 0.8 % C contain mixtures of ferrite and pearlite.

Martensite is formed in quenched steels. The rate of cooling generally has to be rapid to avoid the formation of pearlite. It appears like laths, in steels of up to 0.5 % C, with some triangular symmetry. In steels of medium-carbon content (0.5 %-1.0 % C) it can display an acicular appearance.[8]

Its hardness depends upon the C % and can vary from 200 to over 800 VPH. This is associated with extreme brittleness (a razor blade, for example). Gentle reheating (in the region between 200°C and 600°C) reduces the hardness somewhat but increases the toughness considerably. The preferred modern technique of hardening steels is full-quenching followed by tempering, but this depends upon knowledge of the carbon content and the accurate control of temperature and time. All of these requirements were difficult to meet in Medieval Europe, and so other techniques of heat-treatment were often employed. Delayed or interrupted quenches (generally called "slack-quenches" and not practised today) led to mixtures of martensite and pearlite being formed.

7. V. Buchwald, "Metallurgical study of 12 prehistoric bronzes from Denmark," *Journal of Danish Archaeology* (1990), p. 64-102.

8. R. W. K. Honeycombe, "Steels" (1982) passim.

Results

The sallet

The microstructure consists of alternating bands of ferrite and martensite, with sharp divisions between them. This has evidently been made by forge-welding together at least eight layers of iron and steel of very different composition to form a single sheet.

An attempt has been made to improve the sallet by quenching, and a hardness of 290 VPH has been obtained in the steel. The iron layers have not hardened on quenching.

The most likely explanation of this very unusual microstructure would seem to be the recycling of older metal.

Evidence for recycling

The Regulations of the Nürnberg Armourers' Craft dating from 1478 (with later additions to 1624) have been published by Alexander von Reitzenstein.[9]

The raw material came predominantly from the iron districts of the adjacent Palatinate (Oberpfalz). This metal went first of all to the hammer masters, that operated the trip-hammers powered by water- mills which turned blooms or billets into plates (*Blech*, or *Zeug*) for the armourers. The hammer masters were a craft with their own regulations, that specified that the masters from the hammers at Dutzendteich, Lauffenholz or other places, were to promise to make the metal for the harnesses of "at least half-steel or better" (*zum wenigsten halbstählern oder besser*) and nothing less, and to sign it with the city mark or their own mark. Some armours have a letter (a Gothic *n*) within a circle stamped on the insides of their plates. It has been suggested that such marks were those from the hammer-masters, but this remains to be proved. Whether "half steel" should be taken to mean a mixture of iron and steel is debatable; a low-carbon steel rather than an iron/steel mixture may have been intended, and the metallographic examination of Nürnberg armour does indeed show a low-carbon steel frequently used.[10]

However as well as material supplied by the hammer-masters, the armourers might *also* make use of a material described as *gewellten zeug*, that was also to be at least half steel. The Regulations stated:

> *Doch so mögen die Plattner auch gewellten Zeug machen, der auch auf das mindeste halb stählern sein soll. Der Plattner ist also befugt, das Mischungs-*

9. Alexander von Reitzenstein, "Die Ordnung der Nurnberger Plattner," *Waffen- und Kostümkunde* (München, 1959) New Series, I, p. 54-85. *Idem*. "Der Nürnberg Plattner," *Beitrage zur Wirtschaftsgeschichte Nürnbergs* (Nurnberg, 1967) Band 2, p. 700-725.

10. A. Williams (2003) see chapter 5.10.

verhältnis seines Materials durch das "Wellen" von Stahl und Eisen, die er folglich auch getrennt beziehen kann, in eigener Werkstatt herzustellen.

However so, the Armourer may make also "welded stuff," that should also be at least "half steel." The Armourer therefore is authorized to produce in own workshop, through the "welding together" of steel and iron, that may be obtained from his own materials, and with this ratio.

[my translation – Reitzenstein's rendering of "Wellen" is *Verschweissen* i.e. "welding"].

If it could be made up in the workshop, and did not need a water-hammer, then it must have already been in sheet form, so it may have been pieces of obsolete or damaged armour, or perhaps offcuts from other sheets, that were being recycled.

The upper breastplate

The microstructure of the upper breastplate is a mixture of ferrite and a little pearlite. This is a low carbon steel (0.2 % C) of hardness 200 VPH.

The backplate

The microstructure of the backplate is a fake made from wrought iron of hardness 180 VPH. This was apparently made in the 19th century.

The left gauntlet

Both gauntlets bear the mark of the Treytz family of armourers. The microstructure of both consists of martensite with some ferrite. The average hardness is 440 VPH.

At least three generations of the Treytz family worked as armourers in Innsbruck, which was justly famous for producing the best armour in Europe.

Different members of this family used very similar marks. Konrad Treytz the Elder (d.1469) used a 3-lobed clover leaf. His son Jörg (active to 1499) may have inherited his father's mark. He had two brothers, Adrian (active to 1492) and Christian (active to 1487) one of whom employed a very similar mark to their father, with the stem pointing right instead of left.[11]

In 1530 Sir Robert Wingfield (the English Ambassador to the Imperial court) wrote of a "complete harness which was made for myself at Innsbruck and given to me by the Emperor Maximilian ... whereof I do warrant that a fairer or *a better metal* cannot be found" (my italics).

11. B. Thomas, O. Gamber, "Die Innsbrucker Plattnerkunst" (1954), p. 55.

The left poleyn (knee-defence)

The microstructure consists largely of martensite with average hardness 400 VPH. Such a hardened steel is typical of South Germany in the 16th century. It is certainly not Italian.[12]

The left greave (lower-leg defence)

The microstructure consists of ferrite and pearlite. This was made of a 0.4 % C steel of hardness 220 VPH. If this had been hardened by quenching this would have been of comparable hardness to the poleyn. However, no definite connection can be inferred from this.

The left sabaton (foot defence)

The microstructure of the sabaton is a mixture of ferrite and a little pearlite. This was made of a low carbon (0.2 % C) steel of hardness 150-200 VPH. This is of surprisingly poor quality for 15th century Italian armour.

Conclusions

Allowing for the fact that the backplate was already known to be a 19th century restoration, this armour seems to have been the product of at least *five* workshops. In descending order of quality:

1. The Treytz workshop of Innsbruck.
2. The South German workshop that made the knees.
3. The German workshop that made the lower legs in the 16th century.
4. Another German workshop that made the breastplate in the 15th century.
5. The Italian workshop that made the feet.

12. A. Williams, "The knight and the blast furnace" (Leiden, 2003), p. 215-329.

Fig. 1: Suit of armour A 20; with marks of Rondell, Treytz, and Urban.

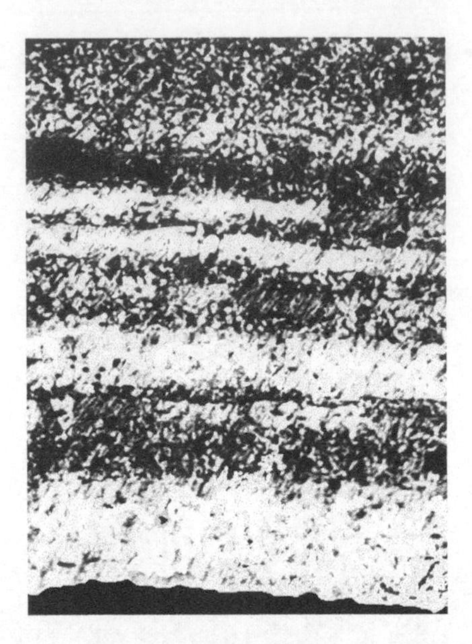

Fig. 2: Microstructure of sallet; ferrite and martensite in bands.

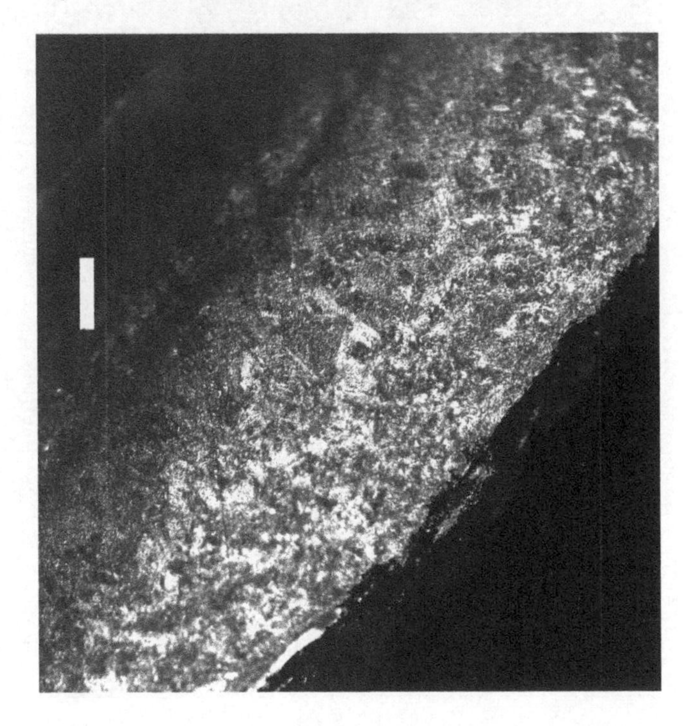

Fig. 3: Microstructure of left gauntlet; tempered martensite.

Fig. 4: Microstructure of left poleyn; martensite.

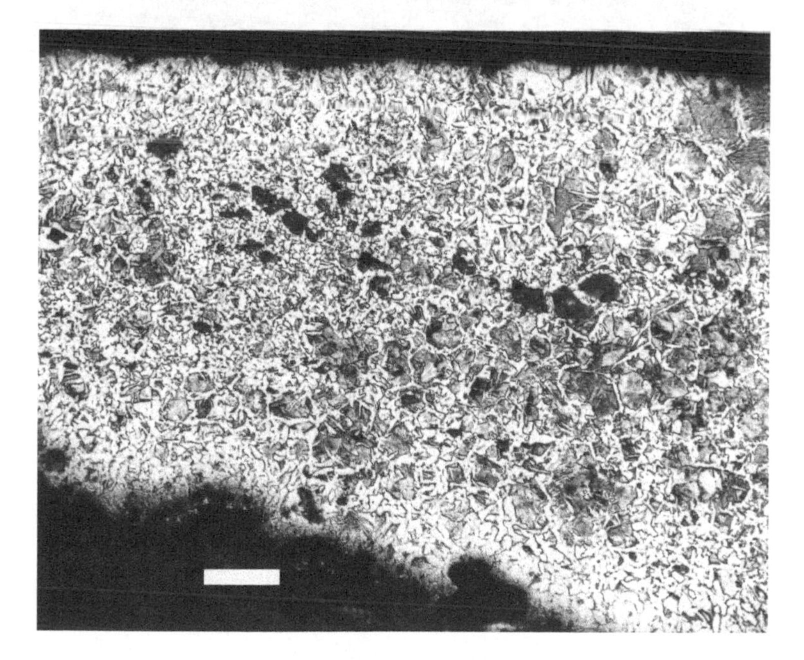

Fig. 5: Microstructure of left greave; pearlite and ferrite, with some slag inclusions.

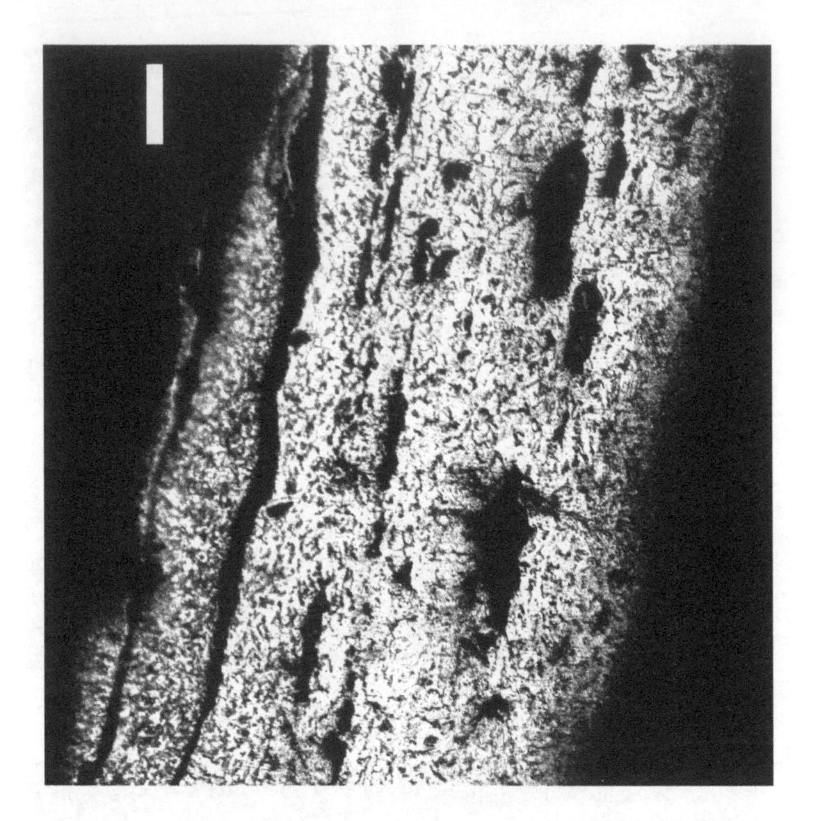

Fig. 6: Microstructure of sabatons; ferrite and a little pearlite, with numerous slag inclusions, and corrosion cavities (scale bar = 50 microns in each case).

THE ETCHING OF IRON BEFORE THE INVENTION OF ETCHED INTAGLIO PRINTING PLATES, 1200-1500

Ad Stijnman

Introduction

It was probably not later than 1500 that the Augsburg armourer Daniel Hopfer (c. 1470-1536) started etching flat iron plates in order to print them like copper engravings. By that time he was accomplished in decorating iron armour by means of etching. How he came to his invention is not known, etching decorations in iron was already practised since the 13th century.[1] A parallel development is the engraving and printing of copper plates practised in Germany from around 1430 and in Italy after the middle of the 15th century.[2] The first copper engravings were printed by hand by rubbing the back of the printing paper.[3] From 1460-65 plates are printed on a roller press, a kind of mangle, in the Upper Rhine region and in Italy (Mantua, Florence) from 1470.[4] Hopfer combined the techniques of etching an image in a flat iron plate with printing this plate with ink on paper by means of a roller press. In that way he invented etched intaglio printmaking. Metals, stone, glass, ivory and bone can be corroded by an etchant, usually a strong mineral acid or else a mixture of various salts in vinegar or water, and occasionally a lye is found. There are three ways to create a design on the object to be etched. The simplest is to paint or draw with the mordant on the surface of the object. This will give some discolouration or matting of the surface. More effective is to paint the design with a resist or ground, such as molten wax or oil paint, on the surface. Next the object is covered with an etchant or placed in a tray with an etching fluid, which corrodes the bare metal. The other method is to completely cover the surface of the object with a ground, draw the design into it with a needle and next let the

1. Harzen 1859; Köhler 1897; Landau & Parshall 1994, p. 323; Williams 1934.
2. Geisberg 1923, p. 26; Levenson 1973, p. XV.
3. Metzger 2009, p. 20-21, 23.
4. Bocquentin 1993, Vol. 1, p. 10-15, 59.

mordant act upon the metal, stone or whatever material. Both these methods create a tangible relief and both techniques are described from early on.

Antecedents

The materials for etching are available around the world and the chemistry is the same everywhere. One can therefore imagine that it is possible to find examples of objects with etched decoration in various corners of the Earth and in various periods.[5] A very early example is the corrosion of semi-precious stones, such as carnelian, agate and chalcedony. The oldest of such objects are from the Middle East and date from before 2,700 BC. Through trade they appeared from the Crimea and West Siberia, to China, to South India later in time, and here was still a production site in what is presently Pakistan in the third quarter of the 19th century.[6] Presently the technique is performed in India again for low-rate work.[7] The technique is simple. A pattern is drawn on the bead with a paste which contains an alkali such as sodium carbonate as its reactive ingredient. The stone is placed in a clay holder, heated for some minutes and a chemical reaction takes place between the paste and the stone. The effect is local discolouration of the surface. There is no relief and layers of tiny white "bubbles" can be seen in cross sections with microscopic enlargement.[8]

From Mesopotamia we move further East to China. Bronze mirrors, swords and spear points from the mid-5th century BC (late Chou and Han periods) show typical, etched patterns. The metal shows a clear, lighter discoloration due to chemical changes at the place of etching, and etched patterns on swords and spear points have a shallow relief. What chemical process was used is not known.[9] Inscriptions in intaglio in a number of late Chou (4th-2nd century) bronzes are of such a character that they cannot be easily explained as to be made by casting or engraving. The edges of the grooves are flat, irregular and undercut, which excludes casting. This suggests an etching technique using a ground and an etchant. Further explanation is waiting, while considering the Chinese of the Han period (2nd century BC-2nd century AD) were versed in chemistry.[10] A similar kind of decoration is observed with Egyptian copper alloy objects from around 2000 BC to Roman times.[11]

Crossing the Pacific Ocean to North America and travelling further across the Rocky Mountains to Arizona we arrive at the Gila River valley, North of

5. Smith 1981, p. 277-281.
6. Beck 1933.
7. Information kindly supplied by Dolores Snel.
8. Francis 1980, p. 24.
9. Chase & Franklin 1979, p. 243-256; Loehr 1956, p. 204-205.
10. Pope 1967-1969, vol. 2, p. 141, 146-147, fig. 182, 183, 187.
11. Ogden 2003, p. 160.

the Gulf of California. Once an agricultural people lived here. They are called the Hohokam by archaeologists and may have inhabited the valley from the 4th century BC until after 1400 AD. The Hohokam seem to have been trading with coastal tribes or travelling there themselves, as many marine shells are found in archaeological diggings. These shells are often decorated by carving, cutting and painting, but occasionally they show a relief which is clearly etched. The etching shows by the irregular edges of the grooves, rounded corners and undercutting of the edges. One such object has remains of pitch, which was used as a resist. The argument was that the chalk of the shell is easily corroded by a mild acid such as fermented juice (vinegar) from the local *saguaro* cactus fruit. A test showed that painting a design with a resist and leaving the shell in this vinegar for three days was enough to produce the same results as observed with the originals.[12] Experimentation by the present author etching a sea shell with strong vinegar produced a relief of 0.2 mm in twelve hours.

Pattern welding

The above examples are incidental and isolated in time and place, there have been no mutual connections and there are no developments towards modern printmaking. The European techniques of "pattern welding" and "copper gilding on iron" therefore come closer. In Europe, during the early middle ages and before, iron sword blades and spear heads were constructed by so-called "pattern welding," a term introduced by Herbert Maryon in 1948. Pattern welding is a metal craft practised in Europe from the 3rd century to the end of the 10th century, gradually disappearing afterwards. The technique is still used in various East and South-East Asian countries, famous are the Japanese sword blade and the Java kris.[13] By the earliest forms of this technique strips of pure and carburized iron were welded into a bar and next folded double once to be forged into a tough blade. Pattern welded weapons are found from the 4th century BC (Etruscan examples) and the 2nd century BC (Celtic examples), developed further from the 2nd century AD onward and came to full bloom during the Merovingian and Carolingian periods, with export of the arms forged in this manner to Eastern Europe in later centuries.[14] Pattern welded swords were in great demand by the Normans invading the empire and by tribes threatening the Eastern borders of Charlemagne's territory. That was reason why the Carolingian administration tried to prohibit export of armour and weapons to these enemy clients.[15] Pattern welding should not be confused with the production

12. Gladwin 1965, p. 148-151; Haury 1967, p. 677, 680.

13. Böhne 1963, p. 227; Meijer 1981, p. 34; Smith 1965, p. 8-9.

14. Aitchison 1961, p. 225-226, 253-257; Anteins 1966; France-Lanord 1964, 321, 326-327; Glosek & Kajzer 1977; Kirpicnikov 1986, p. 15; Neuman 1927; Panseri 1965, p. 37, 39; Sachse 1993, 116-117; Tylecote 1992, p. 66, 68.

15. France-Lanord 1952, p. 417; Salin 1957, p. 279-281.

of Wootz steel, practised in India and the Middle East, and called "damascening" after the city of Damascus where the technique was applied. The technical differences with the latter are the production method of the steel, which is tempered slowly in order not to disturb the crystallisation process, by a higher carbon content and by the presence of traces of the metal Vanadium in the ore which creates a hard and tough alloy. The patterns made by damascening are more complex and refined than the patterns in the earlier pattern welded blades.[16]

In the technique of pattern welding thin bars of hard, carburised (0.3-0.6 % carbon) and soft (pure) iron are welded together to create tough blades which stay sharp. The smith can fold and hammer the iron of the blade in various patterns, which patterns are disclosed by polishing or "etching" the metal.[17] What technique the European medieval smith applied to disclose the welded patterns is unknown, although the general acceptance is that the blades were etched, as this gives best results. A possible reference to etching may be found in a letter Cassiodorus (490-585), as secretary to Theodoric the Ostrogoth, wrote to Thrasamond, King of the Warnorii sometime between 523 and 526. Thrasamond sent two swords as a present to Theodoric. Cassiodorus thanks him for that and describes how shiny they are, how smooth their edges are shaped and that "the centres are hollowed out with beautiful grooves, (that) seem to undulate with worm-like marking" (*media pulchris alveis excavata quibusdam videntur crispari posse vermiculis*).[18] To distinguish between carburized and non-carburized iron differences in shades of colour are enough. Colour hues are created by polishing followed by rubbing the metal with a weak acid, which is the technique for damascene steel.[19]

Experimentation shows that just leaving iron in a bath of vinegar is not effective as the vinegar is not acidic enough. The iron discolorates (micro-etching) but no tangible relief is created. Suspending an iron object above a vinegar bath to expose it to its vapours in combination with oxygen in the air makes it rust quickly. When the rust is brushed off after a week it leaves a shallow (>0.1 mm) relief. It is conceivable that a similar action continued for a month or so would etch the weaker iron in a pattern welded blade more than the harder iron. After cleaning and polishing an agreeable structure will appear. A reference to exposing iron to a mild acid is given by Diodorus Siculus (fl. 60-30 BC). He describes the swords of the Iberian Celts as "two-edged and wrought of excellent iron." According to him, they did bury "plates of iron in the ground and leave them there until the rust has eaten out (περιφαγοντοσ) what is weak in the iron and what is left is only the most unyielding, and of

16. Sachse 1993, p. 36, 94, 162-163, 165; Verhoeven 1998.

17. Böhne 1963, p. 230-233; Jones 1997, p. 1-3; Sachse 1993, 28-29; Smith 1965, p. 4.

18. Cassiodorus, Variarum, V, 1; Salin 1957, p. 273-274, nr. 247; Smith 1965, p. 6-7.

19. Allan 1979, p. 86; Verhoeven 1998, p. (4).

this they then fashion excellent swords." The humic acid in combination with the oxygen present in the soil will have made the iron rust. By next scraping off the rust a piece with a higher proportion of carburised iron is left, which can be forged into a tougher blade. Thus interpreted this passage does not seem to refer to pattern welding of the swords themselves, but to an intermediate stage in the process of preparing a harder quality iron with veins of the softer iron left.[20]

The terminology used in literature on pattern welding may cause misunderstandings. Commonly the term "etching" is used, while the action of the acidic fluids will have been superficial only. The term "micro-etching" would be more appropriate, in the sense of merely changing (matting, discolouring) the surface of the metal in stead of creating a tangible relief by true etching.[21]

Another misunderstanding concerning etching may rise from the observation of iron swords dug up after two millennia, and next cleaned and further prepared. We do not have pristine, polished and patterned blades of that period, so do not know how they looked like originally. A centuries-long stay in a mild acidic environment may create a form of relief due to the differences in chemical behaviour of the two kinds of iron. Subsequent conservation treatment may further cause an aesthetic effect which is away from the object's original appearance.

However, the exception is a kind of decoration practised by Celtic smiths. Iron weapons have been found in Celtic burial mounds and dredged from rivers. The mass found from close to the village of La Tène in the lake of Neufchâtel in the West of Switzerland dates from the 2nd and 1st century BC and is particularly interesting for the present discussion. The sword blades are skilfully decorated with a variety of means among which *etched* decorations of geometric patterns are conspicuously present. Some of the blades show a clear and tangible relief with smooth raised parts while the recessed parts are coarse. This can only have been produced by partially covering the metal with a resist, such as molten wax or pitch, and etching away the iron around it.[22]

It is tempting to think there might be some kind of relation between the Celtic swords with their simple, etched relieves and late mediaeval swords with their elaborately etched decorations. These are similar kinds of objects, decorated by means of acids and resists, and produced in the same regions. However, the distance in time is large spanning a turbulent period in European history and as long as this relation is not properly researched there is a better candidate for the origin of the late mediaeval etching of armour, a form of copper plating, which is used in "gilding" iron.[23]

20. Diodorus Siculus, Book V, XXXIII, 3-4; France-Lanord 1964, p. 321; Oldfather 1952, vol. 3, p. 184-187.

21. My thanks to Allan Williams for elucidating this matter.

22. France-Lanord 1964, p. 323-324, pl. II-VI; Wyss 1968, p. 664-672, Taf. 3-5.

23. Smith & Hawthorne 1974, p. 49, n. 98.

Copper plating

There was a fair knowledge on chemistry available in the middle ages and within the reach of our subject there are dozens of recipes for gilding iron which are corrosive, too. In this kind of gilding process the object is placed in a copper salt solution, and due to electrolytic action between the copper ions in the solution and the iron object metallic copper is plated onto the iron. This serves as a ground upon which the gold was applied. The oldest known recipe for this manner of gilding is in the Lucca manuscript from around 800. It explains to mix equal parts of copper sulphate, alum and common salt (*calcitarim et alumen Asianum equis ponderibus et sal similiter*) and to dissolve this in water. Together with another part of gum tragacanth this makes a gel which can be applied onto the iron object locally, the gel keeping the solution in place.[24] The same recipe can be found in later manuscripts, such as in the Codex Matritensis (c. 1130).[25] Oakeshott shows a 10th century sword blade with what may be an example of this technique of copper plating.[26]

The Heraclius manuscript, originally dating from the 10th century, has a number of gilding recipes. Two of them tell to cover iron with a layer of copper before actual gilding. The first one explains to dissolve three parts of *atramentum* and one part of common salt in vinegar. The iron object is placed in the solution and is covered with a thin layer of copper. The other recipe prescribes alum, *sal gemma* (common salt in a particular crystalline form) and *calcanthum* in vinegar, with the same effect. This is the same recipe as in the Lucca manuscript and the Codex Matritensis. Thereafter the layers are covered with gold dissolved in mercury, heated, the mercury evaporates and the gold is left on the surface.[27]

The *atramentum*, *c(h)alcant(h)um* and *calcithari(u)m* are terms for salts containing various sulphates and more commonly known as "vitriol," it will have been rich in solvable copper(2) sulphate ($CuSO_4$).[28] Experimentation with plain copper sulphate dissolved in water shows immediate copper plating when iron is dipped into the solution or is rubbed with it. Prolonged exposure of iron in the solution shows a minor corrosion after a few days. Adding kitchen salt etches the iron immediately, because the chlorine in the salt propels the chemical reaction, and when there are equal amounts of both the biting goes fast.[29] Copper acetate dissolved in vinegar works slower, but addition of

24. Biblioteca Capitolare di Lucca ms. 490, fol. 221v; Lucca 2003, p. 102-103, 176.

25. Biblioteca Nacional, Madrid, Ms. A.16, recipe nr. (81) fol. 203r, col. a; with thanks to Stefanos Kroustallis and Mark Clarke.

26. Oakeshott 1981, pl. 48.

27. Ilg 1970, p. 64-67, nr. XVII; Merrifield 1849, p. 222-225.

28. Guineau 2005, p. 88-89 ("atramentum"), p. 179-180 ("calcanthe-calcater"), p. 237-238 ("colcothar"), p. 773-775 ("vitriol"); Perego 2005, p. 529 ("Colcotar"); Stijnman 2006, Appendix 3: "Terms for 'vitriol'."

29. I am grateful to Jana Sanyová for elucidating the chemical process.

kitchen salt or sal-armoniac plates the iron in a minute. Etching the shallowest relief with these mixtures takes some hours.

Nitric acid

Another way of etching metals is by means of nitric acid. The oldest description for the distillation of a liquid with a strong dissolving action is in the manuscript *Liber de inventione veritatis* written by "Geber' around 1300. It prescribes dry distillation of one pound of vitriol (Fe_2SO_4 or $CuSO_4$) with half a pound of saltpetre (NH_4Cl) and a quarter of a pound of alum ($K_2SO_4.Al_2(SO_4)_3$). The distillate has a strongly dissolving property and therefore is nitric acid (HNO_3). Further addition of sal-ammoniac (KNO_3) creates *aqua regia.* This is a mixture of one part of concentrated nitric acid and three parts of concentrated hydrochloric acid (HCl), and the only acid capable of dissolving gold. The distillation of nitric acid is also known from Byzantine manuscripts from the end of the 13th century as well as from other 14th century Western manuscripts.[30] The author "Geber" may be identical with the Franciscan lector Paulus de Tarento working in Assisi in the late 13th century and originally from the south of Italy. The name "Geber" was used first by 14th century Western scholars to refer to an Arab scholar and is derived from "Jabir." There are more persons called by that name, but in our case the likely candidate is Jabir Ibn Hayyan (fl. 8th century) who mainly worked in Bagdad. A voluminous compendium of treatises known as the *Corpus Jabirianum* is attributed to him, but most, if not all, of the works are compiled only a century later.[31] It might also contain a description of the distillation of nitric acid.[32] The literary evidence is not strong enough to attribute the actual production of nitric acid to an 8th century Arab scholar, but at least the transition of oriental knowledge to occidental scholars was stimulating and by 1300 nitric acid is known in Europe. The *Corpus* encompasses all ancient sciences, is usually associated with alchemy and astrology, but also contains practical techniques such as the chemistry of salts, production of steel and illumination of manuscripts. Alchemy arose in Hellenistic Egypt, more particular in Alexandria in the 2nd and 3rd century AD and was partly based on the chemical knowledge present there. Knowledge of Alexandrian alchemy was taken over by Islamic scholars from the 8th century. Greek or Syrian texts were translated into Arabic, their contents studied and further developed by them. Arab alchemy was

30. The Liber de inventione veritatis is part of the Summa perfectionis, Bayerische Staatsbibliothek, München, Ms. Lab. 353, end of the 13th century; Beltran 1998, p. 505; Forbes 1970, p. 63, 86; Geber 1922, p. 8, 113-114, 179; Geber 1928, p. 223-224; Karpenko & Norris 2002, p. 1002; Stijnman 2006, p. 68, Appendix 3: "Terms for 'vitriol';" Von Lippmann 1971, p. 175; Walden 1952, p. 6-7.

31. Newman 1985, p. 77, 79, 81, 88, 90; Singer 1948, p. 49-50.

32. Singer 1954-1984, vol. 2, p. 356.

introduced in Southern Europe in the 12th and 13th century by means of translations of Arabic manuscripts into Latin, and scholars in Toledo were particularly active in this field.[33]

With this we have two different methods of corroding metals, coming from two different backgrounds, one from the practical crafts and the other from scholarly research. Etching iron with a salt mixture is easy and the chemicals are readily procurable. Nitric acid is well suited for etching iron and copper, but is more complicated to obtain. Although known in the Western world in the 14th century, a first invention does not mean nitric acid was produced commercially or was available to everyone immediately. Saltpetre, the source of the nitrogen in the nitric acid, is used for a variety of purposes and for the production of gunpowder it is essential. Important is the increase of guns in warfare and as a result nitric acid was produced on a larger scale in Venice from the 15th century. Larger scale production of nitric acid began in France, and possibly also in Germany, in the 16th century.[34]

Etching iron

A resist or ground is missing in these first mentionings. With both methods they just say it corrodes metals, there is no control of where the metal is corroded. Proof of decorating iron, steel and other metals by means of etching a relief of some form comes from a recipe in a 14th century *Secretum Philosophorum*, which is thought to have originated in the 13th century. The term for the acid is *corrosivum*, oil-paint is applied as resist and after etching the grooves are coloured. This comes close to intaglio printmaking, as the next step would be printing this colour onto a sheet of paper.[35] A printed book called *Pro conservanda sanitate* was published in Germany in 1531. The introduction says it is a publication of a manuscript compiled by the French bishop Vitalis de Furno (1260-1327), discovered by the abbot Laurentius in the monastery of Eberbach and entrusted by him to Schoeffer for publication. No copy of this manuscript seems to be known and the attribution to Vitalis de Furno is uncertain.[36] The text on distilling a "water" good to dissolve all metals explains to grind one pound of saltpetre with one pound of vitriol (*corprosse*, corrected to *corporossae*), to mix this with alcohol and distil it twice. This second water coming from the distillation colours wool yellow and dissolves or liquefies all metals, calcined stones and the like, which means it is nitric acid, but no con-

33. Hanegraaff 2005, vol. 1, p. 22, 27-28, 30-32, 35.

34. Agricola 1950, p. 441-442; Biringuccio 1959, p. 183-188; Forbes 1970, p. 86-87; much concerning the early history of acids I owe to my discussions with Basil Hunnisett and especially Mark Stevenson.

35. British Library Ms. Add. 32622, fol. 12r, under the heading "Grammar;" Williams 1935, p. 88, n. 63.

36. De Furno 1531, fol. aij-r; Von Lippmann 1971, p. 175.

nection is made with etching a relief in metal.[37] Such information is found only further on, where the corrosion of iron or steel (*ferrum vel corallum*) is explained in chapter CCLXXI. The mordant is made of plant ashes, copper acetate (*Viridis graeci seu Verdeti*) and urine or vinegar. The iron object is dipped into melted wax to cover it completely, letters and figures are drawn into the wax down to the iron, and the object placed in the etching liquid. It is left in it for some days until one sees the letters and figures are bitten into the iron.[38]

There are more 14th century recipes for distilling nitric acid and "gilding" (copper plating) iron with a salts and vinegar mixture. However, in order to recognize a recipe as being suited for etching a relief in metal objects, it is important that the recipe also describes to use a resist such as wax or oil paint. Ten recipes from five 15th century manuscripts describe how the iron object is either covered with wax, oil-varnish or oil paint and the text or decoration scraped into it with a stylus, where after it is etched. Or, text or decoration is painted onto the iron with oil paint and the metal around it etched away. Additionally a wax wall can be made around the part to be etched to keep the acid, or the etchant is mixed with charcoal powder to an acrid paste which is applied to the part to be etched. Both techniques come in handy for local etching or etching on curved surfaces.

A Spanish sword is our earliest etched iron object, apart from the Celtic examples. It was found in the tomb of King Sancho IV of Castile and Leon (d. 1295) in Toledo and it is dated c. 1290. The inscription on the blade just under the hilt is, according to Oakeshott, "undoubtedly produced by true etching" and the introduction of Islamic science in Toledo in this period (see above) supports the presence of etching. Oakeshott also refers to a possible second Spanish example and shows two Italian blades with etched decoration from the 14th century.[39] Decorating armour by means of etching is not yet well studied, but there is enough material to demonstrate it was practised in Italy, Spain and German countries throughout the 15th century, with an increase in activity in the last decades.[40] Paulerinus, reporting from the 1460s, tells about the *sagittinus* who makes missiles and other weapons. He embellishes these with texts and images, which he bites into the iron by means of a salts solution using red oil paint as a resist (*quas eciam cavat cum sale armoniaco et scribit textus et ymagines in eis et ubi fuerit mineum scriptum cum oleo lini, illis non ledetur ferrum aliqua corrosione*).[41]

37. De Furno 1531, p. 13, addenda; Forbes 1970, p. 63-64, 86; Von Lippmann 1971, p. 179.

38. De Furno 1531, p. 226; Von Lippmann 1971, p. 181.

39. Bruhn Hoffmeyer 1963, p. 15-16; Norman 1964, p. 105, fig. 108; Oakeshott 1981, p. 141-142, pl. 15, 16, 30C, 31.

40. Alexander 1987, p. 24, 25; Gamber 1976-1990, vol. 1, p. 74-75, nr. A49; Mann 1942, p. 19-20, but see Gamber 1976-1990, vol. 1, p. 85; Oakeshott 1981, pl. 42B; Norman 1964, p. 105-107; Tarassuk & Blair 1982, p. 160-161; with thanks to Alheidis von Rohr and to Karen Watts of the Royal Armories in Leeds for references.

41. Biblioteka Jagiello`nska, Cracow, Ms. BJ 257, fol. 188va (Sagittinus); Hadravová 1997, p. 32.

The origin of etching intaglio printing plates is commonly said to have derived from the etching of iron used in the decoration of arms and armour, and the discussion above gives little reason to doubt that. Iron is harder than copper and difficult to engrave, but can be etched with simple mixtures of various salts, commonly containing a copper salt, in water, vinegar or urine. Copper, a semi-precious metal, is corroded less easily. Etching copper can be done with strong nitric acid (20 % or more in volume), but also with a mixture of various salts in vinegar, prepared by boiling it and thereby creating such an acid in the solution. And although there are no historic references to it, experimentation shows that a saturated solution of plain kitchen salt in strong (8 %) vinegar bites a shallow (≈ 0.1 mm) relief in copper in a month time. Etching iron perhaps was not continuous from the late 13th century onward, but at least the knowledge was available and the technique performed at times. Etched decoration of arms and armour became more and more popular with the skills of the armourer flourishing in the 16th and 17th century.[42] In that period etching is used for decorating a variety of objects, such as dishes, locks, beakers and tools, and in a variety of materials such as pewter, bone, ivory, stone, iron, copper and silver.[43] Finally, not to forget, etching is used for making metal printing plates, for both intaglio and relief, up to the day of today.

Bibliography

Agricola 1950 = Herbert Clark Hoover, Lou Henry (eds.), *Georgius Agricola, De re metallica*, New York: Dover, 1950. § Transl. from the first Latin ed. of 1556. Repr. of: London: The Mining Magazine, 1912.

Aitchison 1961 = Leslie Aitchison, *A history of metals*, repr., London: MacDonald & Evans, 1961 2 vols.

Alexander 1987 = D.G. Alexander, "European swords in the collections of Istanbul. Part II," in: *Waffen- und Kostümkunde*, vol. 46 (1987), p. 21-48.

Allan 1979 = James W. Allan, *Persian metal technology, 700-1300 AD*, London: Ithaca, 1979.

Anteins 1966 = A. Anteins, "Im Ostbaltikum gefundene Schwerter mit damaszierten Klingen," in: *Waffen- und Kostümkunde*, (1966), p. 111-125.

Beck 1933 = H.C. Beck, "Etched carnelian beads," in: *The antiquaries journal*, vol. 13, p. 384-398.

Beltran 1998 = Maria Helene Roxo Beltran, "Algumas considerações sobre as origens da preparação de ácido nítrico," in: *Química nova*, vol. 21 (1998), nr. 4, p. 504-507.

Bernt 1939 = Walther Bernt, *Altes Werkzeug*, München: Callwey, 1939.

42. Gamber 1999; Kren 2005, p. 3; Mann 1942; Norman 1964, p. 59-82; Plattnerkunst 1954, p. 41-43, 60-61, 66-67; Scalini 1997; Von Reitzenstein 1964, ill. 14, 15, 43-45.

43. Bernt 1939, p. 20-21, 192-195; Canz 1976, p. 13; Freudenberg & Mondfeld 1982, p. 141; Haedeke 1963, p. 16, 22-23, 179-180, 347-348, 422-424; Impey & MacGregor 1985, fig. 30-31; Iven 1938, p. 186-187; Pechstein 1985, p. 230, 250, 255, 265, 281, 481; Schopers 1981, p. 33, 35; Wegner 1958, p. 178-179, 184; Weixlgärtner 1911, p. 339-345.

Biringuccio 1959 = Derek J. Price (ed.), *The pirotechnia of Vannoccio Biringuccio*, reissue, New York: Basic Books, 1959. § 1st ed.: 1943. Repr.: 1966, 1990.

Bocquentin 1993 = Jacques Bocquentin, *La gravure sur cuivre, ou taille-douce, dans la problematique de l'image au* XVe *siècle*, Paris: Bocquentin, 1993, 2 vol. § Unpublished thesis for the École des Hautes Études en Sciences Sociales, Paris.

Böhne 1963 = Clemens Böhne, "Die Technik der damaszierten Schwerter," in: *Archiv für das Eisenhüttenwesen*, 34. Jrg. (1963), Heft 4 (April), p. 227-234.

Bruhn Hoffmeyer 1963 = Ada Bruhn Hoffmeyer, "From medieval sword to Renaissance rapier," in: *Gladius*, vol. 2 (1963), p. 5-68.

Canz 1976 = Sigrid Canz, *Schlosserkunst*, München: Bayerisches Nationalmuseum, 1976.

Chase & Franklin 1979 = W.T. Chase, Ursula Martius Franklin, "Early Chinese black mirrors and pattern-etched weapons," in: *Ars orientalis*, vol. 11 (1979), p. 215-258.

De Furno 1531 = Vitalis de Furno, *Pro conservanda sanitate, tuendaque prospera valetudine ad totius humani corporis morbos et aegritudines, salutarium remediorum, curationumque liber utillissimus ... (etc.)*, Moguntia (Mainz): Schoeffer, 1531.

Forbes 1970 = R.J. Forbes, *A Short History of the Art of Distillation*, reimp., Leiden: E. J. Brill, 1970.

France-Lanord 1952 = Albert France-Lanord, "Les techniques métallurgiques appliquées à l'archéologie," in: *Revue de metallurgie*, vol. 49 (1952), no. 6, p. 411-422.

France-Lanord 1964 = A. France-Lanord, "La fabrication des épées de fer gauloises," in: *Revue d'histoire de la sidérurgie*, T. 5 (1964), p. 315-327.

Francis 1980 = Pete Francis, "Bead report II: etched beads in Iran," in: *Ornament*, vol. 4 (1980), nr. 2, p. 24-29.

Freudenberg & Mondfeld 1982 = Elisa zu Freudenberg, Wolfram zu Mondfeld, *Altes Zinn aus Niederbayern*, Regensburg: Pustel, 1982, vol. 1.

Gamber 1976-1990 = Ortwin Gamber (*et al.*, eds.), *Katalog der Leibrüstkammer*, Wien: Kunsthistorisches Museum (etc.), 1976, 1990, 2 vol.

Gamber 1999 = Ortwin Gamber, "Der Harnisch im 16. Jahrhundert," in: *Waffen- und Kostümkunde*, vol. 41 (1999), nr. 2, p. 97-120.

Geber 1922 = Ernst Darmstaedter (transl., ed.), *Die Alchemie des Geber*, Berlin: Springer, 1922. § Repr.: 1969.

Geber 1928 = Richard Russell (transl.), E. J. Holmyard (ed.), *The works of Geber*, London: Dent; New York: Dutton, 1928. § Repr. of: 1678.

Geisberg 1923 = Max Geisberg, *Die Anfänge des Kupferstiches*, 2. Aufl., Leipzig: Klinkhardt & Biermann, 1923 (Meister der Graphik; Bd. 2). § 1st ed.: 1909.

Gladwin 1965 = Harold S. Gladwin, Emil W. Haury, E.B. Sayles, Nora Gladwin, *Excavations at Snaketown, material culture*, reprint, Tucson: The University of Arizona Press, 1965.

Glosek & Kajzer 1977 = Marian Glosek und Leszek Kajzer, "Zu den mittelalterlichen Schwertern der Benedictus-Gruppe," in: *Waffen- und Kostümkunde*, (1977), p. 117-128.

Guineau 2005 = Bernard Guineau, *Glossaire des matériaux de la couleur et des termes techniques employés dans les recettes de couleurs anciennes*, Turnhout: Brepols, 2005.

Hadravová 1997 = Alena Hadravová (ed.), *Paulerinus (Pavel <>Žídek) Liber viginti arcium (ff. 185ra-190rb)*, Praha: Koniasch Latin Press, 1997 (Clavis monumentorum litterarum (Regnum bohemiae); nr. 3. Fontes; nr. 2). § Paulerinus (1413-1471) is Pavel <>Zídek.

Haedeke 1963 = Hans Ulrich Haedeke, *Zinn*, Braunschweig: Klinkhardt & Biermann, 1963.

Hanegraaff 2005 = Wouter J. Hanegraaf (ed.), *Dictionary of Gnosis & Western Esotericism*, Leiden, Boston: Brill, 2005, 2 vol.

Harzen 1859 = E. Harzen, "Ueber die Erfindung der Aetzkunst," in: *Archiv für die zeichnenden Künste ... (etc.)*, vol. 5. (1859), p. 119-136.

Haury 1967 = Emil W. Haury, "The Hohokam, first masters of the American dessert," in: *National Geographic*, vol. 131 (1967), p. 670-695.

Ilg 1970 = Albert Ilg (ed.), *Heraclius, von den Farben und Künsten der Römer*, reprint, Osnabrück: Zeller, 1970. § Original ed.: Wien: Braumüller, 1873.

Impey & MacGregor 1985 = Oliver Impey, Arthur MacGregor (eds.), *The origins of museums, the cabinet of curiosities in sixteenth- and seventeenth-century Europe*, Oxford: Clarendon, 1985.

Iven 1938 = Grete Iven, "Zinnarbeiten der Sammlung Clemens im Kölner Kunstgewerbe-Museum," in: *Pantheon*, vol. 21 (1938), (Jan./June), p. 182-187.

Jones 1997 = Lee A. Jones, "The serpent in the sword: pattern-welding in early mediaeval swords," in: *Catalogue of The Fourteenth Park Lane Arms Fair*, Thropton nr. Morpeth (Northumerland): Oliver, 1997.
§ HTML edition (1998), http://www.vikingsword.com/serpent.html (2013).

Karpenko & Norris 2002 = Vladimír Karpenko, John A. Norris, "Vitriol in the history of chemistry," in: *Chemické Listy*, vol. 96 (2002), p. 997-1005.

Kirpicnikov 1986 = A.N. Kirpicnikov, "Russische Waffen des 9-15. Jahrhunderts," in: *Waffen- und Kostümkunde*, Bd. 45 (1986), p. 1-22.

Köhler 1897 = S.R. Köhler, "Über die Erfindung der Ätzkunst," in: *Zeitschrift für bildende Kunst*, new series, vol. 9 (1897-1898), p. 30-35.

Kren 2005 = Peter Kren, *Arms, armour, and fine arts*.
§ Published in: http://www.myarmoury.com/feature_armsarts.html (2013).

Levenson 1973 = Jay A. Levenson, Konrad Oberhuber, Jacquelyn L. Sheehan, *Early Italian engravings from the National Gallery of Art*, Washington: National Gallery of Art, 1973.

Landau & Parshall 1994 = David Landau and Peter Parshall, *The Renaissance print 1470-1550*, New Haven, London: Yale University Press, 1994.

Loehr 1956 = Max Loehr, *Chinese Bronze Age weapons, the Werner Jannings Collection in the Chinese National Palace Museum, Peking*, Ann Arbor: The University of Michigan Press; London: Cumberlege, Oxford University Press, 1956.

Lucca 2003 = Adriano Caffaro (ed., transl.), *Scrivere in oro. Ricettari medievali d'arte e artigianato (secoli IX-XI). Codici di Lucca e Ivrea*, Napoli: Liguori, 2003. § Transcription and Italian translation of the Lucca ms., Ms. 490 in de Biblioteca Capitolare di Lucca, written between 796-816.

Mann 1942 = James G. Mann, "The etched decoration of armour," in: *Proceedings of the British Academy*, vol. 28 (1942), p. 17-44.

Meijer 1981 = *J.J. Meijer*, "Een javaansch handschrift over pamor-motieven," in: *De wapenverzamelaar*, vol. 19 (1981), p. 33-52.

Merrifield 1849 = Mary P. Merrifield, *Original Treatises dating from the XIIth to XVIIIth Centuries (o)n the Arts of Painting* ..., London: John Murray, 1849, 2 vols. § Reprinted with an introduction and glossary by S. M. Alexander (2 vols.) New York: Dover Inc., 1967. Reprinted as *Medieval and Renaissance Treatises on the Arts of Painting, Original Texts with English Translations.* Mineola, NY, Dover, 1999.

Metzger 2009 = Christof Metzger *et al.*, *Daniel Hopfer, ein Augsburger Meister der Renaissance; Eisenradierungen-Holzschnitte-Zeichnungen-Waffenätzungen*, München: Staatliche Graphische Sammlung, Deutscher kunstverlag, 2009.

Neumann 1927 = Bernhard Neumann, "Römischer Damaststahl," in: *Archiv für das Eisenhüttenwesen*, vol. 1. (1927), nr. 3, p. 241-244, pl. 14.

Newman 1985 = William R. Newman, "New Light on the Identity of 'Geber'," in: *Sudhoffs Archiv*, vol. 69 (1985), nr. 1, p. 76-90.

Norman 1964 = Vesey Norman, *Arms and armour*, London: Weidenfeld and Nicolson, 1964. § Vesey = Alexander Vesey Bethune.

Oakeshott 1981 = R. Ewart Oakeshott, *The sword in the age of chivalry*, rev. ed., London: Arms and Armour Press, 1981. § 1st ed.: 1964.

Oldfather 1952 = C.H. Oldfather (transl.), *Diodorus of Sicily, in twelve volumes*, repr., London: Heinemann; Cambridge, Mass.: Harvard, 1952, vol. 3. § Books IV (continued) 59-VIIII. 1st ed.: 1939.

Panseri 1965 = Carlo Panseri, "Damascus steel in legend and in reality," in: *Gladius*, vol. 4 (1965), nr. 5, p. 5-66.

Pechstein 1985 = Klaus Pechstein (ed.), *Wenzel Jamnitzer und die Nürnberger Goldschmiedekunst 1500-1700, Goldschmiedearbeiten - Entwurfe, Modelle, Medaillen, Ornamentstiche, Schmuck, Porträts*, Nürnberg: Germanisches Nationalmuseum, 1985.

Perego 2005 = François Perego, *Dictionnaire des matériaux du peinture*, Paris: Belin, 2005.

Plattnerkunst 1954 = *Die Innsbrucker Plattnerkunst, Katalog*, Innsbruck: Tiroler Landesmuseum, 1954.

Pope 1967-1969 = John Alexander Pope, *The Freer Chinese bronzes*, Washington: Freer Gallery of Art, 1967-1969, 2 vol.

Sachse 1993 = Manfred Sachse, *Alles über Damaszener Stahl*, Bremerhaven: Verlag für neue Wissenschaft, 1993.

Salin 1957 = Édouard Salin, *La civilisation mérovingienne d?apres les sépultures, les textes et le laboratoire: troisième partie, les techniques*, Paris: Picard, 1957.

Scalini 1997 = Mario Scalini, "Il 'giubbotto di ferro cesellato a foggia di colletto trinciato con scarselle' di Guidobaldo della Rovere (1514/1538/1574) e altri resti rovereschi," in: *Waffen- und Kostümkunde*, vol. 39 (1997), nr. 1 & 2, p. 45-50.

Schopers 1981 = Wolfgang Schopers (ed.), *Zinn. Kataloge des Kunstmuseums Düsseldorf*, Düsseldorf Kunstmuseum, 1981.

Singer 1948 = Charles Singer, *The earliest chemical industry, an essay in the historical relations of economics & technology illustrated from the alum trade*, London: The Folio Society, 1948.

Singer 1954-1984 = Charles Singer (ed.), *A history of technology*, Oxford: At the Clarendon Press, 1954-1984, 7 vol.

Smith 1965 = Cyril Stanley Smith, *A history of metallography, the development of ideas on the structure of metals before 1890*, reprint, Chicago, London: the University of Chicago Press, 1965.

Smith 1981 = Cyril Stanley Smith, *A search for structure, selected essays on science, art, and history*, Cambridge (Mass.), London: MIT, 1981.

Smith & Hawthorne 1974 = Cyril Stanley Smith, John G. Hawthorne, "Mappae clavicula, a little key to the world of medieval techniques," in: *Transaction of the American Philosophical Society*, vol. 64 (1974), nr. 4.

Stijnman 2006 = Ad Stijnman, "Historical Iron-gall Inks," in: Jana Kolar, Matija Strlič (eds.), *Iron Gall Inks, on the Manufacture, Characterisation, Degradation and Sterilisation*, Ljubljana: National and University Library, 2006.

Tarassuk & Blair 1982 = Leonid Tarassuk & Claude Blair (eds.), *The complete encyclopaedia of arms & weapons*, London: Batsford, 1982. § 1st ed.: Milan: Mondadori, 1979.

Tylecote 1992 = R.F. Tylecote, *A history of metallurgy*, 2nd ed., London, New York: The Institute of materials, 1992.

Verhoeven 1998 = J.D. Verhoeven, A.H. Pendray, W.E. Dauksch, "The key role of impurities in ancient damascus steel blades," in: *JOM*, vol. 50 (1998), nr. 9, p. 58-64. § Published in: http://www.tms.org/pubs/journals/JOM/9809/Verhoeven-9809.html (2013).

Von Lippmann 1971 = Edmund O. von Lippmann, *Beiträge zur Geschichte der Naturwissenschaften und der Technik*, reprint, Niederwalluf bei Wiesbaden: Sändig, 1971. § Original ed.: 1923, 1953.

Von Reitzenstein 1964 = Alexander Freiherr von Reitzenstein, *Der Waffenschmied, vom Handwerk der Schwertschmiede Plattner und Büchsenmacher*, München: Prestel, 1964.

Walden 1952 = Paul Walden, *Chronologische Übersichtstabellen zur Geschichte der Chemie von den ältesten Zeiten bis zur Gegenwart*, Berlin: Springer, 1952.

Wegner 1958 = Wolfgang Wegner, "Aus der Frühzeit der deutschen Ätzung und Radierung," in: *Philobiblon*, vol. 1 (1958), p. 178-190.

Weixlgärtner 1911 = Arpad Weixlgärtner, "Ungedruckte Stiche. Materialien und Anregungen aus Grenzgebieten der Kupferstichkunde," in: *Jahrbuch der Kunsthistorischen Sammlungen des Allerhöchsten Kaiserhauses*, vol. 29 (1910-1911), nr. 4 (1911).

Williams 1934 = Hermann W. Williams, "The beginnings of etching," in: *Technical studies in the field of the fine arts*, vol. 3 (1934), p. 16-18.

Wyss 1968 = René Wyss, "Belege zur keltischen Schwertschmiedekunst," in: Elisabeth Schmid, Ludwig Berger, Paul Bürgin (eds.), *Provincialia: Festschrift für Rudolf Laur-Belart*, Basel, Stuttgart: Schwabe, 1968, p. 664-681.

Proto-Scientific Revolution or Cookbook Science? Early gunnery manuals in the craft treatise tradition[1]

Steven A. Walton

Scholars who study craft books and books of secrets in the Middle Ages can engage on internal debates about the modes of transmission or interpretations of numerous facets of these important treatises, but they can all agree on two relatively firm facts. First, in the past, technical knowledge was more an art (*techne*) than a science (*scientia*); that is, it was about doing rather than necessarily knowing for knowing's sake. To put it another way, it was decidedly not *philosophia naturalis* with its emphasis on causes but rather *ars practica*, with an emphasis on doing and making. Second, practitioners of these arts tended to disseminate knowledge from one generation to the next more or less directly through the apprenticeship system rather than through written texts. What this means, on both counts, is that in the Middle Ages and early modern period textbooks in the modern sense were generally unavailable, irrelevant, and simply not needed. The craft treatises, which survive then, can only represent a small fraction of the vast craft tradition, and that they were ever written down at all should give us pause to consider by whom and for what purpose.

Considering the craft of gunnery and its preserved treatises also adds another layer of interpretive difficulty: the study of the rise of gunnery overlaps chronologically with the Scientific Revolution and its historiography in a peculiar manner. For the Scientific Revolution, a traditional narrative exists of the increasing rationalization, quantification, and geometrization of nature. For the military use of gunpowder weaponry, the narrative argues that the gunners, seeking ever-more-accurate bombardment of fortifications, turned to exacting understandings of ballistic trajectories to accomplish this task. The conflation of these two narratives is that the gunners became "scientific," that is, using

1. A revised version of this paper was given as "The *Circumstanciae* of gunnery manuals: didactic recipes, ingredients, and readers" in July 2010 at the Middelaldercentret (Medieval Centre) in Nykøbing Falster, Denmark and will be published as a companion to this piece in *Trebuchet to Cannon: Military Technology 1000-1600*, forthcoming.

science in the modern sense, and by some extension, *quantitative natural philosophy* in the historical sense.

Focusing on the contemporary manuals, this conflation is found to be quite unfounded for gunnery before at least the later seventeenth century. The vast majority of gunners' manuals, treatises, notebooks, and manuals before the eighteenth century have much more in common with cookbooks than they do with the *Principia* of Newton or even *The Two New Sciences* of Galileo. Gunnery manuals, in particular, preserve collections of the description of a variety of substances, processes, and mixtures that provided *aides memoire* to the users and perhaps didactic tools for the preservation and transmission of such knowledge. This study analyzes a number of gunners manuals and notebooks for their content, range of coverage, and material knowledge, and then looks in-depth at a number of firework recipes across a number of manuals to understand the gunners' interests, background knowledge, and goals. The material demonstrates their great felicity with material properties and substances, and at the same time a remarkable disinterest in philosophical frameworks in which these materials might have meaning (notably, alchemical connections are entirely absent, and even Scholastic terminology does not appear).

Gunnery Manuals

In the literature of artillery before its "scientification" in the seventeenth century, there is virtually *nothing* about ballistics or trajectories in these works beyond an occasional non-committal and qualitative paragraph. Indeed, these paragraphs typically only appear when the work saw print. The touchstone for the printed ballistics treatise is Niccolo Tartaglia's *Nova Scientia* of 1537, although one could point to small subsections in printed works as far back as the beginning of the sixteenth century that included at least a small mention of the topic.[2] Manuscript treatises, however, survive from the 1450s onward and remain common throughout the sixteenth century and these virtually never discuss ballistic trajectories. Rather, they favor a more cookbook-like approach with memoranda, recipes, and notes on gunnery, sometimes only vaguely organized. Only by 1620 – in English at least, with the publication of *The Complete Gunner* by Richard Norton – does the "doctrine of Projects" (*i.e.*, ballistic trajectories) begin to appear, and even then as a small addendum to

2. N. Tartaglia, *Nova scientia inventa da Nicolo Tartalea*, Vinegia, 1537, and his later *Quesiti et inventioni diverse*, Venetia, 1546, and for an English translation of sections, see S. Drake and I.E. Drabkin, *Mechanics in Sixteenth-Century Italy*, Madison, WI, 1969. See also, Gualfnerum H. Rivium, *Geometrischen Büxenmeisterey*, Nuremberg, 1547; A. Dürer, *Albertus Durerus ... exacte Quatuor his suarum Institutionum Geometricarum libris, lineas, superficies & solida corpora tractavit*, Paris, 1532; and in general, M.J. Henninger-Voss, "How the 'New Science' of Cannons Shook up the Aristotelian Cosmos," *Journal of the History of Ideas*, 63, 2002, pp. 371-397. On a related note for trajectories of other missiles, see S. Anglo, "The Man Who Taught Leonardo Darts: Pietro Monte and His 'Lost' Fencing Book," *Antiquaries Journal*, 69, 1989, pp. 261-78.

this craft- or cookbook-tradition. Much more important to the practice of gunnery was the identification and measurement of cannon, the composition and practice of using gunpowder, and especially recipes for fireworks.

It is my contention that manuscript treatises, particularly those without elaborate dedicatory prefaces, are closer to the practitioners and therefore more representative of practice than published works. This is not to say that the print manuals failed to represent gunnery practice entirely, but they represent a certain abstraction and idealization of what the practice perhaps *ought* to be. The manuscript manuals more likely represent what it was, or at least what it was *thought* to be. By the end of the sixteenth century, as print and technical manuals began to be more and more one and the same object, a sort of watershed or ridgeline appears that suggests how this particular form of technical knowledge in craft treatises served both practitioners and observers.[3]

In the later sixteenth century, gunnery manuals typically contain three or four parts. First there is always a section of cannon identification and statistics:

> "A demiculverin shot of 4¼-in. high and 12-in. circumference and weighs for dyce and leade [*i.e.*, iron cubes cast about with lead] 12-lb. and an iron [ball] 9-lb. The ladell is 8¼-in. broad and 18-in. long. The cartridge is 12-in. broad, 18-in. long, and 9½-lb. of powder discharges the same shot."[4]

The specific numerical data are not perfectly uniform across period and country, but the kind of information recorded – bore, length, shot size and weight, and loading proportions – are, and do not seem to develop much over time. That is, there is no apparent progression towards any more "efficient" measures or use.

Next, there is a section on how to shoot the cannon, often involving the making or use of instruments. In virtually every case the principle instrument gunners record is the quadrant (it is often the only one), an L-shaped instrument with a plumb line and 90° arc graduated in either degrees or eight "points." Many of the instructions in this section have to do with how to use the quadrant to accomplish tasks both quantitatively and tasks not easily done without it:

> "To shoote a good shott upon a Tower by daye, & ye same againe at night … marke by day with a compasse after what manner the Tower lyeth, East, West, North, or South, & when you shoote this shott first take a thred & lay it on ye gunn, and let the thred hang ouer at the mouth of the peece downe to the

3. On the rise of the printed technical treatises in other fields, see a series of works by E. Tebeaux: "Technical Writing in English Renaissance Shipwrightery: Breaching the Shoals of Orality," *Journal of Technical Writing and Communication*, 38, 2008, pp. 3-25; "Technical Writing in Seventeenth-Century England: The Flowering of a Tradition," *Journal of Technical Writing and Communication*, 29, 1999, pp. 209-53; and *The Emergence of a Tradition: Technical Writing in the English Renaissance, 1475-1640*, Amityville, NY, 1997.

4. Richard Wright, MS gunnery manual, 1564 [London, Society of Antiquaries of London, MS 94], fol. 8^{v}.

> ground, & keepe ye sight, & set a marke there with a pricke or pinne, & behind at ye tayle of the pece set amarke in ye ground, y^n by a squadron [quadrant] looke whether your peece stand too lowe or too high at ye one side or ye other or right & iust, if it be iust take a lether table of finger broad & a palme & halfe long hollowed out, & againe sticke it with wax on ye one side & poynte at the other side, before you shoote doe this to be in ye mouth of your gun, & let a lead with a poynty thinge fall downe from above ye gun, this table, ye waxe & ye poynte keepe well, y^t hee goe not out, y^n shoote ye piece off.
>
> If you suppose that it is a good shot remember the sayde [compass] poynte & the poynte in the waxe table, & ye height & lownes upon ye squadron & with the poynte you shal shoote by night againe."[5]

Much of this is very practical and effective instruction, but it is worth noting that it does not rely on any knowledge whatsoever of ballistics.

Third, in most manuals and notebooks a substantial section relates to knowing one's cannon and powder and their goodness, although the types of entries that occur in this section are highly variable across the manuscripts. Cannon barrels must be tested for straightness, for example, and a dispart (sighting adjustment) had to be fashioned individually for each cannon. Similarly, gunners needed to know how to test the goodness of gunpowder to see its strength, and hence adjust their loading to compensate for variation. Thus we get passages such as the following:

> "A rule to know the goodness of all sortes of Gunpowder
>
> Gunpowder may be known by 3 mann*er* of p*ar*ts:
> 1 By tasting of the tongue to know its sharpness
> 2 By faireness of color
> 3^{ly} by the burning.
>
> The tasting of the tongue shows if the powder bee of a high receipe or low … The fairness of colour shows the powder to be good for if it have abundance of moisture and is well wrought it will have a fair colour … [and] by fire you shall know, for if it lack working there will remain after the burning as it were white p*ar*ts of the Master wch signifies evil working … also by the fire you shall know whether the master was well refined or that it be greasy or salt: after the burning there will remain small knots when it was burned & the place will be dankish … Also there is another sort of powder that by burning will lie like pearls white & red as the other did but this burning shall be nothing so quick, & of a dark colour that signifies lack of moisture, & of this powder you may boldly give a piece more by the ninth p*ar*te than of the other."[6]

5. "The Secret of Gunmen" [Oxford, Bodleian Library, Oxford, MS Ashmole 343], fol. 129^r. The author is unknowingly recommending the spherical coordinates for the shot.

6. *Ibid.*, fol. 137^v-138^r.

Finally, manuals invariably contain a section on gunpowder recipes and firework recipes – both for amusement and for incendiary use against the enemy. Fireworks were a large part of Elizabethan pageantry:[7] fireworks accompanied the English entry into the Netherlands' conflict; feasts and processions abounded at which "cannon roared, tar barrels blazed, bells pealed, dragons soared on fiery wings, [and] wreaths of [firework] flowers descended."[8] An example of the importance of these is found in and Elizabethan manuscript entitled, "The Secretes of Gunmen." The relatively short treatise opens with the three things that the gunner *must* know: "you must know good salt-peter from bad whether it be fatt or salte or allum therein" and "after what manner it ought to be put out profitably: & then the salt peter being good & fayre," in order that the gunner may "make good powder thereof for all manner of good shot." Second, the gunner had to know how to operate his artillery: "you must know all peeces measurably to charge or lade them," as well as, "to [dis]parte y^{m} over feild, land, or roades." Finally, "you must know how to make 3 or 4 sortes of fireworkes at least, whether it be by water or land, *if you will get lords wages*."[9] This announced concentration on recipes and on fireworks, then, will come to define a skilled practitioner of gunnery.

Rezeptliteratur and Fireworks

Gunners' manuals fall into the genre of *rezeptliteratur*, or recipe literature, that include individual descriptions of any variety of substances, processes, or mixtures. Recipe books provided *aides memoire* for their users from the Middle Ages through the Renaissance and form the pedigree of modern cookbooks and how-to manuals (referred to hereafter as "do-it-yourself," or DIY, manuals). One crucial distinction between modern DIY manuals and Renaissance recipe books should be borne in mind, however: examples of historic *rezeptliteratur* tend to assume previous knowledge of the field about which they record information; many of today's DIY manuals are designed for the complete novice. For example, a recipe in a modern (culinary) cookbook may ask for a hard-boiled egg; the index – and it is notable that they all have indices – will direct you to a page that will tell you how to hard boil that egg. Renaissance recipe literature will not include that latter step or any pretense to index-

7. For example, fireworks were used in honor of the Queen at Warwick in 1572 and figured prominently at Kenilworth in 1575: "when [the Queen entered the courtyard] after did follo so great a peal of gunz and such lighting by fyr work a long space toogither ...: for indeed the noiz and flame weat heard and séene a twenty mile of;" F.J. Furnivel, ed., *Robert Lancham's Letter: Describing a part of the Entertainment unto Queen Elizabeth at the Castle of Kenilworth in 1575*, London, 1907, pp. 11-12, 18. See also, P. Butterworth, *Theatre of Fire: Special Effects in Early English and Scottish Theatre*, London, 1997.

8. C. Wilson, *Queen Elizabeth and the Revolt of the Netherlands*, London, 1970, pp. 90-91.

9. "The Secret of Gunmen," fol. 128^{r}, emphasis added.

ing.[10] These works are comprised of item after item, sometimes set off with bold headings, which may be confined to a relatively narrow topic – as are our gunners' manuals – or may range freely over domestic science, medicine, prognostication, agriculture, or the search for the Philosopher's Stone. Information may not be entered in any "logical" manner, or at least it may appear in a seemingly random order, but one whose logic may serve a different purpose than the modern cookbook.

Michael McVaugh once remarked that "recipe collections ... spread out tediously over several folios ... in monotonous ... detail [and] encourage the modern reader to dismiss them as neither tractable nor interesting."[11] Seen in that light, they can appear rather opaque and boring. Further, the recipes in this particular subset of gunnery manuals are intractable in the sense that often the recipe components are unclear to modern understanding, or because it would be hazardous to attempt to replicate their recipes for all manner of explosives and incendiaries (even if local fire regulations would permit the experiments). It is not the goal of this research project to understand how these various recipes behaved chemically, but rather to note how the users of the recipes recorded them.

Firework Recipes

Recipes in Tudor gunnery manuals tend to concentrate on one class of product: fireworks. This focus is interesting in that historians of artillery have largely ignored it for more 'scientific' pursuits, namely ballistics.[12] It is true that the mathematicians and scientists of the time (*i.e.*, natural philosophers or those with pretensions to natural philosophy such as Niccolo Tartaglia, Galileo Galilei, or Thomas Harriot) looked to ballistics as the proper *ens* of artillery; the gunners, however, did not. Not only were gunners usually mathematically unprepared to investigate ballistics, and mathematics itself was not yet sufficient even if they had been, but such considerations would have been largely meaningless to their understanding and especially to their operation of great ordnance. Instead, they focused on a product they could control and understand, at least qualitatively. These firework recipes tend to have rather nonde-

10. Well beyond the scope of this paper is the question of how one manages to categorize and organize information in technical treatises. For a modern philosophical look at this important topic, see G.C. Bowker and S.L. Starr, *Sorting Things Out: Classification and its Consequences*, Cambridge, MA, 1999.

11. M. McVaugh, "Two Montpellier Recipe Collections," *Manuscripta,* 20, 1976, p. 175.

12. The classic study is A.R. Hall, *Ballistics in the Seventeenth Century,* Oxford, 1965. One notable recent exception to this lack of attention are articles appearing in the last few years in the *Journal of the Ordnance Society*; *e.g.,* R.R. Brown, "Troncks, Rockets, and Fiery Balls: Military fireworks of the early modern period" *Journal of the Ordnance Society,* 17, 2005, pp. 25-38. More recent work on firework also bypasses the early military use: S. Werrett, *Fireworks: Pyrotechnic Arts and Sciences in European History,* Chicago, 2010.

script names like "trunkes," "ringes," and "balls," and the surviving manuscript recipe texts list recipe after recipe, page after page, often with seemingly small variations from one to the next.

Compared to the printed gunnery manuals, this manuscript emphasis on various recipes seems puzzling. Although Peter Whitehorne had a section on them in his 1560 addendum to Machiavelli's *Arte of Warre*, William Bourne makes little mention of this sort of thing in his *Arte of Shooting in Great Ordenance* of 1578. The comparison between these two is instructive, for Whitehorne was a translator at the very beginning of English gunnery publication. He seems to have been filling in what seemed missing from Machiavelli (indeed, scholars have noticed that he strangely ignored gunpowder artillery in his works), while Bourne was writing at the end of a career as a gunner himself. Manuscripts, on the other hand from the early sixteenth through the late seventeenth century almost invariably include a substantial section on fireworks.

This difference, I argue, is due to a difference in audience. They appear more prominently in the manuscript recipe books not because they are simply interesting or particularly odd, but rather because they were a common element of the gunners repertoire, one that was difficult to remember, and one which was not used or tested nearly as often as other things in the printed manuals. Conversely, printed manuals contain topics that would introduce to non-military tyros (their assumed readership: the "armchair general," for example) to gunnery. Active gunners would not need much of the information in printed treatises after basic training. The regress of assumed readership for these two modes of knowledge transmission is admittedly a problem, but it is fair to say that the manuscripts tend to be lower cost (and therefore status) items and readership for printed books in the sixteenth century was still very much limited to a small percentage of the elites.

Military firework recipe books also show a simultaneous commitment to conservatism and change. In the twelfth-century recipe-book known as the *Mappae Clavicula*, there are a number of incendiary recipes, including one for "the arrow which emits fire."[13] It begins by describing the arrow as "triple-spiked and perforated." The ingredients are given with their various amounts: naphtha, tow, seasoned pitch, "native sulphur," *climatis*, sea salt, olive oil, "raw bird lime," jet stone, some soap made from olive oil, and a woman's milk. After mixing these together you are left with a "fatty" milk in which you dip some flaxen rope which is then fastened to the arrows. When all is ready, "you stretch your bow [and] set the arrow on fire and immediately shoot it where you want a fire to be started." Although the delivery system is old-fashioned by the time of the gunners' manuals, and ingredients are apparently more

13. C.S. Smith and J.G. Hawthorne, *Mappae Clavicula: a Little Key to the World of Medieval Techniques*, Philadelphia, 1974, no. 266. The section on incendiary mixtures continues to no. 278-A.

"magical" than ours,[14] the style of the recipe is similar. Compare it to one from the sixteenth-century Gunmen manuscript:

> "To make ffire arrowes.
>
> Take heter oyle, quicke brimstone, harpoyes, and gunpowder that is good. Put the same in a bason, & set it in a kettle of hott water to drye, & ketle it, when it is keiled [cooled] … take a little vpon an arrowe head at the end, then put it about a linnen cloth with smale ends of Launces, when you wille shoote fire the launces, & when they fire it cannot be quenched."[15]

In essence they amount to the same weapon, although in the details and ingredients they are worlds apart. Both use a linen wicking material, possibly because it provides a better mantle than wool or other cloth, or perhaps because it was cheap and readily accessible. They have no ingredients in common, that is, unless the unknown "harpoys" is identical with the similarly unknown "climatis."[16] The *Mappae Clavicula* specifies specific amounts of all the ingredients, in *solidi* or ounces; the Gunmen MS simply says to melt the four ingredients together, and wipe them on the head of the arrow with linen strips. Perhaps the sixteenth-century recipe is more forgiving than that of the twelfth, but this is unlikely. Rather, the two recipes were written from two very different didactic standpoints: *Mappae Clavicula* is a wide-ranging recipe book that apparently recorded hundreds of recipes for posterity. It was apparently intended for someone unfamiliar with the various recipes and processes and it therefore presents all the amounts and steps to achieve the desired result. The sixteenth-century gunners' manual, on the other hand, is written as if from one confidant to another, both skilled in the various techniques pertaining to incendiaries, but who might not know (or might forget) which ingredients to mix to make the appropriate fire for linen-wrapped arrows.

These firework recipes are also not a far cry from culinary recipes from approximately the same time. Consider the following juxtaposition of (1) a gunners' recipe for wildfire to (2) a cookbook's recipe for a stuffing:

(1) "ffor a Wilde fire mischievous to shoot in a Citty or Towne: [Take] A light gunstone [&] annointe it or dippe it in molten Swanell & hares & sprinkle it with good Gunpowder & y^{n} put it againe in the swanell & hars, then take a cleane

14. For example, why would woman's milk be suggested rather than cow's milk, goat's milk, or the perennial magical favorite, mare's milk? In *Mappae Clavicula* it appears only in these incendiary recipes, although goat's milk is used elsewhere to cut glass (no. 289).

15. "The Secret of Gunmen," fol. 131^{v}.

16. This is another difficulty with *Rezeptliteratur*: nomenclature. Readers, it was assumed, would know what "harpoys" or "climatis" were, so no further explanation was warranted. Bert Hall rightly critiqued Hawthorne and Smith for not providing more information or suggestions on obscure terms in the *Mappae Clavicula*; see his review in *Isis* 67, 1976, pp. 123-124.

cloathe & rowle it in the swanell [&] hares, & y^n sprinkle that with good gun-powder, & shoote where you will have it."[17]

(2) "A Farce [for] all Things: Take a good handfull of Tyme, Isope, Parselye, and three or foure yolkes of Eggs hard Roasted, and choppe them with Hearbes small, then take white bread grated and raw eggs with sweet butter, a few small Raisons, or Barberies, seasoning it with Pepper, Cloves, Mace, Sinamon and Ginger, working it alltogether as Paste, and then may you stufe with it what you will."[18]

Not only is the mode of instruction similar despite comparing Elizabethan napalm to turkey stuffing, even the ending valediction is similar, inviting the practitioners to use either concoction as they see fit.

This then again returns our attention to an important facet of the entire realm of *Rezeptliteratur*: its intended audience. Textbooks lead you through each logical step towards the desired result where one rarely learns to *do* anything, but presumably learns *how* the topic functions. In gunnery, William Bourne provided the first attempt at a textbook treatment in English in 1578, with his *The Arte of Shooting in Great Ordnaunce*, but it was not until 1628, with Norton's *The Gunner* that a full textbook treatment of great artillery was available. Recipe books for various tasks, on the other hand, have very different readers in mind when they are written. In the modern world, the historical distance can be difficult to achieve when there is a software manual on every desk, a cookbook in every kitchen, and a DVD manual stuffed in a drawer. But all these manuals have one thing in common: even if they are rarely used, they are designed for looking up a specific task when the need arises (How do I reformat a paragraph? Or roast a chicken? Or set the DVR to record a show?); manuals offer learning that has or need little framework and does not claim to bring the learner to a particular state of knowledge at the end. Recipe books list individual tasks in a boiled-down format (the culinary pun is intentional) for someone who already knows how to perform the task and just needs the ingredients and perhaps instructions on some particular step or condition in the process that might be difficult. So, when you look up the recipe for a lemon meringue pie, you are told to beat the egg whites until stiff and glossy; it is assumed that you know that you need a mixer, bowl, how to separate eggs, and what condition "stiff and glossy" refers to.

For one last instructive investigation in the published literature of gunnery recipes in the sixteenth century, consider how Machiavelli's *Art of War* arrived into English. In 1560 Peter Whitehorne published a reasonably faithful translation of the Italian original, but then in 1562 appended to it a lengthy work entitled *Certaine Waies for the Orderyng of Souldiers in Battelray*, which contained not only the topic of its title, but also a long section on fortification, and

17. "The Secret of Gunmen," fol. 131^r.
18. T. Dawson, *The Second Part of the Good Hus-wives Jewell*, London, 1597, pp. 13-14.

one on “howe to make Saltpeter, Gunpowder, and diuerse sortes of Fireworkes or wilde Fyre, with other things appertaining to the warres.” In this treatise, Whitehorne adds about two dozen recipes for gunpowder. The recipes themselves vary in overall quantity of ingredients used, but if one reduces the proportions of the three principle ingredients – saltpeter, charcoal, and sulphur – then the range of recipes is striking (fig. 1). We might expect slight variations here and there or distinct groupings for different types of powder, but the spread is phenomenal: from an even 1:1:1 mixture [A] to another over 83 % charcoal [B]. The three principle components vary by factors of 2½ [33.3-83.3 %], nearly 5 [6.83-33.3 %], and over 4 [8.3-33.3 %], respectively and although there is a slight grouping around the “ideal” (*i.e.*, most effective) powder with a proportion of about 8:1:1 [C], it seems clear that zeroing in on that ideal was of little concern for Whitehorne or his source(s).

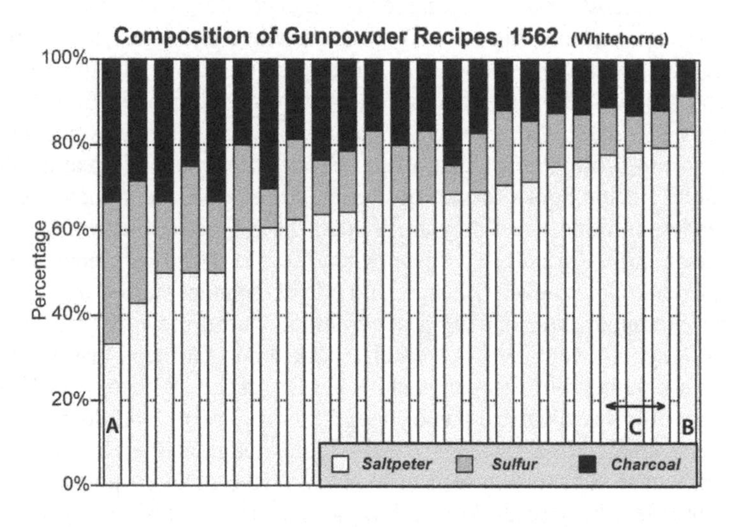

Fig. 1.

One striking feature, although obscured by this method of graphing, is that in these recipes Whitehorne shows a commitment to ingredients mixtures in whole number ratios (*e.g.* 8:1:2 or 5:4:3). This should remind us of the mental framework for numeracy in which these recipes are even contemplated as well as executed. While it is not possible at this time to say how many of these recipes simply would not work (*i.e.*, explode for use as a propellant or even sustain burning for incendiaries), we do know that many would be difficult to ignite – particularly those with a saltpeter content of less than 60 %, or certainly less than 50 % – and would likely not produce usable effects under normal circumstances. The range of recipes he presents does at least suggest that he may not have tried (m)any of the these recipes and implies that the purpose of recipes in this type of craft treatise was not strictly functional.

Further confirmation of this comes from a series of recipes given by Cyprian Lucar in his *Three Bookes of Colloquies concerning the Arte of Shooting in great and small peeces of Artillerie* (1588), itself a free English translation and adaptation of Niccolo Tartaglia's *Quesiti et Inventioni Diverse* (1546), in which he provides 16 modern and 5 ancient powder recipes (fig. 2) as well as 22 recipes for different uses in use in his day (fig. 3). In doing so, he confirms that Renaissance gunners were experimenting with numerous composition recipes for gunpowder. Figure 2 does show a "progression" from the early, low-saltpeter recipes to the then modern high-saltpeter, *if* one chooses to read it that way. However, it also shows that even his modern powders had a great range of variability – not the sort of optimizing behavior one would expect from "science," but rather the variations one would expect in cookbooks (that many of the recipes fall in a straight line is the result of their whole-number mixing ratios). Figure 3, which differentiates his contemporary powders by use, also shows a great deal of variation in the recipes, but it also belies the fact that a modern "scientific" understanding of types of powder was as yet unknown, at least to him. While we know that there were indeed different grades of powder used for different weapons, Lucar's recipes do not fall into any discernible groupings (with the possible exception of pistol powder) and all groups considerably overlap each other. My interpretation here is that the variations in the recipe ingredients was of much more interest to the gunners and recipe-makers than was the actual performance characteristics of the resulting mixtures. Thus, they were not interested in the composition of gunpowder "scientifically," but rather as craft practitioners.

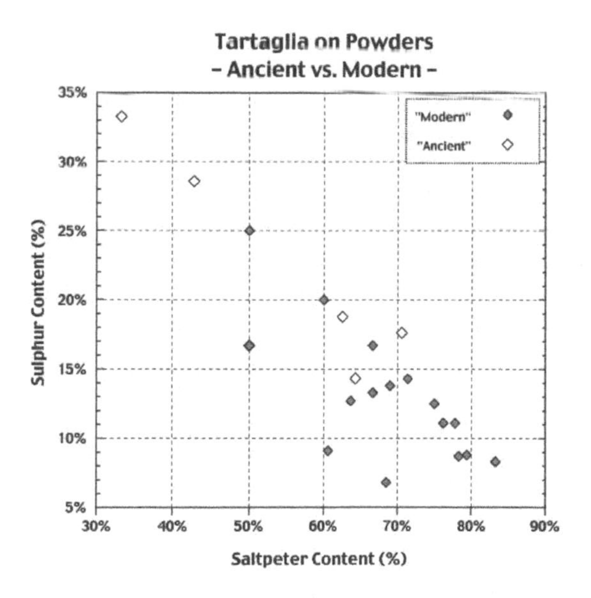

Fig. 2.

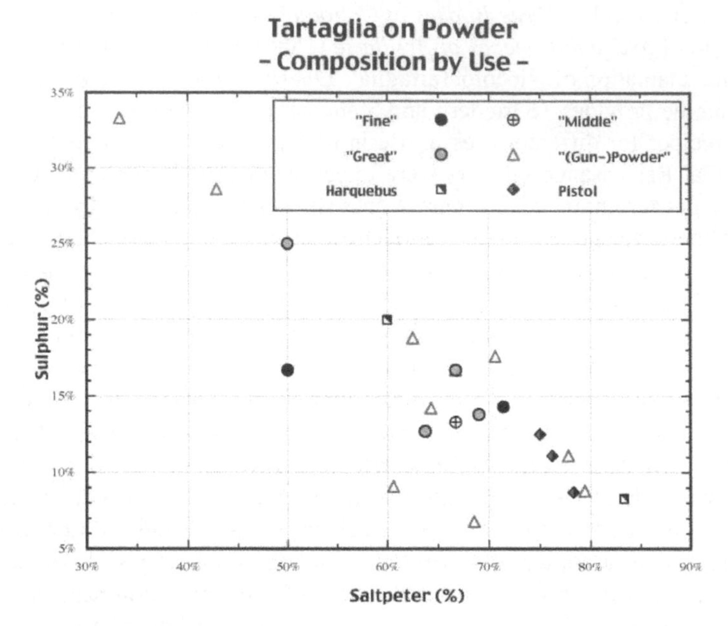

Fig. 3.

Gunners' Manual Recipes

Turning from the audience to the reader, manuscript treatises on gunnery shed further light on how technical information was transmitted in a textual form.[19] Since these manuals were not usually written for outside eyes, some of the information contained in them presents the modern reader with difficulty in terms of the nuances of the art left out by the authors. First, there is the matter of the literacy of the author and its relation to the format of the manuals. Clearly since these were written manuals, we can assume a reasonable level of literacy for the readers. Still, the evidence indicates that they did not have an exceptional level of mastery of language: they were, in fact, what I would call "paraliterate." I use this term to imply a person who could read and write but without extensive secondary schooling. They were outside ("para-") and relatively subordinate to the people typically thought of as "literate;" and would tend to lack the Scholastic or Humanist frames of reference of a fully literate individual of the day. If we accept a paraliterate status for the authors, then we

19. For a recent survey of the mode of transmission, see L. Hilaire-Pérez and C. Verna, "Dissemination of Technical Knowledge in the Middle Ages and Early Modern Era: New Approaches and Methodological Issues," *Technology & Culture*, 43, 2006, pp. 536-65.

may ask where this sort of knowledge fit into society and, conversely, what cultural constructs the authors would have brought to the study of gunnery.

The Wright MS provides one answer in the formulaic and somewhat enigmatic entry after a number of its entries: *probatum est*. This Latin phrase, meaning simply "it is proven" or "it is tested" is added at the end of the various recipes or hypotheses. At first glance this suggests that the various items had been proposed and then later tested for validity. We might therefore assume that the elements which do not have *probatum est* added were either untested, had failed the test, or even were not considered to have required testing. We cannot reach any clear conclusion about whether Wright himself tested any of these recipes, but it would not be incongruous to assume that if the Wright MS is indeed a fair second-generation copy (produced in the evenings after class, as it were), then the items which are marked *probatum est* might well have been demonstrated to or tried by him directly.

The items that garner *probatum est* fall into two categories: generalized rules of operation and specific recipes (*i.e.*, fireworks). The generalized rule so annotated relates the strength of powder to its temperature (*e.g.* a query on whether a hot cannon shoots differently than a cold one): "A question whether it is the piece that shall make the shot to mont after twice or thrice shooting or that it is the powder by reason of the heat of the pes" (fol. 5). The solution to this query suggests a positive correlation: "Reason: If that when the powder is in the hot peice and the powder growing hot also makes the powder stronger by reason of the heat that dries it, for powder being hot burneth stronger then [when it] is coulde." Wright then also noted the proof of this observation: "the proof: take … powder out of a barrel and … warme [it] in a pan over the fire and when it is hot, then burn the cold and the warm both together asunder and you shall find the warm powder [is] the stronger." This "proof" – if in fact it is true – suggests that Wright learned this concept from his instructor, possibly by demonstration or possibly by assertion, and recorded the results in his manual duly noted as "proven."

In another case of *probatum est*, Wright relates the charge of powder to the diameter of the bore, specifically, a 3:1 ratio: "Item from a saker upward, take the height of the mouth of the piece then give him iij times his height to his charge of the piece." Wright does not say that for a certain diameter bore, use a certain weight of powder. And although the tables in later works will list powder charges in pounds and parts thereof, Wright's usage, also echoed in other manuals, indicates how the gunners themselves would have dealt with these charges. Scales were few and far between and relatively difficulty to use in battle conditions; instead, for certain bore White instructs gunners to use a certain *length* of powder. How this could have been "proven" is unclear; that Wright thought it was, however, is.

When one considers gunnery in the texts as well as in the illustrations and surviving materials associated with it, one cannot escape the impression that gunnery was above all else a *numerate* profession. In parallel with their paraliterate status, we may also now speak of these practitioners as "paranumerate" as well: he could count and do simple sums and products, but higher maths such as conic sections or (later) differential calculus would be beyond them. Gunners are constantly told to measure the barrel length, diameter, and offset; the ball's circumference, weight, and windage; the ranges, distances, and elevations to the target; and then compile them into nice compact tables or scales – sometimes even inscribing those tables or scales onto wooden or brass instruments. In all these cases, standard measures in inches, feet, and yards are given, but when one looks at the many recipes, they appears to us to use the "wrong" unit, as in this example of a "length" of powder.

Although this cavalier use of incongruous (if not quite incommensurate) units strikes modern eyes as odd, we must not confuse a *functional* unit with a *tangible* unit. Tangible units may be considered the unit that measures a fundamental property of a substance and which directly captures its quantity without ambiguity. Thus, for a powder of any sort, either weight or volume might do; temperature or time would not. Functional units, on the other hand are units we use to describe the materials in particular, defined situations. The most common example would be a thermometer where we measure temperature using length or angle (for thermometers with linear or dial faces, respectively). That there might be alternate functional units for the same situation – consider that American cooking recipes typically specifying flour or butter by volume (cups) while British recipes use weight (grams) – should pose no difficulty, as long as the user knows what s/he is expected to do. Graduated cylinders in a chemistry lab or measuring cups in the kitchen use a similar logic in that although they are marked in milliliters (a volume), the scale is actually a linear one and in effect, one is measuring the height of the column of liquid and assuming a constant cross-sectional area. Renaissance cannoneers went one step further. Since the cannon barrel was more-or-less constant, they too were measuring volume by specifying length, but volume only provides an intermediate (unarticulated) value between the functionally measured length and the tangibly prescribed weight, which is itself yet one more intermediate step from the needed value, namely the explosive power (pressure) generated by the powder.

This methodology would find good use shortly after Wright in the construction of standardized ladles matched to each class of cannon by which a proper shot was charged with a specific number of ladlefuls, all measurements whether functional or tangible having been dropped for ease of operation by paranumerate gunners. In essence, this is an early form of gauging and can be seen as yet another example in the long line of military necessities surrounding gunpowder weaponry which led to standardization and, ultimately, to interchangeable parts. The other side of the development is seen in the profusion of instrumentation that developed for gunnery in the middle of the sixteenth century. Quadrants, sectors, theodolites, and engraved linear scales known as "gunners' rules" appear in myriad forms from this point onward and become badges of both honor and proficiency for the gunners themselves.[20] Instruments with engraved scales abound, with versions made in wood, brass, ivory, and even gilt if the customer could afford it. That they survive in all these materials shows that potential users of all classes *did* afford them, and presumably put them to good use, whether on the battlefield, shipboard, the classroom, or even only in the princely study.

Finally, it is worth considering gunners' knowledge of one last component of their craft: the ingredients in their recipes. If they needed to bypass basic units to make gunnery less daunting to potentially paranumerate practitioners, then we might well ask what sort of understanding they would have had of the myriad ingredients in their recipes, especially those for fireworks. The principle ingredients found in most firework recipes were saltpeter, charcoal (sometimes specified as either willow or birch), sulphur, rosin, turpentine, and various oils (often linseed and possibly others such as "sweet," rape, camphor, "debaye," or benedict). But the range of trace ingredients is indeed mind-boggling. Some are fairly common: "smyths dust" (presumably iron filings), verdigris, *aqua vitae* (alcohol), forms of mercury (liquid and "subley," or sublimate), arsenic, red lead, "unwrought" (unslaked) lime, red and white vinegar, powdered glass ("deme glase"), sawdust, tallow, and oakum. Others are either rare, unidentifiable, or both: asafetida, "detratiuan" or "deteatary," "mysket," "callemare" (perhaps calomel [L. *calomelas*, mercurous chloride]), "combuste," or the mysterious "Exodus."

Even though many of the substances may have been relatively mysterious to the gunners of the time (as well as to us today), many also had currency in contemporary alchemical investigations. In gunnery treatises, these connections – specifically the ascription of causes to ingredients or any macro-microcosmic significance – are left unexplored. Instead, recipes are treated as very straightforward processes that produce (relatively) specific results. In this sense, A.R. Hall's belief that these ingredients "show the impact of chemical ... science upon the traditional art of war"[21] cannot be sustained, unless "chemical science" also be allowed an impact on alchemy and folk medicine or other processes that worked or seemed to work but the understanding of which eluded the practitioners. Far from being a science, the gunners in the sixteenth and into the seventeenth century and later cared not so much for the science of their ingredients but for the art of their mixture.

20. S.A. Walton, "Mathematical Instruments and the Creation of the Scientific Military Gentleman," in S.A. Walton, ed., *Instrumental in War: Science, Research, and Instruments Between Knowledge and the World,* Leiden, 2005, pp. 17-46. K. Watson, "Notes on Cataneo's Scale, Some Gunners' Daggers and a Gunners' Rule," *Arms & Armour*, 8, 2011, pp. 106-116.

21. A.R. Hall, *Ballistics in the Seventeenth Century*, Oxford, 1965, p. 23.

Of course, it is possible that the fireworks recipes were taught to the gunners and were never used in the field. Such potentially superfluous instruction is not uncommon in virtually any technical field, and serves as a means to engage and maintain the interest of the student above and beyond the more mundane information. This is not to say that this information is entirely useless, but merely to make the rather obvious point that in any occupation, 95 % of an employee's time is occupied by the mundane tasks. The gunners might not have had to make them often, but were fireworks ordered, the gunners would have been the men who provided the entertainment – and earn "lord's wages" in the process.

Conclusion

It is in this intersection of both paraliterate and paranumerate gunners – they are paraliterate and yet fixated on recipes for gunpowder and fireworks without any clear sense of functional differentiation, and they are paranumerate and yet fascinated by numerical instruments and obsessed with measuring their technologies – that we can discern the early modern gunner and his craft treatises. Gunnery manuals are not part of the proto-Scientific Revolution as is usually conceived.[22] Rather they are the proud descendants of centuries of craft treatises and recipe literature that shows a practitioner population more interested in cookbooks than in treatises on natural philosophical causes. The gunners created useful and beautiful things such as fireworks, they used similar recipes to alternately entertain the masses or burn down enemy villages and ships, and they also occasionally lobbed iron and stone balls onto a target. What they seemingly did not do, however, was contemplate the mechanics of motion that got those balls to their target. These men were more like chefs and sou-chefs than they were even like engineers.[23] While I do not claim that these gunners were ignorant of *scientia* (although I am willing to say they did not know *philosophia naturalis*), I do claim that they were firmly in touch with the *techné* of their art. Thus, if I do not overstep my bounds, and I may be permitted to end with a single claim, for what it is worth: *probatum est*.

22. See G. Freeland and A. Corones, eds., *1543 and All That: Image and Word, Change and Continuity in the Proto-Scientific Revolution*, Dordrecht, 2000.

23. Although many of them were indeed engineers or at least proto-engineers as well, and their profession would develop into the modern field of engineering, it might be well at this juncture to remind the reader that engineers themselves never really became "scientific" until the late eighteenth and really the nineteenth century.

Italian Renaissance Bronze Casting Technology: The Written Record

Gertrude M. Helms

The writings of Biringuccio, Vasari and Cellini are an invaluable resource for students of the technology of Italian Renaissance bronze sculpture. Recently, their works were used in the interpretation of the technical studies of fifty-two small bronzes in the Kunsthistorisches Museum in Vienna. The researchers concluded that although there was considerable variation in the details of production, for example, in armature construction or chaplet type and insertion, there was a strong correlation between descriptions of bronze methodologies in the treatises and actual workshop procedure.[1] This study is one of many in which their writings have been taken as the standards that technical studies are compared with. Rarely have these accounts been questioned. In this essay I will try to answer three fundamental questions about their writings. How standard were the procedures they described? Did they reflect only contemporary practice? And which sculptors might have provided technical information that they drew on?

Biringuccio's *The Pirotechnia* was probably written around ca. 1535 and published in 1540.[2] It is a ten book treatise on minerals, the practice of smelting or casting metal as well as all related subjects. Biringuccio, while working for the Sienese ruling family, the Petrucci, traveled to Germany as well as many parts of Italy. He was appointed to the Sienese armory in 1513, the Sienese mint in 1514 and became architect and director of Siena Cathedral in 1535 before taking over the Papal Foundry and munitions in 1536. His discussion of bronze sculpture was included in the book on casting which focused primarily

1. B. M. Weisman, C. L. Reedy, "Technical Studies on Renaissance Bronzes," *Materials Issues in Art and Archaeology VI*, 712, 2002, pp. 483-495.

2. V. Birincuccio. *De la pirotechnia*, ed. A. Carugo, Venice, 1540; facsimile, Milan, 1977. V. Biringuccio, *The Pirotechnia of Vannoccio Biringuccio,* trans. and ed. C. S. Smith and M. T. Gnudi, New York, 1959, reprint 1990. Both of these volumes have short biographies of Biringuccio. The most pertinent sections for bronze sculpture are Biringuccio, 1990, pp. 218-221; 228-234; 248-255.

on guns and bells. GiorgioVasari, a Tuscan painter and architect, first published his *Lives of the Artists*, in 1550.[3] The technical introduction to this work, divided into sections on architecture, sculpture and painting, comprised most of the technical information needed for constructing and decorating a building. Benvenuto Cellini, a Florentine goldsmith and sculptor, published his *Treatise on Sculpture* in 1568,[4] while his autobiography, probably written in about 1558, was known only in manuscript form until it was first published in 1728.[5] Although all three were Tuscans, each had traveled extensively throughout Italy. Vasari and Cellini most likely knew Biringuccio's treatise and they certainly knew each other. From comments in all three treatises it is also evident that they were interested in the technical knowledge of other artists.

Biringuccio's, Vasari's and Cellini's accounts of bronze sculpture methodologies have technological differences that are both subtle and distinctive, particularly their descriptions of the materials needed. In order to consider the questions listed above, I have analyzed the surviving documents, both published and unpublished, of over eighty fifteenth and sixteenth century Italian large bronze sculpture projects. Most of the documents I have studied were payment records, although I have also consulted commissioning documents when they included a list of materials, and, where possible, I have also transcribed some of the sculptors' own notebooks. The level of detail in commissioning and payment records varies greatly depending on patron, project and material. However, large bronze sculptural projects, being both expensive and lengthy, were usually run through the patron's account books with the patron responsible for the payments of salaries and for the purchase of the materials needed for the project. Some projects only have one or two surviving records concerning materials, many have a substantial list, while a few projects are so fully documented that it is possible to speculatively recreate the technology being used. These documents provide an overview of practice that gives a context to the writings of Biringuccio, Vasari and Cellini and an insight into their sources.

In interpreting the documents I have made the following assumptions: (1) If a material was ordered and purchased it must have been used. (2) Materials

3. G. Vasari, *La Tecnica dell'arte negli scritti di Giorgio Vasari,* ed. R. Panichi, Florence, 1991; G. Vasari, *Vasari on Technique,* trans. L. Maclehose and ed. G. Baldwin Brown, 1907, reprint, New York, 1960. Both of these are of the 1568 edition. The section on bronze sculpture in Vasari, 1960, is pp. 158-166. G. Vasari, *Lives of the Painters, Sculptors, and Architects,* trans. G. du C. de Vere, London, 1912, reprint London, 1996.

4. B. Cellini, *I Trattati dell'Oreficeria e della Scultura,* ed. C. Milanesi, Florence, 1857, reprint, Florence, 1994. B. Cellini, *The Treatises of Benvenuto Cellini on Goldsmithing and Sculpture,* trans. C. R. Ashbee, London, 1888, reprint, New York, 1967. Both of these are based on a codex in the Marciana rather than the 1568 edition. B. Cellini, *Due Trattati uno intorno alle otto principali arti dell'oreficeria*, Florence, 1568. *The making of the Modern Economy,* Thomson Gale, 2006. The section on bronze sculpture in Cellini, 1967, is pp. 111-126.

5. B. Cellini, *Vita,* Milan, 1985 and B. Cellini, *The Autobiography of Benvenuto Cellini,* trans. J. A. Symonds, New York, 1963.

were generally obtained in the sequence in which they were needed. There seems to have been little stockpiling of materials. It was the norm for materials to be reordered as needed. (3) The lack of a material doesn't necessarily mean it wasn't used if there was other evidence to suggest it was. (4) And finally, comments made in the documents about materials are valid.

The descriptions of bronze casting in Biringuccio, Vasari and Cellini can be understood as a series of steps that can be broken down into the materials needed for each step. As both their accounts and the documents are sequential, it is useful to consider each material in order. The simplest method, direct casting, was called "the ordinary way" by Biringuccio[6] and used by Cellini for the Perseus and omitted by Vasari. In direct casting, the wax sculpture is modeled directly over a core and is therefore both the original and full scale model. If something goes wrong with the casting the original is lost.

The first step in direct casting is modeling a clay core over an iron armature. Both Biringuccio and Cellini emphasized the importance of choosing good clay. As experienced casters themselves, they were both aware of the importance of choosing the best raw materials available for a project and preparing them properly. Biringuccio and Cellini describe preparing molding clay by beating cloth clippings into clay. It took Ghiberti's assistants for the St. Matthew two days to beat 100 pounds of cloth clippings into the clay.[7] Vasari described using horse dung and hair to make the clay mixture. Biringuccio suggested that those materials were ones that could be used if cloth clippings weren't available or when cloth clippings made the core too strong to be removed easily, but he preferred cloth clippings. When describing making the clay, Cellini divulged his secret, he let the clay and cloth clippings mixture decompose for four months or more.[8] He may have done this, but I found no evidence of this in practice. Since the same clay and cloth clippings mixture was used for both cores an outer molds there were usually several purchases of both throughout a project. Clay was sometimes left out of the materials purchased for a project, but its use can be inferred by the purchase of cloth clippings. For example, there are several purchases of cloth clippings but not of clay for the bronzes Donatello made for the High Altar of S. Antonio, Padua, and a mention of clay for his contemporary equestrian monument to the Gattamelata, Padua.[9]

6. Biringuccio, 1990, p. 228.

7. A. Doren, "Das Aktenbuch für Ghibertis Mattäus-Statue An Or San Michele zu Florenz," *Italienische Forschungen,* I, 1906, pp. 3-58.

8. Florence, Biblioteca Riccardiana, Codice Riccardiana, 2787. B. Cellini, *Libro segnato A.* It is not clear from Cellini's accounts whether he actually practiced this or not.

9. A. Sartori, *Archivio Sartori. Documenti di storia e arte Francescana.* Vol. I, Padua, 1983. Santo High Altar, pp. 215-222; Gattamelata, pp. 850-854.

The core was carefully finished. Cellini suggested finishing it to within a finger's thickness of the bronze while Biringuccio said that it should be as large as one wants the hollow in the finished bronze to be. The core is then dried. Once the core is ready, wax is applied. Cellini's method was to add that missing finger's thickness of wax to achieve the final bronze, while Biringuccio describes adding or thickening the wax as much as you want the bronze to be. At this point, the sculptor would have a molded core with a coat of wax that had the aspect and thickness of the final bronze.

The sculptor could also reach this point of a fully modeled wax over a core by indirect casting, a process that involves making a gesso mold of the original. Cellini rejected this method for the Perseus because indirect casting is much more time consuming than direct casting. It does have the important advantage, however, that the original model is retained. As Cellini noted, the model can then act as a guide for an assistant or another sculptor to finish the bronze. He blamed the time required and the result of the chasing of the Perseus on the fact that it was directly cast, which meant there was no fine model to finish the work from. In their treatises, Biringuccio, Vasari and Cellini all describe versions of indirect casting, Biringuccio and Cellini in some detail and Vasari more summarily.

For Biringuccio, indirect casting was a useful method because the model could be bronze, marble or terracotta. He suggested that if the model was large it would need to be greased with tallow, pig fat or oil or covered with tin foil or beaten gold or silver before the gesso mold was made. Although both Vasari and Cellini describe using a clay model, only Cellini explained how to make an isolating coat for the clay by brushing the model with warmed wax mixed with turpentine before applying a layer of tin foil that was then well oiled.

I have found only one project in which the tin ordered might have been tin foil rather than tin for making bronze. Purchases for the bronzes that Donatello made for the High Altar of S. Antonio, Padua, include one for "wax, resin, turpentine from larches and tin to make the molds over the design of the altar."[10] This tin may have been for making the bronze alloy as the merchant was listed as providing some tin in the metal account. However, a later payment for "turpentine from larches, varnish, tin had master Donato to mold the figures of the altar" was made on the same day and account page for payments for copper delivered to the coppersmith who received the metal and cast the bronzes for the altar.[11] There are a number of purchases of tin such as the 21 pounds to temper copper[12] and 40 pounds to cast the sculpture of the Virgin[13] that are clearly identified.

10. Sartori, 1983, p. 217, doc. 56.
11. Sartori, 1983, p. 220, doc. 110.
12. Sartori, 1983, p. 220, doc. 108.
13. Sartori, 1983, p. 221, doc. 112.

Once the model was chosen or finished, a gesso piece mold was made. Although the documents are unclear on this point, it seems reasonable to conclude that every artist probably had their own method for casting, marking and assembling the gesso pieces and molds. It is unclear when the use of indirect casting and gesso piece molds for large bronzes became widespread. It has been suggested that 15th century large bronzes were directly cast. Most 15th century contracts were similar to Ghiberti's contract for the St. Matthew which stipulated that he was to receive "clay, ironwork to assemble the figure, wax, brass, charcoal, wood and all other things occurring and necessary to the figure."[14] By the mid 16th century many contracts, such as that for the grates for the Arca of St. Anthony, S. Antonio, Padua, 1543, included gesso in the standard list of materials. The artists Danese Cattaneo and Tiziano Aspetti agreed, "here to put the wax the gesso molds the ironwork the metal the furnace the charcoal ..."[15]

I did find two possible uses of gesso in the fifteenth century. For Donatello's Santo bronzes not only was there a purchase of tin but also a curious purchase of lime, "calzina."[16] There was no reason for Donatello to need lime because he did not cast the bronzes nor do the masonry work for the altar. So it may be that "calzina" was being used as a generic in the same way that Biringuccio said that plaster of Paris was a variety of "calcine."[17] The Santo project was also unusual in that Donatello and his assistants were paid individually for chasing specific evangelist and angel relief's which suggests that they may have had the original Donatello models to follow.[18] Recently, the technical examination of the Amor Atys, a sculpture attributed to Donatello, has revealed that it was indirectly cast.[19] This supports the possibility that Donatello indirectly cast the Santo bronzes.

The other mention of gesso is in the inventory of goods in Verrocchio's Venetian workshop after his death. There was a quantity of stones to make gesso that was valuable enough to have already been sold on for quite a high price.[20] This could be gesso for the possible indirect casting of the Colleoni or as Vasari recounts in his Life of Verrocchio the special gesso that Verrocchio used for casting natural objects such as hands and feet so that he could imitate them, with greater convenience.[21]

14. Doren, 1906, pp. 26-30.

15. Sartori, 1983, pp. 370-371, doc. 563.

16. Sartori, 1983, p. 220, doc. 106.

17. Biringuccio, 1990, p. 401.

18. Sartori, 1983, p. 216, doc. 51 for the contract. The angels had already been cast and the evangelists were in wax covered with clay. For individual payments see p. 218, docs. 75-91.

19. S. Siano, "Considerazioni tecniche," *Il ritorno d'Amore: L'Attis di Donatello restaurato*, B. Paolozzi Strozzi, ed, Florence, 2005, pp. 122-135.

20. D. Covi, "Four New Documents Concerning Andrea del Verrocchio," *Art Bulletin*, XLVIII, 1966, 97-103, doc. IV.

21. Vasari, 1996, Vol. I, p. 555.

Once the gesso mold is made, Biringuccio, Vasari and Cellini each described slightly different methods of proceeding in terms of making the wax layer and producing the core. Their variations and others involving armatures, the size and placement of chaplets, etc. are among the details so useful in the technical analysis of small bronzes. The result is, however, the same in that one has a wax positive sculpture over a core, a retained original and in some cases a reusable plaster mold. After piece molding, the wax was cleaned to remove casting fins and other artifacts.

Wax was so indispensable that it was almost always listed in commissioning documents and often listed on a separate wax page in account books. Cellini simply specified wax for the wax layer while Biringuccio said that wax or wax and tallow could be used. Vasari advised using yellow wax softened with a little turpentine and tallow. Wax was ordered in different grades: white, yellow and green and sometimes from specific places. Given the large amount of wax needed for a project, it was often recycled from sources like the drips and drops from candles or from broken wax sculptures. In some projects other materials were purchased to mix with wax; turpentine to mix with wax appeares in both 15th and 16th century accounts. Sometimes, however, there were more unusual purchases like the tallow, black pitch, and tar to mix with wax and black to tint wax that Maso di Bartolomeo bought for a monumental candelabra for SS. Annunziata, 1447,[22] or the large quantity of tar that Donatello bought for an unspecified 1456 project.[23] The Maso mixture was similar to the mixture of wax, a little animal fat, turpentine and black pitch that Vasari suggested for making wax models for marble sculptures[24] and the ingredients, (tallow, turpentine, and colophony), that Cellini bought to alloy the wax of models.[25] Biringuccio in his section on molding various kinds of reliefs recommends several molding compounds to replace plaster including one of wax and Greek pitch or ship's tar.[26] There is no evidence that Maso di Bartolomeo and Donatello were making models or molding reliefs with these compounds.

Once the wax layer is finished then the sculpture is prepared for casting. Chaplets are added. A fine coat of thinned clay or other fine material is dusted on to the wax before successive layers of the clay and cloth clippings mixture are applied to form the outer mold. Small molds were reinforced with iron wire and larger molds with iron bands, rods and/or plates as well as iron wire. Mold reinforcement was the most important use of iron mentioned in the account books. Other uses for iron were for the founding furnace, tools and the arma-

22. Prato, Biblioteca Roncioniana, Roncioniana 388. Maso di Bartolomeo, *Ricordi*, c. 2 right and c. 3 right.

23. G. Corti and F. Hartt, "New Documents Concerning Donatello, Luca and Andrea della Robbia, Desiderio, Mino, Uccello, Pollaiuolo, Filippo Lippi, Baldovinetti and Others," *Art Bulletin*, XLIV, 1962, 155-167, doc. 20 a.

24. Vasari, 1960, p. 148.

25. Cellini, Riccardiana 2787, c. 11 1/2 left and c. 29 right.

26. Biringuccio, 1990, pp. 329-332.

ture of the core. There are only a few instances where it is possible to identify iron as made for internal armatures.[27] Due to its importance, ironwork is usually listed in commissioning documents. Although it was usually purchased from ironsmiths some sculptors probably had the capability to do some of their own smithing. For example, there were records of blacksmithing as well as bronze working tools in the Ghiberti workshop.[28]

It was at this point that both Biringuccio and Vasari describe making channels and vents for the mold. Cellini, when discussing indirect casting, suggests that they were added after touching up the wax. In his autobiography, Cellini described his channels and vent method "which differed very considerably from that of all the other masters in the trade"[29] and in his treatise he elaborated on it. He designed the vents to point down to make it easier to remove wax. In the casting ditch he used little terra cotta water pipes to bring these downward vents up through the ditch fill to the surface. His was the only purchase of these pipes that I have found.[30]

After being reinforced, the mold is gently heated to remove the wax. This was very important as wax left in the mold could be catastrophic for the project. Biringuccio suggested using wood and or charcoal for this step. After this first "drying" the mold has to be "baked," a very important step left out by Vasari. In order to bake the mold an oven or furnace of brick or brick ends is assembled a few finger lengths away from the mold. Biringuccio suggested firing it with wood and charcoal or wood only. Cellini was adamant in his treatise that only soft wood, alder, lime, beech or twigs, should be used and that green wood like oak should be avoided and charcoal never used.

Bronze sculpture projects were very fuel intensive, because fuel was needed to dry and bake the molds and alloy and prepare metal for casting. Most projects had many purchases of fuel, usually of both wood and charcoal. Sometimes the use of the fuel is stipulated, for example, Maso's purchase of large wood to bake the molds of the Annunziata candelabra[31] or Donatello's cart of charcoal to ready four saints[32] for the High Altar of S. Antonio and two carts of wood and sedge to fire the figures of the altar.[33] Ammanati, in 1559, used charcoal to dry the mold of the Hercules and Antaeus and oak and charcoal to bake the molds.[34] In his autobiography, Cellini said that he used pine-

27. Florence, Biblioteca Riccardiana 2788. Benvenuto Cellini, *Giornale segnjato A*, c. 5 right, for iron armatures.

28. T. Krautheimer-Hess, "More Ghibertiana," *Art Bulletin*, XVLVI, 1964, pp. 307-321.

29. Cellini, 1963, p. 369.

30. Cellini, Riccardiana 2787, c. 11 left, 17 left, and 18 right, Appendix E.

31. Maso di Bartolomeo, Roncioniana 388, c. 3 left.

32. Sartori, p. 220, doc. 107.

33. Sartori, p. 220, doc. 108

34. M. Fossi, "Note documentaire sul gruppo di Ettore e Anteo dell'Ammannati e sulla villa Ambrogiana," *Architettura e politica da Cosimo I a Ferdinando I*, ed. L. Olschiki, Florence, 1977, pp. 463-479.

wood to dry the Perseus mold and implied that he also used pine to bake the mold.[35] He did buy pinewood from the Frescobaldi pine forest but didn't identify its use.[36] However, charcoal, pine, oak and alder were obtained to found bronze and metal.[37]

After baking, Biringuccio recommended inspecting the mold for cracks before casting. He then suggested filling any cracks with a plaster made of egg white, crushed brick and quicklime or instead of brick, baked crushed mold clay.[38] I have found only one reference to this. In the payments for Francesco di Giorgio's Angels for Siena Cathedral, 1483-90, there is an entry for a large amount of wood to bake the molds and twelve eggs to plaster.[39]

Once ready, the mold is lowered into the casting ditch, which should be deep enough for the sculpture to stand upright. Earth is then rammed in around it. This ditch was dug near the metal melting furnace so that the molten metal would not have far to flow. As described by Biringuccio[40] and Cellini[41] this furnace was constructed of bricks and or refractory stone, tiles and ironwork. Although Biringuccio and Cellini use the term furnace, during the 15th century the words oven and furnace seem to have been used interchangeably. Founding furnaces were not permanent; the materials for one were often included in the list of materials for a project. For example, for the St. Matthew, Ghiberti purchased refractory stones, "ironwork, bricks, lime, wood and other things to make the oven in order to cast the said figure."[42] Ghiberti's list of materials needed is very similar to the materials that Cellini purchased for making his own melting or founding furnace.[43] Unless the use of bricks is identified, it is not always possible to tell whether the bricks are for ovens/furnaces to bake the molds or ovens/furnaces to melt metal. Francesco di Giorgio bought common bricks to make ovens to bake molds and to found metal for the Siena Duomo Angels.[44] Ammanati ordered 800 tiles to make an oven and 300 small pots to put over the furnace for his Hercules and Antaeus project, 1559-60.[45]

Metal was the most important and expensive material purchased for a bronze project. It came in many forms, as various types of ingots and in vary-

35. Cellini, 1963, p. 368.

36. Cellini, Riccardiana 2787, c. 8 right.

37. Cellini, Riccardiana 2787, c. 8 left and 30 left.

38. Biringuccio, 1990, p. 253.

39. C. Zarrilli, "Francesco di Giorgio pittore e scultore nelle fonti archivistiche senesi," *Francesco di Giorgio e il Rinascimento a Siena 1450-1500*, ed. L. Bellosi, Milan, 1993, pp. 530-538, doc. 25.

40. Biringuccio, 1990, pp. 281-288.

41. Cellini, 1967, pp. 127-133.

42. Doren, 1906, p. 34.

43. Cellini, Riccardiana 2787, for example, the expenses on c. 7 left, 7 right, 9 left, 9 right, 14 left, and 14 right.

44. Zarrilli, 1993, doc. 45.

45. Fossi, 1977, pp. 465-466.

ing degrees of purity. During the "metal drought" of the fifteenth century, it must have been quite difficult to amass sufficient metal. Sculptors resorted to buying it from scrap merchants, pawnbrokers, and bric brac dealers. There are mentions of broken basins, old candlesticks and even metal that used to be the statue of a lion. It is clear that it was not necessary to buy metal in proper ingots as best practice suggested that the alloy be prepared before casting, a straightforward process for the founder. Metal had a different price per pound or hundredweight depending on its quality and sculptors bought metals of varying quality for the same project. Both Biringuccio and Vasari discussed the appropriate alloys of copper and tin for casting sculpture. According to Vasari, statuary metal was a combination of two-thirds copper and one-third brass[46] while Biringuccio said that bronze was a compound material "of copper or tin, of brass or lead." Biringuccio noted that "metal" and "bronze" are the same thing.[47] In the fifteenth century metal, bronze and brass are used interchangeably even for projects when copper and tin were ordered separately. Even though lead has been found in the alloys of some sculptures, for example in some of the alloys of Donatello's Judith and Holofernes,[48] there are few specific purchases of it. One was, however, for "four pounds of lead in order to cast" for Donatello's unrealized doors for the Siena Duomo, 1457.[49]

Once a sculpture was cast, it needed to be cleaned. In any bronze project the major expenses were metal and manpower. Assistants were needed not only to help prepare the sculpture for casting but for the many man days of work necessary to finish a bronze. As Biringuccio said, after a gun or statue is freed from clay, "do not be dismayed if you see the castings so rough and in a form that scarcely shows what they are."[50] Although the ideal, according to Cellini, was to produce a casting that was perfect enough not to need surface working, this was rarely achieved and some bronzes took years to finish because they, like Maso di Bartolomeo's grate for a chapel in Prato Cathedral, needed to be "... worked, sculpted, filed, polished and burnished as is seen in similar work."[51]

With few exceptions, the descriptions of direct and indirect casting in the treatises can be used to explain the purchase documents for the materials of bronze sculpture projects. Conversely, an analysis of the documents suggests that the treatises can be used as the standards for ordinary direct and indirect

46. Vasari, 1960, pp. 163-164.

47. Biringuccio, 1990, pp. 299-300.

48. Centro Ricerche Europa Metalli, L.M.I., "Ricerche su leghe e indagini metallografiche," *Donatello e il restauro della Giuditta*, ed. L. Dolcini, Florence, 1988, pp. 58-63.

49. V. Herzner, "Donatello in Siena," *Mitteilungen des Kunsthistorischen Institutes in Florenz*, XV, 1971, pp. 161-186.

50. Biringuccio, 1990, p. 307, doc 12c.

51. G. Marchini, *Il Tesoro el Duomo di Prato*, Prato, 1963, p. 106.

casting as long as one is aware that they do reflect individual practice as well as their milieu.

Biringuccio as an artillery maker wrote about metallurgy and making cannons from personal experience. Like Cellini, he described direct and indirect molding of sculptures as well as a type of "slush molding" similar to the one Vasari mentioned. He also described two other ways of making sculptures. One of these was a form of direct casting where the clay sculpture is finished to the exact dimensions of the bronze and then, after baking, the surface is carved away and replaced by wax. This was a way to make sure that the statue would be hollow and the bronze equally thin all over. This laborious, if careful, method used the same materials as a direct cast. The other method for statues is a type of indirect casting. A statue of tow and paste is modeled on an iron armature then clothed with thick or thin canvas covered with glue that is then evened up with wax and tallow mixed with turpentine. Biringuccio said that it was like a painter's model. A mold of two, three of four pieces is made over this model. This mold is baked so that the model burns up leaving a negative mold. A layer of wax is spread in this mold and chaplets are inserted through the wax into the mold, after which the procedure for creating a core for an indirect cast is followed. Although Biringuccio doesn't say where he learned of this technique, it was probably as a child in Siena where his father worked with Francesco di Giorgio on several Sienese projects including commissions in 1491.[52] In 1488-90, Francesco di Giorgio made two 125 cm. bronze angels for the Siena Duomo. Some of the materials purchased for them were quite unusual. Besides the customary materials, such as: metal, wax, clay, wood, bricks, charcoal, and ironwork, he bought a "chest of iron to cast" at the beginning of the project, a hanging of fine linen cloth in order to make the shifts to the molds of the angels, flour and tow for the cores of the angels, 20 arm lengths of cloth in one fine sheet to make the cloth of the angels, a sheet of hemp, cheese in order to make glue, a round trunk, and gesso to fill the molds.[53] Biringuccio may have been only eight when Francesco di Giorgio began the angels but he was seventeen when there was a controversy between Francesco di Giorgio and the Duomo over payment and Francesco di Giorgio submitted quite detailed payment claims based on his own account book. Biringuccio may even have known it. It has recently been suggested that Donatello's Judith and Holofernes, 1460s, was produced using this method described by Biringuccio.[54] Donatello worked in Siena in the 1450s so he may

52. G. Chironi, "Appendice documentaria," *Francesco di Giorgio architetto*, eds. F. Fiore and M. Tafuri, Milan, 1993, p. 405.

53. Zarrilli, 1993, pp. 532-535, docs. 25-49.

54. R. Stone, "A New Interpretation of the Casting of Donatello's *Judith and Holofernes*," *Small Bronzes in the Renaissance*, Studies in the History of Art 62, Center for Advanced Study in the Visual Arts, Symposium Papers XXXIX, National Gallery of Art, ed. D. Pincus, Washington, 2001, pp. 55-69.

have either learned the technique there or introduced it. Unfortunately, there are no certain documents for the Judith or for bronze sculptures in Siena in the second half of the fifteenth century earlier than Francesco di Giorgio's Angels. Further research in Siena and the other cities where Biringuccio worked may find other projects that directly relate to techniques described in his treatise.

Vasari's technical introduction to the Lives should be analyzed in conjunction with the Lives, because he included quite a lot of technical information while also making assessments of the technical abilities of various artists. He was very complimentary about the Florentine sculptor, Jacopo Sansovino, who at that time was working in Venice. In the accounts for Sansovino's Sacristy door for San Marco, 1545-1546, there are payments for the facture of wax of the reliefs; wax and turpentine; gesso to make the molds; payments for Master Gasparo for making the molds or reliefs, figures, putti and heads; and a payment to a Master Andrea gesso worker for having cast the figures of wax in said molds.[55] One of Sansovino's pupils, il Tribolo, was a friend of both Vasari and Cellini, with whom he had taken a trip to Venice to see Sansovino. At the time of his death in 1550 he was employed by the Medici and was designing the gardens and fountains for the Medici Villa of Castello. Many of the water pipes and bronzes for Castello were cast by Zanobi Portighari. His account book for that work, which was begun in 1547, is entitled "Account of the iron of wax of lead of copper of tin of money had from the fabric at Castello on the order of the sculptor Tribolo." Among the payments listed are ones for wax to cast a figure and for ironwork for a figure. For a satyr he was paid for making the wax, making the mold and casting the bronze.[56] On Tribolo's death, the great fountain was left incomplete until 1559 when Ammanati made the Hercules and Antaeus. The account book lists payments for: gesso; wax and cinnabar; iron; 2 flasks of oil to oil the two pieces of the mold – something mentioned by Vasari; wool clippings; tiles and furnace makers to make the oven; charcoal; white clay; charcoal to bake the figures; iron wire to tie the figures, etc.[57] These sculptors were part of Vasari's milieu so it would not have been surprising for him to turn to them for practical technical information and then to have written only about indirect casting as it seems to have been the norm for these artists.

Cellini's description of bronze casting in his *Treatise on Sculpture*, as he pointed out, was based on his own recollection of the direct casting of the great lunette of the Porte Doreé for the Chateau of Fontainbleau and the Perseus and

55. B. Boucher, *The Sculpture of Jacopo Sansovino*, New Haven, 1991, vol. I, pp. 201-202, doc. 112.

56. Florence, Archivio di Stato di Firenze, Corporazioni religiose soppresse dal Governo Francese, serie 103, pezzo 328. Z. Portighiani, "Contte di fero di cera di pionbo di reame edi stagio e di danari autti dala fabrica di e castelo per ordine del tribolo scultore," Zanobi Portighiani helped Cellini in the indirect casting of the bust of Cosimo I; see Cellini, Riccardiana 2788, C. 2 reverse.

57. Fossi, 1963, pp. 463-479.

Medusa for Florence. His account books that include the expenses for the Perseus and Medusa as well as the indirectly cast monumental head of Cosimo I survive and I believe that he was following his own workshop procedure quite closely in the treatise. His account of the indirect casting of large bronzes may also have been based in part on his experience in France where he made his first large bronze sculptures. Although he said that the lunette was directly cast there are notarial documents for master founders who were to make, mold, found, join, assemble and dress his great jamb figures, victories and other sections. This suggests that at least some elements of the door were to be indirectly cast.[58] While he was in France bronze sculptures were being cast from molds taken from antique sculptures in the Vatican.[59] Cellini was well aware of the technology being used in France and its excellent reputation, although he did think that his methods were superior. Perhaps it was molding after the antique in Rome that Biringuccio was thinking of when he said that models of bronze, marble or terra cotta needed a different method from the ordinary and then described piece molding.

In conclusion, with few exceptions, the descriptions of direct and indirect casting in the treatises can be used to explain the materials purchased for both fifteenth and sixteenth century bronze sculpture projects. The documents also provide the basis for accepting the treatises as standards and as well as provide insights into the sources of Biringuccio, Vasari and Cellini.

58. C. Grodecki, "Le Séjour de Benvenuto Cellini a l'Hôtel de Nesle et la fonte de la Nymphe de Fontainebleau d'après les acts des notaries parisiens," *Bulletin de la Société de l'Histoire de Paris et de L'Ile-de-France*, 98, 1971, pp. 76-78, doc. XII.

59. L. de Laborde, *Les Comptes des Bâtiments du Roi (1528-1571)*, vol. I, Paris, 1877, pp. 191-204.

De la recette à la pratique :
l'exemple du *lutum sapientiae* des alchimistes

Nicolas Thomas

Dans les textes alchimiques, il est dit qu'il faut « luter » les vases avant utilisation[1]. Le mot peut avoir plusieurs significations qui dépendent du contexte : il peut s'agir de recouvrir la panse d'un pot avant de le passer au feu, de fermer hermétiquement un récipient ou encore de solidariser deux vases pour former un appareil (figure 1).

Ces trois fonctions du lut ne s'excluent pas mutuellement : on peut vouloir rendre hermétique le joint entre un pot et son couvercle tout en solidarisant les deux parties. Autre fonction : le lut peut être simplement utilisé pour fermer un vase afin de conserver un produit. L'opération est commune et pourtant les auteurs accordent une grande importance à ces compositions au point d'en donner les ingrédients permettant de les réaliser. Il s'agit de recettes, c'est-à-dire de la transcription d'une chaîne opératoire plus ou moins complète et explicite mais que l'on ne peut dissocier de son contexte technique et culturel qui l'a produite. Bien sûr, on trouve surtout des recettes de lut au détour d'autres recettes, par exemple au milieu d'une recette de distillation, de sublimation ou encore de cémentation : les recettes de lut sont donc des recettes contenues dans d'autres recettes[2]. En dépit de cette remarque, il demeure que de nombreux traités fournissent des recettes de lut isolées. Luter apparaît alors étrangement comme une opération autonome.

La fréquence des mentions, comme la variété des procédés et des compositions, témoignent donc d'une préoccupation constante des alchimistes pour une technique qui apparaît somme toute banale. Enfin, le vocabulaire employé rend

1. Le mot vient du latin *lutum* : boue, en référence à l'enduit utilisé, le plus souvent une terre argileuse.

2. N. Thomas, « Prendre de l'acier pour de l'or : imaginaire et procédés métallurgiques du Moyen Âge au XVIIIe siècle », dans *Hypothèses 2005*, Paris, 2006, p. 175-186 ; N. Thomas, « L'alambic dans la cuisine », dans Ravoire F., Dietrich A., dir., *La cuisine et la table dans la France de la fin du Moyen-Âge, Contenus et contenants du XIVe au XVIe siècle, Actes du colloque de Sens (2004)*, Caen, 2009, p. 35-50. N. Thomas, Claude C., « Les vases à fond percé : pratique de la distillation *per descensum* au bas Moyen Âge en Île-de-France », *Revue archéologique d'Île-de-France*, 4, 2012, à paraître.

compte de ce paradoxe en survalorisant cette action triviale de luter : l'enduit destiné à recouvrir les vases est souvent appelé *lutum sapientiae*. L'étude de la littérature technique et alchimique, associée à l'analyse des compositions et des procédés de mise en œuvre permet d'apprécier la distance entre la recette et sa réalisation, mais elle fournit également des hypothèses pour expliquer cette quasi obsession dont témoignent les textes à propos du lut.

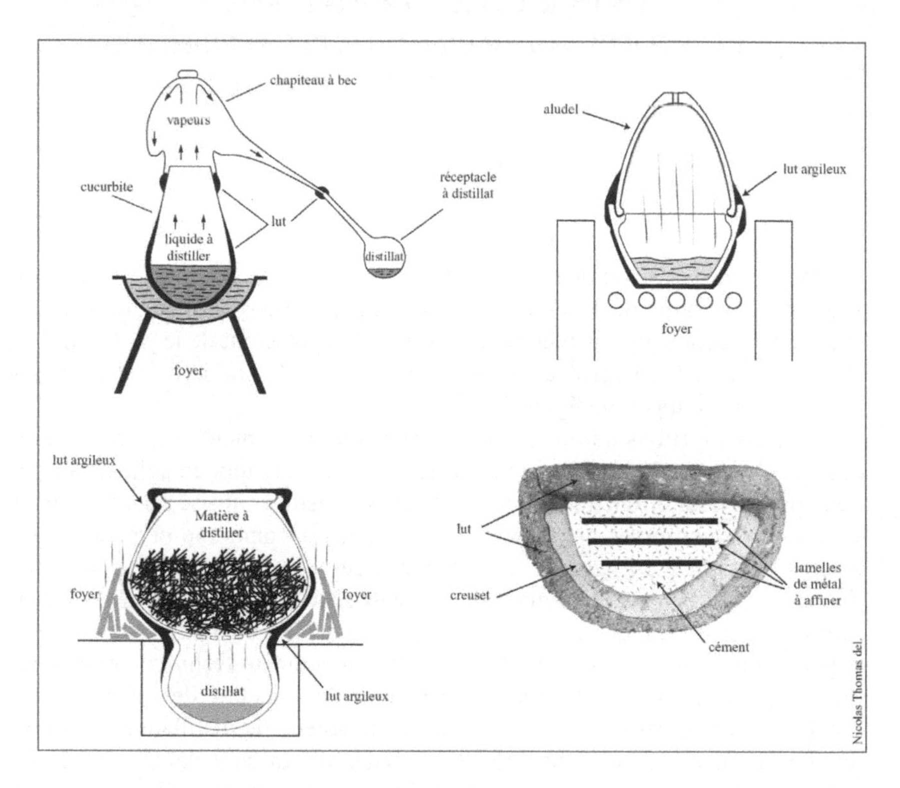

Figure 1 : schémas de la distillation *pers ascensum*, de la sublimation dans un aludel, de la distillation *per descensum*, de l'affinage de l'or dans un creuset par cémentation, position du lut (les 4 schémas ne sont pas à l'échelle).

Les recettes et la composition des enduits

Dans les traités techniques les plus anciens, la composition de l'enduit est déjà donnée. L'alchimiste Zosime de Panopolis mentionne ce procédé à de nombreuses reprises[3]. Il s'agit le plus souvent de colmater les joints entre dif-

3. M. Mertens, *Zosime de Panopolis, Mémoires authentiques*, Paris, 1995, p. CXV (Les alchimistes grecs ; 4, 1).

férents vases constituant un appareil afin d'en assurer son étanchéité : « scellez tout autour les jointures de l'instrument, de façon sûre, afin que l'odeur soit gardée »[4]. Chez Zosime, le lut est à base de farine et d'eau, de graisse mélangée à du gypse, de la propolis[5] ou encore de la chaux huilée[6]. Le même auteur préconise également d'enduire un vaisseau de terre cuite avec de l'argile dite « réfractaire » avant sa mise au bain de fumier[7].

Dans les recettes alchimiques syriaques ou arabes[8], on trouve le plâtre (ou le gypse)[9], la cire[10] et de l'œuf mêlé à de l'argile et à du crin[11]. L'argile est plus souvent mentionnée, elle est utilisée seule[12] ou mêlée à du crin[13]. De nouveaux ingrédients apparaissent comme le naphte blanc, un bitume, mêlé à du sel[14] ou le vinaigre[15]. Le sel peut aussi être cité avec du plâtre ou du blanc d'œuf[16]. Les vases peuvent être fermés simplement avec des chiffons ou du

4. *Idem*, p. 31.

5. *Ibid.*, p. 15, la cire est aussi utilisée comme lut, s'agit-il dans ce cas de cire d'abeille ? Nous trouvons la graisse dans *Le papyrus de Stockholm* au sein d'une recette de fabrication d'émeraude : R. Halleux, *Papyrus de Leyde, Papyrus de Stockholm, Recettes*, Paris, 1981, p. 121 (Les alchimistes grecs ; 1).

6. *Ibid.*, p. 31 : ἐλαιοκονία est traduit par M. Mertens : « chaux huilée ». Κονία désigne indifféremment la poussière, la chaux, la cendre ou le sable. Le matériau étant souvent désigné par son aspect, la confusion est ici possible. M. Berthelot, traduit par « cendre mélangée d'huile », M. Berthelot, C. E. Ruelle, *Collection des anciens alchimistes grecs*, Paris, 1888-1889, III, p. 143. La chaux comme la cendre sont également utilisées dans l'alchimie occidentale pour la préparation de luts. La chaux hydratée $Ca(OH)_2$ se présente sous la forme d'une poudre blanche, ce qui peut entretenir cette confusion avec la cendre. D'après le lexique syriaque de Bar Bahloul : « *Conion*, selon les artisans, la chaux. Dans un manuscrit, c'est la cendre – Gabriel a dit : c'est le nom de la lessive de cendre ; en grec, κονία », M. Berthelot, R. Duval, « L'alchimie syriaque au Moyen Âge », dans M. Berthelot, *La chimie au Moyen Âge*, Paris, 1893, II, p. 137.

7. Πυρίμαχος πηλός, M. Mertens, *op. cit.*, p. 32.

8. Ce *corpus* serait daté du VIIe au Xe siècle. Il serait composé à la fois de traductions de textes grecs et de commentaires plus récents, M. Berthelot, R. Duval, *op. cit.*, p. I *sqq.*

9. *Doctrine de Démocrite le philosophe* : M. Berthelot, R. Duval, *op. cit.*, p. 53 et 65.

10. Seuls les joints sont lutés avec de la cire, l'enduit, recouvrant le récipient et destiné à le protéger du feu, est une terre argileuse : « lute les joints avec de la cire. Enduis-le ensuite avec de la boue » *Ibidem.*, p. 38 ; et p. 98 : afin de conserver l'eau de sel ammoniac, on la met dans une fiole fermée hermétiquement au moyen d'un linge et de cire. On retrouve ce procédé utilisé jusqu'à aujourd'hui.

11. Ici l'œuf est entier et incorporé à de l'argile mêlée de crin, *Ibid.*, p. 94.

12. *Ibid.*, p. 38 et 53.

13. *Ibid.*, p. 45, 46 et 94.

14. « Enduis bien avec du naphte blanc et du sel ». Il s'agit sans doute du bitume de Judée qui possède la propriété de blanchir à la lumière, *Ibid.*, p. 68.

15. « Prends la terre que tu as blanchie, place-la dans un vase d'argile, lute-le avec le lut des philosophes qui est fait de [mot corrompu] et de vinaigre. Sèche le vase que tu as ainsi enduit, et introduis-le avec précaution dans le fourneau de calcination », *Ibid.*, p. 39.

16. « L'eau qui aura monté, mets-en la moitié dans une fiole. Lutes-en l'orifice avec du plâtre ou du sel. Laisse sécher [cet enduit]. Prends une marmite d'argile ; place au milieu la fiole qui a été enduite de sel. Recouvre-la avec du sel et lute la marmite avec de la terre », *Ibid.*, p. 53. Nous ne comprenons pas très bien l'intérêt d'enduire la fiole de sel. Peut-être faut-il lire « Lutes-en l'orifice avec du plâtre et du sel » ou s'agit-il du « lut blanc » préconisé dans l'alchimie arabe, lut fait d'un mélange de sel calciné et de blanc d'œuf. Pour la deuxième partie de la citation, le sel à l'intérieur de la marmite semble contribuer à maintenir la fiole au centre du vase.

parchemin[17]. Ailleurs, ce peut être du coton[18]. Les indications sur la composition du lut apparaissent souvent accessoires : elles se présentent sous la forme de mentions au hasard d'une recette sur l'or, la distillation, etc. Toutefois, le lut peut devenir lui-même l'objet de la recette, ce « plus petit élément en quoi un texte se laisse ultimement décomposer »[19] :

> « Prends de la terre de Cimole, rouge ou blanche, grasse, qui soit exempte de pierres. [...] projette dessus de l'eau salée, [...] Laisse-la sécher. Ensuite pile-la avec un bâton, comme la terre à jarres ; puis passe-la dans un crible fin, et pile-la une seconde fois dans un mortier ; puis passe-la dans un tamis de crins rudes. Ensuite humecte-la avec de l'eau, dans laquelle tu auras fait macérer du son de riz, séparé de la farine, un jour et une nuit ; puis réduis-la bien en pâte ; couvre-la un jour et une nuit. Passe du fumier propre dans un crible fin et ensuite dans un tamis de crins rudes. Tu le broieras bien dans un mortier, et tu le mêleras avec la terre, en quantité égale. Tu mettras, pour chaque livre de terre, dix drachmes de sel de cuisine et trois livres de poterie, broyée et tamisée au tamis de crins rudes, ainsi qu'une poignée de poils de bête hachés. Fais un feu supportable ; ramollis un peu la matière et traite-la convenablement. Sers-t'en pour luter : ceci est le meilleur lut des philosophes »[20].

Outre le fait que l'on rencontre pour la première fois dans le corpus arabe l'expression « lut des philosophes », la référence dans ce texte à « la terre à jarre » montre la proximité des alchimistes avec les artisans potiers. Ces derniers doivent être les fournisseurs d'appareils et de matière première, mais ils sont également en des lieux où l'on observe et interprète les techniques et la transformation des matériaux. Encore discrets dans les plus anciens textes grecs, les luts faits d'un mélange de terre et de dégraissants organiques deviennent fréquents dans l'alchimie arabe. Rāzī préconise une argile fermentée mélangée avec des excréments broyés, des poils d'animaux et du sel[21].

17. *Ibid.*, p. 64. On retrouvera cette technique utilisée par les apothicaires pendant très longtemps. Elle explique entre autre la forme de la lèvre des *albarelli* et de la plupart des pots de pharmacie : « fermés par un couvercle ou une peau serrée d'un lien autour du col », J. Mouliérac, « Objets de pharmacie sur quelques miniatures arabes », dans *Archéologie et Médecine*, VIIe rencontres internationales d'archéologie et d'histoire d'Antibes (Octobre 1986), Juan-les-Pins, 1987, p. 436.

18. Dans un traité : « Mets sur l'orifice de la bouteille un bouchon de coton et dispose-la dans le fourneau », *Ibid.*, p. 141.

19. R. Halleux, *Les textes alchimiques*, Turnhout, 1979, p. 74 (Typologie des sources du Moyen Âge occidental ; 32).

20. M. Berthelot, R. Duval, *op. cit.*, p. 152.

21. Ms. Rampur publié par H. E. Stapleton, R. F. Azo, « Alchemical Equipment in the Eleventh Century A.D. », *Memoirs of the Asiatic Society of Bengal*, 1, 4, 1905, p. 47-71, en particulier p. 63. M. A. Maqbul Ahmad, « A persian translation of the eleventh century arabic alchemical treatise : *'Ain as-San'ah wa 'Aun as-Sana'ah* », *Memoirs of the Asiatic Society of Bengal*, 8, 7, p. 419-460, en particulier p. 425. Voir également J. Ruska, « Al-Râzî's Buch Geheimnis der Geheimnisse », *Quellen und Studien zur Geschichte der Naturwissenschaften und der Medizin*, 6, 1937, p. 96. Dans la version latine du *De aluminibus et salibus*, traité anonyme, l'expression *lutum sapientiae* est très fréquente, s'il n'y a pas de recettes explicites, on trouve tout de même la mention *et pone in vase vitreato* [et] *eum circumluta argilla albleversi*, qui fait référence à une terre argileuse. La terre est parfois mêlée à du crin ou des cheveux coupés comme le signifie l'expression *lutum capillarum* : J. Ruska, *Das Buch der Alaune und Salze*, Berlin, 1935, p. 27 pour les commentaires et p. 82 pour le texte latin. Voir aussi les mentions dans la version du ms BNF lat. 6514, le seul attribuant ce texte à Rāzī : R. Steele, « Practical chemistry in the twelfth century – Rasis de aluminibus et salibus », *Isis*, 12, 1929, en particulier p. 20.

Dans les textes du bas Moyen Âge, écrits en latin, en grec, ou en langue vernaculaire, rares sont les auteurs à ne pas donner leurs recettes[22]. Les luts argileux sont nettement majoritaires. Ils semblent toutefois surtout liés à des opérations métallurgiques ou du moins à hautes températures. Dans les plus anciens traités en langue latine, comme la *Mappae Clavicula*, on trouve l'argile mentionnée à de nombreuses reprises[23]. La terre est utilisée seule, ainsi que le plâtre[24]. Dans le *De diversis artibus* du moine Théophile, l'auteur recommande de recouvrir l'intérieur et l'extérieur des fourneaux d'une argile fortement pétrie mêlée à du fumier ; il en est de même des vases en fer nécessaires à la purification du cuivre[25]. Pour l'affinage de l'or par cémentation dans un creuset fermé, cette terre particulière des alchimistes est le *lutum sapientiae* d'après Albert le Grand[26]. Dans un autre texte dont l'attribution à Albert est discutée, les luts argileux, mais également ceux à base de farine et de blanc d'œuf sont préconisés pour la sublimation et la distillation, mais ce sont des recettes ajoutées lors des premières éditions et absentes des versions plus anciennes manuscrites[27]. Indistinctement, les luts argileux sont donc utilisés pour les opérations à basses ou à hautes températures, comme ceux composé de plâtre, de gypse ou de terre très calcaire[28].

22. S'il n'y a pas de recettes de lut dans la *Summa* du pseudo-Geber, peut-être le Franciscain Paul de Tarente, on en trouve plusieurs dans sa *Theorica et practica*, comme dans le *De investigatione*. Pour la *Summa*, voir W. R. Newman, *The « Summa perfectionis » of Pseudo-Geber a critical edition, translation and study*, Leiden, 1991, IV-785 p. (Collection de travaux de l'Académie internationale d'Histoire des Sciences ; 35). Pour la *Theorica et practica* : W. R. Newman, *The « Summa perfectionis » and late medieval alchemy : a study of chemical traditions, techniques and theories in thirteenth century Italy*, Thesis, Harvard University, Cambridge, 1986, III-1, p. 136-137. Pour le *De investigatione*, il y a un chapitre sur les luts et les vases : *Ibid*, III-2, p. 248-250 et de nombreuses références à des luts argileux déposés sur la panse des appareils, par exemple, p. 14, 100, 105, 182, 181, 188, 189, 190, 192, 195 et 226.

23. C. S. Smith, J. G. Hawthorne, « *Mappae Clavicula*. A little key to the world of medieval techniques », *Transactions of the american philosophical society*, new series, 64, 4, Philadelphie, 1974, n° 15-16, p. 32, n° 74, p. 38, n° 139, p. 47, n° 209, p. 59.

24. *Idem*, n° 11 et 12, p. 31.

25. C.R. Dodwell, éd., *Theophilus, The various arts, De diversis artibus*, Oxford, 1961, p. 121 et 125 ; Cte de l'Escalopier, éd., *Théophile, prêtre et moine, Essai sur divers arts, diversarum artium schedula*, Nogent-le-Roi, 1977, p. 221 et 225-226.

26. « *Et tenaci luto, quod sapientiae lutum vocant alchimici* », dans le *De mineralibus*. A. Borgnet, éd., *B. Alberti Magni, Ratisbonensis episcopi, ordinis praedicatorum Opera omnia*, Paris, 1890, V, p. 93.

27. Les recettes sont très explicites dans l'édition du *Libellus de Alchimia* d'Albert le Grand : *Additio.* [...] *Et si sublimatorium fuerit de vitro, fiat mediantibus cineribus, & lutum fit de creta pulverizata cum farina distemperata cum albumine ovorum : Et si sublimatorium fuerit terreum fiat super carbones, & fit lutum de argilla, calce viva cum stercore equorum, cum aqua salis & papyro optime madefacto*, et plus loin : *Lutum quo juncturae claudantur fit ex cinere, argilla, sale communi dissolutis in urina. Aliqui cum albumine ovi & calce viva. Idem*, XXXVII, p. 561 et 563 ; *Theatrum chemicum, praecipuos selectorum auctorum tractatus ... continens*, Ursel, 1602, L. Zetzner, Strasbourg, 1613-1661, II, p. 507 et 510.

28. « Ensuite lutez la jointure au sommet, à l'exception du fond où est l'œuf, avec de la craie de forgeron bien mêlée avec de la rouille et de la laine brute hachée finement. Ensuite séchez bien au soleil. Après dessiccation couvrez la jointure et le couvercle avec un chiffon de lin qui retient la craie ... » Ms. de Palerme, 4Qq A10, f° 273v°-274r°, transcrit et traduit partiellement dans C. Opsomer, R. Halleux, « L'alchimie de Théophile et l'abbaye de Stavelot », dans *Comprendre et maîtriser la nature au Moyen Âge : Mélanges d'histoire des sciences offerts à Guy Beaujouan*, Paris, 1994, p. 449.

En revanche, les luts à base de farine et de blanc d'œuf sont destinés uniquement aux basses températures et plus particulièrement à la distillation afin de fermer le joint entre la cucurbite et le chapiteau. La cire est utilisée seulement pour fermer hermétiquement un vase pour la conservation des produits contenus à l'intérieur. Marcus Graecus, dans le *Liber ignium ad comburendos hostes*, lute l'alambic avec un mélange de farine de froment et de blanc d'œuf[29] et l'orifice d'un ballon de verre à l'aide de cire et de sel calciné[30]. Pour le moine franciscain Jean de Rupescissa (XIVe siècle), « le lut de sapience » se fait au moyen de farine de froment « très subtile », de blanc d'œuf et de papier humide « diligemment charpiné », c'est-à-dire coupé finement[31]. Plus loin, une fiole contenant de l'eau-forte doit être bouchée à l'aide de cire[32].

Dans les textes du bas Moyen Âge et après, les recettes de lut ne sont pas seulement incluses dans d'autres recettes. Ce sont également des recettes autonomes dans les traités techniques : *Se tu vol fare el luto de sapientia, togli terra bene tenace e seccala bene. Poi fane polvere suttile e mettilo in un poco d'acqua. Poi fa polvere de sterco de cavallo e mesteca con chiara d'ovo e sbatti bene. Poi mesteca con quell'altre cose questo se domanda loto de sapientia col quale intonica l'ampolla*[33].

Les recettes sont très nombreuses dans les textes alchimiques du bas Moyen Âge, mais aussi, en plus grand nombre encore, dans les traités imprimés de la Renaissance et jusque dans les traités techniques ou de chimie du XIXe siècle. Entre le XVIe et le XVIIIe siècle, les recettes de lut font l'objet de chapitres entiers dans certains ouvrages d'alchimie, de chimie ou dans les collections de secrets[34]. Cette pratique très répandue dans les textes imprimés est toutefois

29. M. Berthelot, *La chimie au Moyen Âge ..., op. cit.*, I, p. 126, d'après le ms. 197 de Munich.

30. ... *orificio lutato cerugi et sale combusto bene recluso et in fimo ..., Idem*, p. 111, d'après le ms. BNF latin 7156 datant du XIIIe siècle. Voir également le texte établi et publié par F. Hoefer, *Histoire de la chimie depuis les temps les plus reculés jusqu'à notre époque*, Paris, 1866-1869, I, p. 517 *sqq*. La transcription est quelque peu différente : ... *orificio lutato cera graeca et sale combusto bene recluso et in fimo ...*

31. J. Rupescissa, *La vertu et propriété de la quinte essence de toutes choses*, Lyon, 1549, p. 25. Les deux mêmes luts se trouvent dans *L'obratge dels philosophes* écrits en langue d'Oc au début du XVe siècle : *Pregnes doncas aquest esperit et gardas lo dedins .I. fiola de veyre tapatda am cera blanca communa* et : *Pueys aprop metes vostre alembic desus et luctas la junctura am pasta facha de farina distrenpratda am album de huou* [Puis mettez votre alambic sur le vase et bouchez la jointure d'une pâte faite de farine trempée de blanc d'œuf], S. Thiolier-Méjean, *L'alchimie médiévale, L'obratge dels philosophes, La Soma et les manuscrits d'Oïls*, Paris, 1999, p. 164 à 167. Mêmes mentions dans le *Testamentum* du pseudo-Lulle : M. Pereira, B. Spaggiari, *Il « Testamentum » alchemico attribuito a Raimond Lullo*, Firenze, 1999, p. 316, 320, 322, et 336.

32. J. Rupescissa, *op. cit.*, p. 31.

33. Laur. Ash. 349, f° 20v° (XVe siècle), Bibliothèque Laurenziana-Medicea (Florence). Transcription réalisée par Ricardo Córdoba que je remercie pour les recettes de lut communiquées.

34. Il y a une nette inflation du chapitre sur les luts entre les deux éditions des secrets d'Alexis, de un folio dans l'édition de 1559, on passe à plus de 3 pages dans celle de 1573 : Alexis le Piémontois, *Les secrets du Seigneur Alexis Piemontois ...*, Anvers, 1559, f° 54r° ; Alexis le Piémontois, *Les secrets du Seigneur Alexis Piemontois ...*, Paris, 1573, p. 330 à 333. Voir également les luts dans C. Gesner, *Quatre Livres des secrets de médecine et de la philosophie chymique ...*, Lyon, 1593, fos 23r° à 25r° ; J. Brouaut, *Traité de l'eau-de-vie ou anatomie théorique et pratique du vin ...*, Paris, 1646, p. 75-76 ; J. R. Glauber, *La description des nouveaux fourneaux philosophiques ou art distillatoire ...*, Paris, 1659, p. 8-12 ; C. Glaser, *Traité de la Chymie ...*, Paris, 1668, p. 57-61 ; N. Lémery, *Cours de chymie ...*, Paris, 1675, p. 22-24.

plus ancienne car déjà au XIVe siècle, Guillaume Sedacer, dans la *Sedacina totius artis alkimie*, consacre un chapitre entier à cette question avec pas moins de 12 recettes différentes[35].

Au XIXe siècle, les ingrédients de base sont toujours les mêmes. Mermet, dans ces *Manipulations de chimie*, indique comment préparer les luts. Il utilise de l'argile grasse, une pâte épaisse à base de plâtre ou dextrine et d'eau, deux tubes en verre se trouvent lutés à leur jointure au moyen de bandes de papier et de plâtre[36]. Un mélange de farine de froment et d'eau était aussi utilisé au XIXe siècle par les distillateurs de grains et de pommes de terre pour luter la jonction des appareils[37]. Enfin aujourd'hui, c'est surtout en cuisine que l'on utilise toujours des luts à base de farine et de blanc d'œuf[38].

Les matrices et le nombre d'ingrédients ajoutés

Pour cette étude, ont été rassemblées plus de 120 compositions différentes de lut sur une très longue période depuis les premiers textes grecs, soit le *Papyrus de Stockholm*, jusqu'aux recettes contenues dans la seconde édition du Cours de Nicolas Lémery et celles de l'Encyclopédie[39]. Ce recueil n'a évidemment rien d'exhaustif et sont exclues les simples mentions ou encore les recettes insuffisamment explicites pour former un ensemble cohérent et représentatif. La répartition par période est bien sûr liée aux sources disponibles, ce qui explique le faible effectif des époques les plus anciennes (figure 2).

35. *Capitulum tricesimum sextum et ultimum huius primi libri de lutis requisitis in hac arte quibus crucibuli et vasa componuntur et sigillantur* : P. Barthélemy, *La Sedacina ou l'œuvre au crible. L'alchimie de Guillaume Sedacer, carme catalan de la fin du XIVe siècle*, Paris, 2002, II, p. 274-281. Il est à noter que certains de ces luts correspondent bien à des enduits destinés à fermer les appareils ou à les protéger du feu, mais d'autres sont destinés à fabriquer des vases. Il demeure que les recettes ne diffèrent pas beaucoup de celles rencontrées ailleurs pour le *lutum sapientiae*.

36. A. Mermet, *Manipulations de chimie - métalloïdes*, Paris, 1885, p. 8, 502, 621 et 661. Voir également dans la pharmacie : F. L. M. Dorvault, *L'officine ou répertoire général de pharmacie pratique*, 6e éd., Paris, 1866, p. 1200.

37. A. P. Dubrunfaut, *Traité complet de l'art de la distillation*, Paris, 1824, p. 240. Quelques distillateurs du Cognaçais d'un certain âge m'ont assuré avoir vu leurs pères utiliser des préparations à base de farine pour luter les joints de leur alambic à la place de pièces de caoutchouc devenues défectueuses.

38. Pour la préparation du baeckeoffe en Alsace. Voir par exemple : P. Montagné, *Larousse gastronomique par Prosper Montagné, Maître cuisinier avec la collaboration du docteur Gottschalk*, Paris, 1937-1938, p. 654 : « Lut : enduit, qui durcit en desséchant, servant à assurer la fermeture hermétique d'un récipient. Le lut d'amandes se fait en mélangeant du tourteau d'amandes pulvérisé avec de la colle d'amidon. En cuisine on se sert habituellement d'une pâte de farine appelée aussi *repère* ou *repaire* ». Voir aussi, plus récente, la recette de la « poule faisane aux endives en cocotte lutée » dans P. Wells, *Le meilleur et le plus simple de Robuchon, 130 recettes présentées par P. Wells*, Paris, 1992, p. 137 *sqq.*

39. Il est difficile de répartir le corpus par aires chronologiques ou culturelles de manière satisfaisante. Dans le corpus grec, entre les IVe et Ve siècles, nous avons placé uniquement le *papyrus de Stockholm* et Zosime de Panopolis. Pour la troisième période, il s'agit de textes écrits en latin, en langue vernaculaire, ou en grec, même si certaines recettes sont plus anciennes, comme la *Mappae clavicula* (IXe-XIe s.), la plupart de ces textes sont à situer entre les XIIe et XVe siècles.

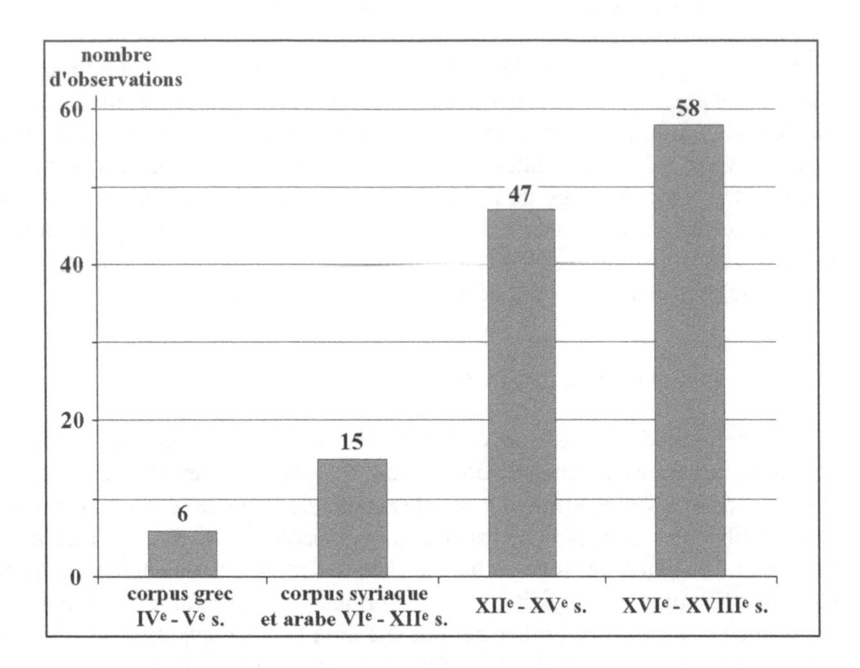

Figure 2 : répartition des recettes de lut (en nombre d'individus).

L'analyse de la composition des recettes permet de distinguer un élément principal, appelé ici matrice, des ingrédients secondaires qui sont des dégraissants, des fondants ou des agrégats en fonction de la nature de la matrice, une ou plusieurs substances humidifiant la matrice et enfin des éléments structurants. Les principaux matériaux constituant la matrice sont[40] :

– les terres ou argiles,
– les pâtes à base de farine,
– les ciments, à base de plâtre, gypse, craie ou chaux (vive ou éteinte),
– les bitumes naturels, poix (d'origine végétale), graisses et cire …

40. Seulement 16 recettes ne citent pas ces substances. Comme autres matrices utilisées seules, sont mentionnés : la cendre (3 occurrences), l'empois (3), le sel (2), le verre pilé (2), le blanc d'œuf seul (2) et enfin l'utilisation d'une vessie de bœuf ou de porc appliquée autour de la jonction cucurbite/chapiteau d'alambic (4).

La matrice ne varie guère : tous ces matériaux, à l'exception de la farine absente des textes arabes consultés, sont présents pour toutes les périodes. Chez les premiers alchimistes grecs, les mentions des quatre matrices sont également réparties. En revanche, dans les textes arabes, les luts réalisés avec une matrice argileuse sont nettement majoritaires (figure 3). S'il est possible d'émettre des réserves sur ces résultats à cause d'effectifs relativement faibles, il n'y a plus de doute concernant les luts utilisés dans le monde latin à partir du XIIe siècle : la terre est le matériau le plus souvent cité, viennent ensuite les luts à base de chaux puis ceux réalisés avec de la farine. Dans 86 % des cas, il n'y a qu'une seule matrice à laquelle sont ajoutées des matières secondaires (figure 4). Dans les 14 % restant, la matrice est constituée de deux, voire trois matériaux mélangés, toutefois ces mentions se rencontrent surtout pour les périodes les plus récentes : un quart de ces cas est médiéval tandis que les trois autres quarts sont datés entre le XVIe et le XVIIIe siècle.

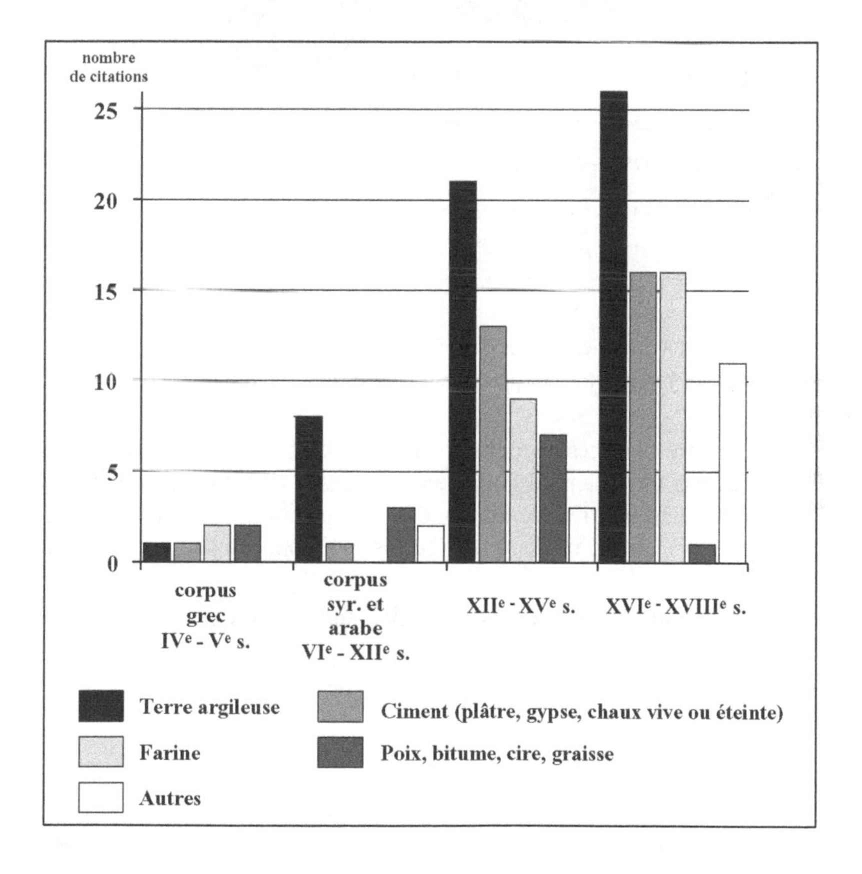

Figure 3 : matrices citées dans les recettes par périodes (en nombre d'individus).

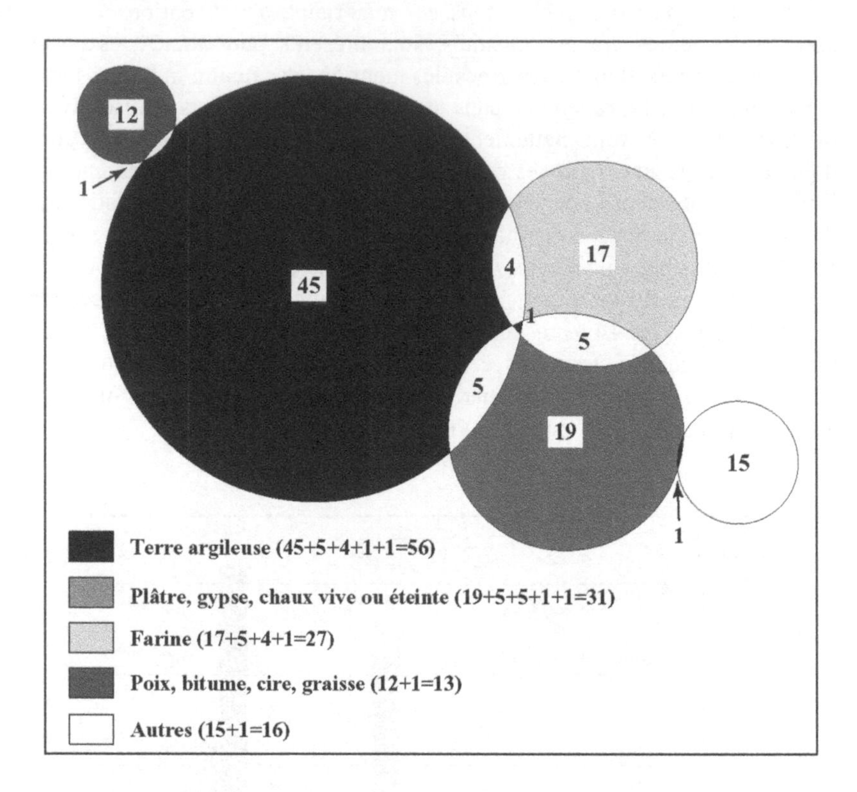

Figure 4 : matrices citées dans les recettes pour toutes les périodes, et les recettes mêlant plusieurs matrices (en nombre d'individus).

Si les mêmes matrices sont citées entre le IVe et le XVIIIe siècle, les recettes ne forment pas un ensemble homogène au regard du nombre de constituants. Ce dernier peut-être très variable : des compositions ne citent qu'un seul constituant, la matrice, tandis que d'autres mentionnent jusqu'à 8 ingrédients (figure 5).

Nous remarquons clairement une inflation surtout entre le XIIe et le XVIIIe siècle. Pour les périodes les plus anciennes, les recettes ont plutôt un nombre faible de constituants. Entre le XVIe et le XVIIIe siècle, la moyenne du nombre de constituants est autour de 4. Entre le XIIe et le XVe siècle, la situation est intermédiaire et un passage progressif sur toute la période peut être remarqué. De même, la variabilité des ingrédients peut être montrée. Cette augmentation du nombre de constituants n'est pas seulement liée aux faibles effectifs des périodes anciennes. De nouveaux produits apparaissent constamment et leurs mentions passent de 12 à 34 sur toute la période. Il est légitime de s'interroger sur l'efficacité réelle d'un lut avec 8 matériaux différents dont certains pourraient paraî-

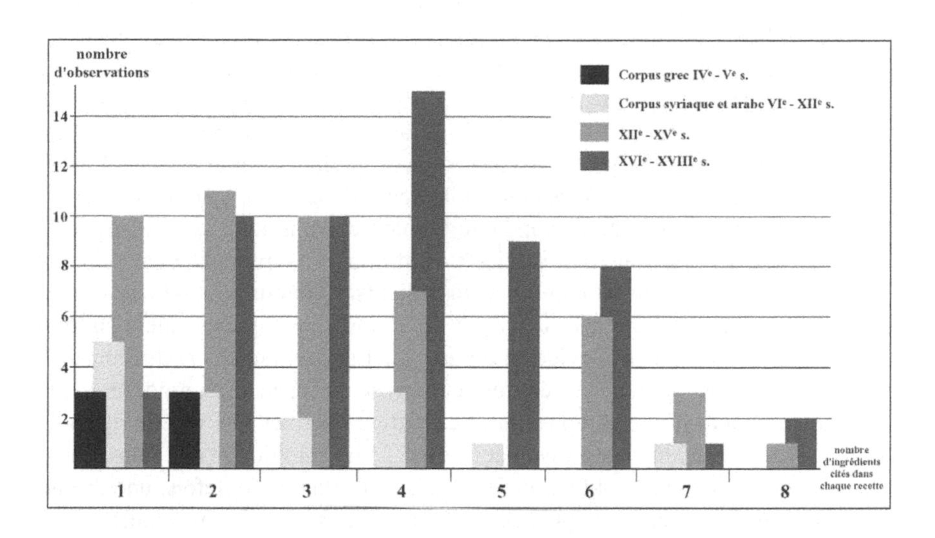

Figure 5 : distribution des recettes en fonction du nombre total d'ingrédients cités.

tre inutiles comme la poudre de soufre[41], la céruse[42] ou encore du vermillon[43]. Farfelues sembleraient les mentions de fromage ou de foie de lièvre[44] ? La question peut ne pas être posée uniquement en termes d'efficacité technique,

41. *pulveris sulphuris*, P. Barthélemy, *op. cit.*, p. 279.

42. *Recipe vitri partem I, calcis vivae partem I, tegularis pulveris partem I, cerusae partem I. Et haec omnia pulverisentur optime et impastentur cum albumine ovi.* Ms. BNF lat. n° 7147, f° 70v°, XVIe siècle (1535-1537).

43. C. Gesner, *op. cit.*, f° 24 v°.

44. « Pour faire bon lut de sapience : Recipe chaulx vive et aubin de eufz et de jus de mauves, et ung peu de foie de lièvre et les empastes ensamble très bien et destrampes de eaue de sel. Aultre leut : Recipe fromaige froys, de chaux visve trible, ensemble destrampée de gleres d'eufz et de froumaige de Moulyn, celle que est volative ; c'est bon lut ». Ms. BNF lat. n° 12993, fos 56r°-v°, cité par P. Cézard, « L'alchimie et les recettes techniques », *Métaux et Civilisations*, I, 2, 1945, p. 44. La recette cite deux fois le fromage. L'expression « froumaige de Moulyn » est curieuse. En fait, peut-être faut-il y voir une erreur du copiste. Il nous semble qu'il s'agit en réalité de « froument » (confondu avec « froumenge »), c'est-à-dire de farine de froment provenant d'un moulin, ce qui s'accorde avec la suite « qui est volatile », ce qui signifie « subtile », « fine ». En revanche, la première mention de fromage n'est pas étonnante. En effet, le fromage et la caséine rentrent dans la composition de nombreuses colles ou mastics. Pour s'en convaincre, il suffit de consulter Théophile, chap. XVII, qui fournit une recette à base de fromage et de chaux vive pour coller le bois, Cte de l'Escalopier, *op. cit.*, p. 31, C.R. Dodwell, *op. cit.*, p. 16-17, ou encore, plus récent, le recueil de recettes de F. Margival, *Colles, mastics, luts et ciments*, Paris, 1914, p. 80 *sqq.* Par exemple, p. 88, on trouve la recette suivante très proche du lut cité : « Mastic pour raccommodage de porcelaine, se prépare avec : fromage blanc frais, lavé et pressuré 120 g, blancs d'œufs 2 g, gousses d'ail 6 g, chaux vive pulvérisée 20 g ». Il y a une parenté entre la recette citée par P. Cézard et le *De investigatione* publié par W. R. Newman où on retrouve du foie animal, de la mauve, et du blanc d'œuf, toutefois, ici la matrice est une terre argileuse, et non de la chaux, et le foie est réduit en cendre : *Accipe hepatis alicuius animalis combusti et pulverizati, scorie ferri, amborum ana partem unam, argille seu crete rubee bone partes .2. Conficiantur cum succo malve et bismalve, et per noctem unam dimittantur insimul fermentari. Deinde informa vasa ad tuum libitum et sicca ea. Deinde ferruginem et bolum trita et misce cum albuminibus ovorum. Inlinias intus ipsa vase, et fac ea decoqui ; et durabunt in igne perpetua, op. cit.*, III-2, p. 249.

aspect abordé dans la dernière partie de cette étude. Ainsi, le nombre d'ingrédients et leur variabilité doivent également être liés aux préoccupations et à la représentation que l'on se fait du lut. La terminologie adoptée, comme *lutum sapientiae*, trahit une survalorisation de cette opération. Cette expression apparaît pour la première fois dans l'alchimie arabe et elle est reprise telle quelle dans les textes latins et grecs postérieurs avec quelques variantes[45]. À l'aspect technique du lut, qui est de retenir les vapeurs dans les opérations de distillation ou de sublimation, correspond une idéologie alchimique dans les représentations que l'on se fait des esprits[46]. Enfin, il y a probablement, dans le phénomène d'inflation du nombre d'ingrédients mentionnés et de leur diversité, une surenchère à associer à ce que nous pourrions appeler une « stratégie éditoriale », bien sûr plus évidente à partir de l'apparition des textes imprimés et des recueils de *Secrets*[47]. L'écriture et la mise en forme d'un procédé, et plus particulièrement d'une composition, permettent de fixer une recette qui pourrait apparaître comme normative par rapport à une tradition et une transmission uniquement orale qui serait par définition plus flottante. Toutefois, une des formes de certains manuscrits puis de livres imprimés sont des accumulations de recettes. Il faut compiler les recettes anciennes et en ajouter de nouvelles[48].

45. En arabe, on trouve : ṭīn al-ḥikma, dans les textes latins on trouve : *lutum sapientiae, lutum philosophorum* et *lutum magisterium*. Voir chez Michel Scot (XIIe siècle) : *et luta eam undique luto sapientie* ou *fenestras omnes claude cum luto philosoforum* : H. Thomson, « The texts of Michael Scot's Ars Alchemie », *Osiris*, 5, 1938, p. 551 et 556. La première expression est toutefois la plus fréquente dans cette édition du *Liber dedali* : J. Wood Brown, *Life and legend of Michael Scot*, Edimburgh, 1897, p. 249, 253, 257 et 259. Voir aussi la traduction latine du *Livre des soixante-dix*, de Jâbir ibn Hayyân, le *Liber septuaginta* : *Sume vas unum rotundum et involve totum ex luto magisterii*, M. Berthelot, *Archéologie et histoire des sciences*, Paris, 1906, p. 347. (Mémoires de l'Académie des sciences de l'Institut de France ; 49). Dans les textes grecs du bas Moyen Âge, on a les traductions : πηλὸς σοφιστικός, πηλὸς φιλοσοφικόσ, πηλὸς διδασκαλικὸς, ou encore πηλὸς τῆς σοφίας : A. Colinet, *L'anonyme de Zuretti ou l'art sacré et divin de la chrysopée par un anonyme*, Paris, 2000, p. 13 (Les alchimistes grecs ; 10) ; A. Colinet, *Recettes alchimiques (par. gr. 2419 ; Holkhamicus 109), Cosmas le Hiéromoine, Chrysopée*, Paris, 2010, p. 18 et 20 (Les alchimistes grecs ; 11).

46. Il faut retenir les vapeurs, car dans celles-ci se trouve ce qui est précieux : « Et quand tu ouvriras le dit pertuis, si tu sens une si merveilleuse odeur [...] donques sera la dite quintessence parfaite ». J. Rupescissa, *op. cit.*, p. 25. Savonarole s'inspirant de Rupescissa, confirme cette réputation faite à la quintessence : « En effet, quand tu estimeras l'avoir confectionnée après un certain nombre de jours, tu la retiendras enfermée avec le lut de sapience et après, l'orifice ouvert, toute la maison embaumera, de telle sorte qu'aucun parfum du monde ne pourra lui être comparé, comme il n'y a aucune douceur semblable à l'émission du sperme », cité par D. Jacquart, Médecine et alchimie chez Michel Savonarole (1385-1466), dans J.-C. Margolin, S. Matton, *Alchimie et philosophie à la Renaissance*, Paris, 1993, p. 113. Sur la quintessence et ses représentations voir : S. Colnort-Bodet, *Le code alchimique dévoilé, Distillateurs, alchimistes et symbolistes*, Paris, 1989, 388 p.

47. Voir note 40.

48. Par exemple Conrad Gesner : « Prenés craye ou argille ou terre, à laquelle, si d'avanture est trop grasse & gluante, meslés quelque peu de sable ou arene, adjoustez aussi des filaments de drappeaux, & fien de cheval, incorporez le tout ensemble à ce qu'il soit d'une consistence plus mollastre que dure ou seiche. Leonard Fieraventi » (C. Gesner , *op. cit.*, f° 23r°) cite dans ce passage un de ses contemporains, Leonardo Fioravanti, médecin bolonais qui fréquenta un groupe d'alchimistes en Espagne, et plus précisément le chapitre XLIII du Livre 3 de ses *Caprices* : *Si piglia creta, et se è tenace, o viscosa, vi si mette dentro un poco di arena, over sabbione, & ui si mette cimatura di panni, et sterco di cavallo* ..., L. Fioravanti, *De'Capricci medicinali dell'excellent. Medico* ..., Venezia, 1568, f°s 200r°-v°.

Outre les erreurs de copie ou de traduction, il s'agit parfois de se distinguer en apportant une « touche personnelle »[49], à l'exemple aussi de ce qui peut se passer dans le domaine de la cuisine où les recettes sont promptes à s'ajuster à des goûts et à des modes changeants[50]. Par ailleurs, il est aisé d'observer le même phénomène d'inflation du nombre d'ingrédients pour d'autres recettes alchimiques. Autre exemple de variabilité puisé dans la littérature des recettes : pour la fabrication du cinabre ou sulfure de mercure rouge par sublimation, recette que l'on peut suivre depuis les *Compositiones* de Lucques jusqu'aux textes du XVIIe ou même du XVIIIe siècle, les ingrédients ne changent évidemment pas : il faut du mercure et du soufre ...[51] Il existe toutefois une part mouvante, ou en quelque sorte un espace de liberté, qui ne se trouve pas dans les ingrédients, mais dans les proportions entre les deux matériaux, dans l'appareil à sublimer, dans le mode opératoire ou encore dans la manière de le luter. Même dans la sphère technique où le « goût du siècle et la mode du temps » n'ont pas la même importance que dans la gastronomie, en réalité, la recette écrite est loin de fixer une composition et un procédé : sa variabilité n'est pas seulement à associer directement à des critères d'efficacité.

Les ingrédients ajoutés et l'efficacité technique

L'augmentation des constituants entrant dans les compositions, le mélange de plusieurs matrices autrefois utilisées seules, le vocabulaire employé, tous ces éléments tendent à montrer que la réussite d'un bon enduit résistant au feu et aux vapeurs en demeurant relativement étanche n'est peut-être pas si évidente. La préoccupation du praticien rejoint celle de l'alchimiste, quand bien même ces deux hommes seraient séparés ... Poser la question en terme d'efficacité c'est poser la question des matériaux préconisés et évaluer leur adéquation avec les qualités recherchées. Différentes matrices sont proposées dans les textes car le lut peut avoir des fonctions très différentes. Il est utilisé dans des opérations pratiquées à des températures très variables et pour des durées iné-

49. « Plusieurs ont écrit des luts des vaisseaux, et les ont composés de tant de drogues, que huit jours ne suffiraient pas pour les faire ; pour moi, suivant ma simplicité ordinaire, voici ce que je vous conseille. Prenez de la terre à potier sèche et réduite en poudre subtile, que vous délayerez avec des blancs d'œufs bien battus ; un peu de bourre ouverte, de la limaille de fer bien déliée, ou du sable, et un peu d'urine, pétrissez le tout ensemble en consistance de pâte molle, et en lutez vos cornues et matras, ou autres vaisseaux que vous laisserez sécher doucement à l'air sans feu, ni soleil. Ce lut résiste au feu. Pour refaire les vaisseaux cassés, vous réduirez la chaux vive en poudre, et délayerez avec du blanc d'œuf ; vous tremperez un linge bien délié dedans, et l'appliquerez promptement sur les cassures. Pour luter les récipients et les courges avec leurs chapiteaux, il ne faut que de l'empois et du papier », J. Jacques, *Marie Meurdrac – La chymie charitable & facile, en faveur des Dames – 1666*, Paris, 1999, p. 45-46.

50. B. Laurioux, *Le règne de Taillevent, Livres et pratiques culinaires à la fin du Moyen Âge*, Paris, 1997, p. 215-264. A. R. Girard, « Du manuscrit à l'imprimé : le livre de cuisine en Europe aux XVe et XVIe siècles », dans J.-C. Margolin, R. Sauzet, *Pratiques et discours alimentaires à la Renaissance*, Paris, 1982, p. 107-117.

51. H. Hedfors, *Compositiones ad tingenda musiva*, Uppsala, 1932, p. 32.

gales. Pour la distillation *per ascensum*, les températures ne dépassent guère une centaine de degré, pour une sublimation, elles peuvent largement atteindre quelques centaines de degré, dans les opérations métallurgiques qui entraînent la fusion du métal, elles peuvent être supérieures à 1100 °C pour des durées relativement longues parfois répétées. Dans certains cas, le lut sert à fermer hermétiquement un appareil, dans d'autres, appliqué sur toute la panse d'un vase, il est destiné à lui conférer de meilleures propriétés réfractaires. Les enduits composés de chaux, de gypse ou de plâtre risquent de subir des transformations importantes à hautes températures et de complètement se désagréger au-delà de 800 °C[52]. Ils peuvent recouvrir la panse d'un appareil, mais pour des opérations à des températures modestes seulement, comme les sublimations ou certaines distillations *per descensum*. Appliqués sur la panse, et surtout pour des températures importantes, sont toujours préférés les luts argileux qui sont des matériaux céramiques réfractaires. Dans le cas de la distillation, pour fermer le joint entre le chapiteau et la cucurbite, un lut à base de farine ou de chaux est largement suffisant et peut être très efficace. Les produits organiques ajoutés permettent de réaliser des colles tenaces qui rendent les appareils hermétiques. En outre, si l'obsession est de conserver les vapeurs, dans la réalité un lut parfaitement étanche poserait des difficultés de pressions internes : l'astuce décrite par G. Agricola pour palier les problèmes de pression témoigne indirectement d'une réelle efficacité des luts utilisés[53]. Au-delà de la question de la matrice et de l'utilisation du lut, se pose donc la question des ingrédients ajoutés en fonction des fins poursuivies. La multiplicité des composants rend difficilement prédictible le résultat final. Une substance minérale ou organique ajoutée peut avoir des comportements très différents en fonction non seulement de la température, mais également des autres corps en présence dans la pâte. Il peut se comporter comme un dégraissant, puis comme un fondant au-delà d'une certaine température.

En considérant les ingrédients cités plus de cinq fois dans l'ensemble des recettes du corpus, 14 groupes différents se dessinent (figure 6). En dehors de la question de l'humidifiant, qui est le plus souvent de l'eau afin d'obtenir une

52. Pour les luts à base de chaux vive (CaO), il y a également une difficulté pratique à les mettre en œuvre, comme nous l'avons testé : c'est la réaction exothermique en présence d'eau : le dégagement de chaleur est très important ! De plus, tous les matériaux cités dans les recettes comme la calcite, la craie ou encore le marbre perdent leur CO_2 aux alentours de 825 °C et au cours du refroidissement le CaO va se recarbonnater lentement. Au moment du départ du CO_2, il y a donc une forte augmentation de volume ou foisonnement qui réduit en poussière le lut, il est donc peu efficace à haute température. De même, il faut un séchage très lent afin que le départ de l'eau ne soit pas trop violent avant d'utiliser un tel lut. Pour le plâtre et la chaux, voir également les expérimentations dans H. de Contenson, L. Courtois, « à propos des vases en chaux, recherches sur leur fabrication et leur origine », *Paléorient*, 5, 1979, p. 177-181.

53. Agricola livre ce que nous pouvons appeler un « truc » ou un tour de main : au niveau de la jointure entre le bec de l'alambic et le réceptacle à distillat, il fixe un petit clou de fer (ou une cheville de bois) à peine plus gros qu'une épingle à travers le lut. L'opérateur doit le retirer quand trop de vapeurs se sont rassemblées dans la partie supérieure de l'alambic. Hoover H. C., Hoover L. H., éd., *Georgius Agricola, De re metallica, translated from the first latin edition of 1556*, New York, 1950, p. 441.

pâte plastique, et du structurant sur lequel nous reviendrons, l'élément le plus cité est le blanc d'œuf, un très bon liant qui a des propriétés collantes excellentes. Le lut doit adhérer sur son support et durcir après séchage. Le blanc d'œuf est utilisé associé à différentes matrices. Si son association à la farine est bien connue du fait de son utilisation en cuisine, en revanche le mélange de blanc d'œuf et d'un carbonate de calcium ou d'une argile est moins évident[54]. Ce sont là également les propriétés collantes qui sont désirées. Le sang,

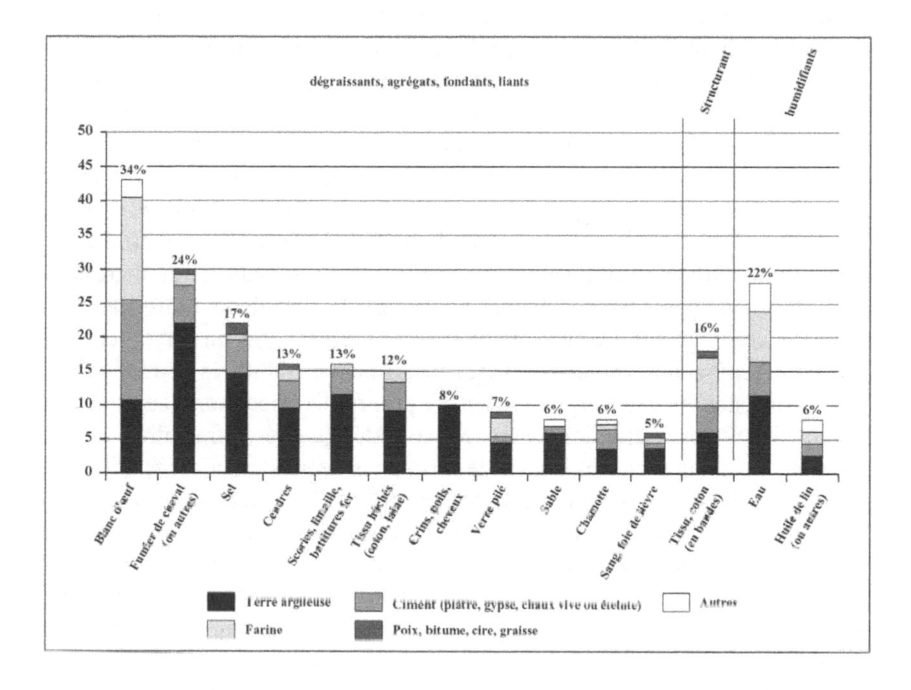

Figure 6 : les ingrédients ajoutés à la matrice cités plus de 5 fois dans tous le corpus (en nombre d'individus et en pourcentage), et matrice associée (en pourcentage).

54. Le blanc d'œuf et la chaux donnent une colle efficace d'après F. Margival, *op. cit.*, p. 98 : « le blanc d'œuf battu en neige et mieux le liquide provenant du dépôt de ce blanc pendant vingt-quatre heures à la cave est mélangé avec de la chaux vive pulvérisée. La pâte ainsi obtenue forme une colle très énergique et très rapide ». Cette colle est utilisée pour réparer les vases en verre : « On colle le verre, la porcelaine avec un lut obtenu en mélangeant vivement de la chaux éteinte avec du blanc d'œuf ; ce lut durcit très rapidement, aussi doit-il être employé aussitôt qu'il est préparé », E. Bouant, *Dictionnaire manuel illustré des connaissances pratiques*, 4e éd., Paris, 1901, à l'article « lut ». Voir également J. Chardin, *Voyages en Perse*, Paris, 1811, IV, p. 147 (1re éd. 1671-1681) : « Je ne dois pas oublier qu'ils ont en Perse l'art de recoudre le verre fort adroitement [...] pourvu que les morceaux ne soient pas plus petits que l'ongle, ils le cousent ensemble avec du fil d'archal, et passent par-dessus la couture du blanc de plomb ou de la chaux calcinée, avec du blanc d'œuf ; ce qui fait que l'eau ne sauroit du tout passer au travers » (le fil d'archal est du fil de laiton). En Égypte, cette pratique a également été observée : « Enfin nous avons remarqué qu'ils ont l'art de recoudre le verre avec du fil d'archal et de couvrir cette suture de blanc de plomb ou de chaux, l'un ou l'autre délayé par le blanc d'œuf ». J.-P. Boudet, « Notice historique sur l'art de la verrerie, né en Égypte », dans E. Jomard, *Description de l'Égypte ou recueil des observations et des recherches qui ont été faites en Égypte pendant l'expédition de l'armée française* ..., Paris, 1813-1822, XXII, p. 27.

comme l'huile de lin qui dans les textes apparaît comme un humidifiant, sont certainement employés pour les mêmes raisons. Même s'ils peuvent surprendre, certains produits mentionnés ne sont donc pas inappropriés, bien au contraire. Ils permettent de fabriquer des colles parfaitement adaptées à des températures modestes.

Les luts en terre, les plus fréquemment cités et les plus affectés par une variabilité des composants, sont le résultat d'une convergence d'intentions multiples. Celles-ci ne sont pas tout à fait les mêmes que pour un potier. Il est évident que le lut doit, avant tout, résister le mieux possible à des températures parfois élevées pendant des temps de cuisson longs. Pour sa mise en œuvre, la pâte doit présenter de bonnes aptitudes au façonnage. Lors de l'utilisation de l'appareil luté, les fissures et le décollement du lut doivent être évités, car la contraction en volume est à l'origine de contraintes dans le joint entre le lut et la surface du vase. Ces contraintes sont liées à deux phénomènes : la perte naturelle en eau de l'enduit lors du séchage et les variations thermiques au cours de la cuisson. Certaines argiles remplissent aisément la première condition, à savoir, posséder une bonne résistance à la chaleur. Mais l'argile est déposée crue sur une surface relativement stable : verre ou terre déjà cuite. Quand la température s'élève, le vase, en lui-même, subit très peu de variation de volume, tandis que l'argile crue perd progressivement son eau et voit son volume diminuer. Plasticité, adhérence, solidité, qualités réfractaires, un minimum de retrait au séchage, facilité de mise en œuvre avant et après utilisation : voilà les principales qualités recherchées du lut. Sa composition est donc le résultat de compromis entre ces diverses caractéristiques. Les ingrédients comme la chamotte ou le sable sont des dégraissants reconnus pour réaliser un bon réfractaire[55]. Les dégraissants non plastiques limitent le retrait de la matrice et l'apparition de fissures. De même, les dégraissants organiques comme le fumier de cheval, le crin ou les tissus hachés, vont limiter les fissures lors du séchage[56]. Tout en atténuant le retrait, ils peuvent avoir d'autres

55. L'adjonction de céramique déjà cuite, ou chamotte, à une argile crue est très répandue. L'explication en est donnée par Albert le Grand dans le *De Mineralibus : Est autem durissimum illud caementum : quia quotiescumque decoquitur optesi siccum terrestre, calcinatur et iterato optesi supra modum durum efficitur, quasi inconsumptibile per ignem. Hoc autem in operibus probatur artis, ubi testa contrita et calcinata, et iterum humido commixta in testam formata et optesi coagulata durissima efficitur et inconsumptibilis ab igne : propter quod taliter facta vasa quaerunt artifices cum fundunt metalla.* A. Borgnet, *op. cit.*, p. 21. Pour la fabrication des creusets, le fait est largement attesté par l'archéologie, notamment au bas Moyen Âge : I. Katona, D. Bourgarit, N. Thomas, A. Bouquillon, « From furnace to casting moulds : an exceptional 14th century copper-metallurgy workshop studied in the light of refractory ceramic materials », dans *Proceedings of the 8th European Meeting on Ancient Ceramics (EMAC '05)*, Lyon 25-29 octobre 2005, Maison de l'Orient et de la Méditerranée, Lyon, sous presse. I. C. Freestone, « Refractory Materials and their procurement », dans A. Hauptmann, E. Pernicka, G. A. Wagner, éd., *Archäometallurgie der Alten Welt/Old World Archaeometallurgy, Der Anschnitt*, Beiheft 7, Deutsches Bergau-Museum, Bochum, 1989, p. 155-162.Voir également J.-C. Échallier, *Éléments de technologie céramique et d'analyse des terres cuites archéologiques*, Lambesc, 1984, p. 13 *sq.* (Documents d'archéologie méridionale, série méthodes et techniques ; 3).

56. Le fumier de cheval, séché et tamisé, est finalement de l'herbe très finement hachée.

conséquences comme un enrichissement de la pâte en carbone et surtout une augmentation de la porosité. En effet, les éléments végétaux vont brûler lors de l'utilisation, laissant des vides et des empreintes caractéristiques lors de la cuisson. Ces matériaux organiques sont abondamment utilisés en métallurgie pour fabriquer des moules de fonderie, mais parfois aussi en céramique[57]. D'ailleurs, le médecin Fioraventi fait lui-même un rapprochement explicite entre le *lutum sapientiae* et la terre utilisée par les fondeurs pour fabriquer leurs moules[58]. Or, la porosité d'un matériau céramique est inversement proportionnelle à sa conductibilité thermique[59]. Le lut enrichi en matières végétales devient un isolant thermique. Déposé sur la panse d'un vase, il l'isole du contact de la source de chaleur et répartit les gradients thermiques tout autour de celui-ci. Il s'agit sans doute d'une des propriétés recherchées par les praticiens, du moins intuitivement. Le lut permet de limiter les chocs thermiques en homogénéisant la température autour du vase. De plus, la porosité fait baisser la résistance à la traction lors d'un choc thermique en favorisant l'apparition de fissures[60], mais limite dans le même temps leur propagation et donc une rupture catastrophique, alors que le vase est un matériau plus dense, moins sujet à l'apparition de fissures, mais si une fissure survient, elle se propage rapidement entraînant la rupture de l'appareil. Ces deux résistances, l'apparition de fissures et leur propagation, sont des propriétés qui s'opposent dans une céramique. Le lut apparaît donc comme un moyen de résoudre ces difficultés.

L'ajout de fondant dans les recettes de lut argileux semble plus problématique. Les fondants vont abaisser la température de fusion du matériau à une certaine température, alors qu'une autre des propriétés recherchées d'un réfractaire est sa résistance au fluage, c'est-à-dire à la fusion du matériau. Les fondants très couramment cités sont le sel, le verre pilé mais également la cendre et peut-être aussi le fer sous forme de battitures par exemple. Pourtant, les

57. I. Katona *et al.*, *op. cit.* ; M. S. Tite, V. Kilikoglou, G. Vekinis, « Strength, toughness and thermal shock resistance of ancient ceramics, and their influence on technological choice », *Archaeometry*, 43, 3, 2001, p. 301-324. Pour la céramique scandinave (XIIe-XIVe siècle), mise au jour en Écosse. D. R. M. Gaimster, « Dung-Tempering ? A late Norse case study from Caithness », *Medieval Ceramics*, 10, 1986, p. 43-47. Le même auteur donne également un exemple ethnographique du centre de l'Inde.

58. « ... il faut premierement faire les moules de terre, d'un tel artifice qu'ils ne crevent, & soient fermes contre la chaleur du métal fondu : la terre pour faire les moules s'accoustre en cette maniere, l'on prend de la terre craye seiche on la pile & estant pilée on la passe : puis on la destrempe avec eau, y adjoustant de la fiente de cheval, & cimature, la meslant tresbien ensemble. Et cela s'appelle bouë de sapience, dont on fait les moules des cloches, des artilleries, des mortiers & autres grosses choses. Et puis l'escorce de dessus se faict de quelques sortes de terre plus menuë, comme de craye destrempée avec glaire d'œufs, ou pierre ponce bruslée, & mise en paste ou destrempeé, pour jetter choses subtiles & délicates ». L. Fioravanti, *Miroir universel des Arts et Sciences ..., mis en François par Gab. Chappuys, Tourangeau ...*, Paris, 1586, p. 107. « Jeter » signifie ici couler l'alliage en fusion.

59. G. Aliprandi, dir., *Matériaux réfractaires et céramiques techniques, I. Éléments de céramurgie et de technologie*, Paris, 1979, p. 122.

60. Les vides laissés dans la pâte, par la carbonisation et la volatilisation des éléments organiques, sont autant de fissures préexistantes dans le matériau. Ces dégraissants limitent donc l'apparition de fissures lors du séchage, mais en augmentent considérablement le nombre au cours de l'utilisation.

mentions sont récurrentes et l'on ne peut y voir des ajouts fantaisistes des rédacteurs. Est-ce qu'un état pâteux à hautes températures, à la surface du lut, ne serait pas désiré ? En effet, la vitrification de la surface externe du lut évite le décollement du matériau du vase ou l'effritement. La cohésion peut être ainsi assurée à très hautes températures dans les opérations métallurgiques, pour peu que le lut ne devienne pas totalement liquide et ne s'écoule pas complètement dans le foyer. Cette vitrification doit être contrôlée et limitée seulement à la surface.

Enfin, dans les recettes est très souvent préconisé l'ajout de tissu ou de papier que l'on enroule autour du vase et que l'on recouvre de l'enduit. Ce structurant à l'échelle macroscopique permet de maintenir le lut et d'éviter son décollement. Il est utilisé avec toutes les matrices citées. Dans le cas des opérations métallurgiques menées à hautes températures, le structurant va se carboniser lentement laissant à la fin des vides et des empreintes. Toutefois, son action est surtout nécessaire lors du séchage et de la cuisson du matériau, donc au début de l'opération.

La mise en œuvre

Les textes ne fournissent pas uniquement des compositions. Ici et là, se trouvent des informations sur le mode de mise en œuvre et sur l'épaisseur des luts. Il est parfois indiqué de plonger directement le récipient dans la pâte liquide, de le recouvrir de plusieurs couches d'enduit en laissant sécher entre chaque opération. Quelquefois, des bandes de tissu ou de papier sont enroulées autour du vase après avoir été trempées dans la pâte ou dans du blanc d'œuf. Ainsi, dans l'anonyme de Zuretti, la description du processus accompagne une recette de sublimation, ce sont effectivement les propriétés collantes du blanc d'œuf qui sont recherchées dans ce lut argileux :

> « Le lut de sapience se fait comme suit. Prenez de l'argile à pots, pure de cailloux, de l'argile mélangée blanche et rouge ; ajoutez-y une part de sel et de cheveux coupés avec des ciseaux ainsi que de la cendre et des charbons bien pilés, pour obtenir de la colle. Avec l'argile, enduisez trois fois l'extérieur du vaisseau et laisser sécher. [...] Lorsque vous voulez serrer les jointures de l'aludel, faites auparavant un peu de lut de sapience dans le bas et laissez sécher. Prenez ensuite de la toile de lin, enduisez-la de colle de l'art et collez de l'argile au-dessus pour entourer l'aludel ou la coupelle une fois ou deux. Et si vous n'avez pas de tissu, mettez du papier à la place. Quant aux orifices des urinaux ou des ballons de verre, bouchez-les d'abord avec du coton, ensuite avec de la colle de chaux et de blanc d'œuf ... »[61].

61. A. Colinet, *op. cit.*, p. 65. La recette de *La colle de l'art* n'est pas explicitement donnée dans toute l'œuvre. Toutefois, il pourrait probablement s'agir d'une préparation à base de blanc d'œuf et de plâtre.

Avant utilisation, il est souvent recommandé de laisser sécher le lut, comme le préconise explicitement le *De aluminibus et salibus* : *Postea circumlinias illud plumbum luto sapientiae et permitte siccari*[62]. Luter un vase correctement s'inscrit donc dans une chaîne opératoire complexe et longue, surtout quand il faut plusieurs couches et laisser sécher entre chaque application. Agricola conseille d'enduire la surface d'un pot d'argile liquide huit ou dix fois de suite en laissant sécher après chaque couche. Chaque pellicule de terre doit avoir l'épaisseur d'une lame de couteau, l'épaisseur totale atteignant le travers d'un pouce[63]. Blaise de Vigenère demande de recouvrir la surface d'un chapiteau d'alambic d'un enduit d'un « bon pouce » d'épaisseur [64]. À la fin du XVIIIe siècle, en Égypte, la fabrication du sel ammoniac par sublimation est pratiquée dans des ballons de verre lutés au moyen d'une argile mêlée de laine hachée : quatre couches de 3 à 4 cm d'épaisseur au total sont nécessaires[65].

L'iconographie des appareils de chimie est très abondante, tant dans les manuscrits médiévaux que dans les recueils imprimés[66]. Toutefois, le *lutum sapientiae* n'est guère illustratif : aucune ne semble lui accorder de place, à l'exception de la *Pyrotechnie* de Biringuccio où deux gravures y sont consacrées (figures 7 et 8)[67]. Sur la première, un homme prépare les ingrédients en les passant au crible tandis qu'un autre bat le mélange fortement avec un bâton sur une table. Un troisième applique l'enduit sur la surface de plusieurs pots, tandis que le dernier vérifie très certainement la bonne et égale application du lut sur l'ensemble du pot en regardant à l'intérieur. Sur la seconde gravure, les pots lutés sont mis à sécher au soleil, retournés sur un banc conçu à cet effet. Au XIXe siècle, Mermet indique aux étudiants en chimie comment luter un creuset. Le joint entre le creuset et son couvercle est luté au moyen d'un boudin d'argile appliqué fermement au niveau de la jointure (figure 9)[68].

62. J. Ruska, *op. cit.*, p. 61. [...] *et luta enim lute saphientie, et sicca ipsum, et pone ovum in igne stercoris vacce* ..., R. Steele, *op. cit.*, p. 33.

63. Hoover H. C., Hoover L. H., *op. cit.*, p. 441. Un traité arabe donne une indication d'épaisseur, mais la traduction est loin d'être claire : *Half the pot should be luted with the 'Clay of Wisdom' to the thickness of a thumb, and on the other half there should be made for it* [4 handles] *of clay. Dry the sides of the Aludel* ..., M. A. Maqbul Ahmad, *op. cit.*, p. 426. La moitié de l'aludel est enduite sur une épaisseur d'un pouce, l'autre le serait sur l'épaisseur de 4 paumes ?

64. B. de Vigenère, *Traicté du Feu et et du Sel* ..., Paris, 1618, p. 94.

65. « Les ballons [fabriqués spécialement], avant d'être remplis de la suie [produite par combustion de fiente de bétail] dont on retire le sel ammoniac par sublimation, sont couverts d'un enduit de terre mêlée de laine hachée, de 3 à 4 centimètres d'épaisseur. Cet enduit est formé de quatre couches successives, dont chacune est desséchée au soleil en y restant exposée pendant deux jours », E. Jomard, *op. cit.*, VIII, p. 611 *sqq*.

66. J. Van Lennep, *Alchimie, Contribution à l'histoire de l'art alchimique*, Bruxelles, 1985.

67. C. S. Smith, M. T. Gnudi, *Biringuccio. De la Pirotechnia 1540*, New York, 1990, p. 184 et 197.

68. A. Mermet, *op. cit.*, p. 622.

Figures 7 et 8 : manière de luter des cucurbites
pour la distillation des acides d'après Biringuccio, *De la Pirotechnia*, 1540.

Figure 9 : manière de luter un creuset d'après Mermet, *Manipulations de chimie*, 1885.

Conclusion

Le lut n'est évidemment pas au centre des problématiques posées en histoire des sciences et des techniques. Jusqu'à présent, il n'avait que peu suscité de curiosité auprès des spécialistes n'étant qu'un moment dans une chaîne opératoire qui se trouve en marge de l'opération et des fins de l'utilisateur. De plus, on peut penser qu'il suffit de prendre de l'argile dans un coin de l'atelier et d'en enduire les pots avant utilisation. Le procédé peut sembler trivial, donc sans intérêt ... Et pourtant, les textes montrent que le lut peut faire l'objet de préoccupations importantes. L'abondance des mentions et des recettes témoigne, en fait, de la difficulté des procédés. Les compositions appliquées sur des jointures sont différentes de celles posées sur une panse au contact du feu. Les constituants varient parfois en fonction des produits utilisés. Les techniques de mise en œuvre sont complexes. La fréquentation des textes suggère aussi l'existence de tours de main, sous la forme d'un ajout dans la pâte ou d'une astuce pour la mise en place. Ces références mettent au jour les problèmes et les enjeux. Ces derniers sont assez évidents, du lut dépend la réussite d'une opération. Un échec a un coût que nous ne pouvons certes pas évaluer, mais que l'on songe au prix d'appareils qui sont commandés et fabriqués spécialement pour certains usages.

Le *lutum sapientiae* évoque bien sûr l'idéologie alchimique, mais aussi tout l'intérêt porté à ces produits dont certains sont des réfractaires. Ce sont ces compositions soumises aux plus hautes températures qui semblent poser le plus de difficultés si l'on en juge par le nombre de mentions et la variété des recettes. Toutefois, nous aurions tort de croire que le lut est seulement destiné à palier la mauvaise tenue au feu de vases ou de creusets. De même, le lut ne concerne pas seulement les vases en verre, ce dernier matériau pouvant être moins résistant que la céramique. En réalité, ce sont les conditions d'utilisation qui motivent le procédé et non la qualité des appareils[69]. Le lut n'est pas, comme on pourrait le penser, l'indicateur d'une technologie peu élaborée. Ce n'est pas un produit simple à fabriquer et son efficacité dépend de son comportement thermomécanique. Celui-ci varie en fonction de nombreux paramètres comme sa composition et évidemment la température d'utilisation. La composition d'un lut dépend aussi de l'opération projetée. L'élaboration du mélange nécessite donc un savoir-faire et une connaissance intime des matériaux.

Ces remarques suggèrent un mode d'acquisition particulier des connaissances. Celui-ci est évidemment empirique. La composition d'un enduit dépend

69. Par exemple, l'étude de creusets recouverts de lut découverts à Melle et à Niort montre que la terre utilisée pour la fabrication des creusets est beaucoup plus réfractaire que celle du lut. D'autre part, dans les mêmes contextes archéologiques, des creusets sont lutés et d'autres non, ce qui tend à montrer une utilisation particulière. N. Thomas, « Quand Melle enterrait ses métallurgistes – Apports et limites de l'étude de creusets lutés découverts à Melle et à Niort en contexte funéraire (Deux-Sèvres, France) », *Archéosciences – Revue d'archéométrie*, 30, 2006, p. 45-59.

de l'accord de ses constituants, or l'expérience apparaît comme l'unique moyen de parvenir aux équilibres recherchés. La transmission des procédés devait probablement avoir pour cadre l'atelier ou le laboratoire. L'apprentissage des gestes, des savoir-faire et finalement du savoir se faisait donc au contact du feu. Il faut certainement minimiser le rôle joué par les recueils de recettes dans les mécanismes de transmissions. Il peut s'agir d'aide-mémoire destinés au rédacteur lui-même ou à ses confrères. Le texte est le complément d'enseignements oraux et de travaux pratiques. L'inflation des recettes et la variabilité des substances ajoutées, suggèrent des tentatives nombreuses, des compositions acquises au moyen de l'expérimentation, même si le désir de posséder sa recette personnelle est certainement tangible à l'analyse des textes. Dans tous les cas et même à partir de la diffusion des recueils imprimés, la pratique demeure irremplaçable[70]. Dans le cas des réfractaires, si aujourd'hui nous avons la connaissance théorique des phénomènes physico-chimiques, il n'en demeure pas moins que toute élaboration d'un matériau passe par le mode expérimental.

Les alchimistes ne nous disent pas pourquoi et comment le lut parvient à rendre plus réfractaire un appareil. Les lois régissant les comportements thermiques des matériaux ne sont évidemment pas connues. Toutefois, dans le domaine de la physique, les artisans ont eu l'intuition de phénomènes très complexes. Le lut, quand il ne sert pas à rendre hermétique, est un moyen d'homogénéiser les flux de chaleur, de limiter les chocs thermiques et d'éviter les fissurations catastrophiques.

Loin de représenter des processus simples, le lut apparaît finalement comme une prouesse technique. Tout en se plaçant en marge des opérations alchimiques et métallurgiques, ces procédés témoignent donc de savoir-faire très évolués.

70. C'est ce que reconnaît par exemple Glaser dans son épître à « Messire Antoine Vallot », quand il écrit : « Il y a quelque temps que je fis mettre sous la presse un petit Traité de Chymie pour la commodité de ceux qui assistent aux Leçons que j'en fais tous les ans par vos ordres au Jardin du Roy ... » C. Glaser, *op. cit.*

Late Medieval Italian Recipes for Leather Tanning[1]

Ricardo Córdoba de la Llave

Among medieval texts containing descriptions of industrial recipes, there is a small group devoted to hide tanning. Most of them are brief instructions found in miscellaneous texts, and medical, cosmetic or dietetic recipe books, among whose pages there are references to the tanning and staining of hides. But two of these texts, dating from the 15th and 16th centuries, are veritable handbooks on this subject as they compile a good number of recipes on the same topic and insert them in chapters or parts of a work exclusively devoted to working the hide. All the texts we have selected originate from Italy; dating from the 14th, 15th and 16th centuries they focus on the alum tanning of leather, a modality which was used, among other applications, for making clothes, gloves and apparel accessories.

In the National Library of Florence we find the most copious section of technical manuscripts, the so-called Fondo Palatino, among which are texts on tanning in numbers 867, 916, 934, 941 and 1026, all of them dating from 1450 and 1550. Gabriella Pomaro transcribed some of these recipes in her well-known study on the technical manuscripts of the library's collection (to be specific, those from manuscripts 916, 934, 941 and 1026), while the ones included in 867 remain unpublished.[2] The National Library texts are complemented with some others kept in the Riccardiana Library (MS 1243, 1247, 3052 and 3053) and in the Medicea-Laurenziana (Ricettario del Fondo Antinori, no. 14) which had not been published up to now either. Among the latter, the MSs 1243 and 1247 from the Riccardiana date, the same as those kept in the

1. The present work has been conducted within the framework of project HAR2012-37357, *Il conocimiento cientifico y tècnico en la Peninsula Ibèrica (siglos XIII-XVI)) : producción, difusióny aplicaciones*, funded by the Ministry of Economy and Competitivity. The references taken from unpublished manuscripts conserved in Florentine libraries have been obtained thanks to the carrying out of the project PR2004-0187, *Literatura técnica en la Italia bajomedieval (siglos XIII-XVI)*, financied by the Ministry of Education and Science.

2. G. Pomaro, *I ricettari del Fondo Palatino della Biblioteca Nazionale Centrale di Firenze*, Giunta Regionale Toscana, Firenze, 1991. Cap. Lavorazione della pelle, pp. 139-142.

National Library, from the 15th and 16th centuries, but no. 3052 from the same Library and the Antinori 14 from the Laurenziana Library originate from the 14th century and are therefore the earliest recipes of those studied in this work.

However, already in the first half of the 15th century we find a text in which there are a large number of recipes on working hides; this is the famous *Bolonia's Manuscript* also called *Segreti per Colori*, published in the 19th century by Mary Merrifield.[3] This is a small volume with paper pages conserved in the library of San Salvador de Bolonia, written in Latin and the Italian of Lombardy (northern Italy). The text, compiled by an anonymous author, is ordered by recipe themes probably taken from diverse sources and origins, which were systemized by chapters; at the end of each chapter there are usually some recipes written in a different hand, which were possibly added after the death of the first compilator and half a century later. It is a system treatise which, throughout its first five books, deals with the preparation of pigments and the application of colours, although many other subjects are approached in the following books. The chapters devoted to leather working are found at the end of the treatise, once all the questions related to the preparation of dyes and colours for textiles, the illumination of manuscripts, the composition of glassware and pottery, etc., have been elucidated, so that they form a sort of final chapter of the handbook. They consist of 33 chapters structured in two large sections as the first 17 refer to dyeing techniques and the remaining 16 to alum tanning methods used to make chamois, but in no case is there any information on vegetable tanning.

Already the 16th century, specifically in 1548, saw the appearance of the last of the handbooks studied by us. This was the well-known handbook by Gioanventura Rosetti, *Plictho dell'arte dei tintori che insegna tenger pani, telle, banbasi et sede si per l'arthe magiore come per la comune*, a treatise describing the dyeing of silk, wool, cotton, cloth and hides.[4] Rosetti's work was the first printed handbook devoted entirely to cloth and hides; that is why its importance in the history of technology can be compared, according to Franco Brunillo, to that of the handbooks of Biringuccio and Agricola in the field of metallurgy. The word *Plictho* has originated many discussions and there are

3. M. P. Merrifield, *Original treatises dating from the Twelfth to the Eighteenth Centuries on the Arts of Painting*, London, 1849 (New York, 1967, Dover Publications Inc.), vol. 2, pp. 340-600; introductory study pp. 327-339. Also published by O. Guerrini, C. Ricci, *Il Libro dei Colori. Segreti del secolo XV*, Bologna, Comisione per i testi di lingua, 1969.

4. First edition of *Plictho* in Venice 1548, also published at the beginning of 20th century by I. Guareschi, *Storia della chimica. Sui colori degli antichi*, Turín, 2 vols., 1905-1907. The most useful version is the english edition, with the original italian, by Sidney M. Edelstein & Hector C. Borghetty, *The Plictho of Gioanventura Rosetti. Instructions of the Art of the Dyer*, Cambridge Mass., MIT, 1969; in this book are studied the processes of tanning and dyeing related with this text, and in the paper "Dyeing and Tanning Leather in the Sixteenth Century," *American Dyestuff Reporter*, 54, 1965, pp. 48-52. Chapters on hide dyeing and tanning studied and transcript by F. Brunello, *Concia e tintura delle pelli nel Veneto dal XIII al XVI secolo*, Vicenza, Ente Fiera, 1977, pp. 63-81.

those who have considered that it was an old Greek or Venetian term with an unknown meaning or that it was the name of the author. The term *Plictho* is the forerunner of the Italian word *plico* which means a collection of recipes or instructions. The *Plictho* was published in the Venice of the Cinquecento and was written by an employee at the Venetian Arsenal, probably born in Venice itself, who, in about 1530, worked in the shipyards, where he remained at least until the middle of the century. The only known work published by him apart from the *Plictho* was entitled *Notandissimi secreti de l'arte Profumatoria*, in 1540. As laid down in the prologue of the *Plictho*, it collected dyeing recipes and processes used in his times in Venice, Genoa, Florence and other Italian cities, a work which took him 16 years. His handbook was divided into four sections; the first two on the dyeing of wool, cotton and linen with a great variety of colours; the third on the dyeing of silks, the fourth on the dyeing and tanning of leather.

Rosetti's *Plictho* is the work which provides us with the most detailed information, although very much in line with that given in the *Bolonia's Manuscript* as it should be remembered that both texts come from the north of Italy and are relatively close in time. As in the previous case, the information about leather work is found at the end of the work, making up an independent chapter, the fourth, exclusively devoted to this subject, beginning with the affirmation that "this book teaches the art of tanning hides, making chamois of them and dyeing them from colour to colour, as shown by the whole art." This fourth section is formed by 50 chapters, 11 of which refer to hide tanning procedures and 33 to methods for dyeing them. Although the subject matter is very similar to that of the *Bolonia's Manuscript* (here, also, only mineral tanning with alum is mentioned and never vegetable tanning), the chapter arrangement is less regular and those devoted to dyeing are mixed up with those on tanning, although the latter are concentrated in the last pages of the handbook. Brunello points out the details in Rosetti's recipes, from the risks involved to the hide from incomplete drying, to the baths to smooth the hide which should be given with semolina or flour.[5]

There is no doubt about the connection between these recipe and handbooks and their insertion in a literary tradition orientated towards the dissemination of the industrial techniques of the era and towards the training of the workers. Some recipes are very similar, the same tanning substances are applied, and the description of the work processes are very alike, with some variants which are more the result of the different possibilities existing for each task rather than the carrying out of different work in each of the places where the texts were written. Otherwise, as we shall see in what follows, the treatment processes used in those times by tanners and curriers were similar. That is to say, we find

5. F. Brunello, *Storia del cuoio e dell'arte conciaria*, Vicenza, Fenice, 1991, p. 142.

ourselves with some texts which are of a possibly practical use and which include authentic information, with recipes which would be undoubtably used by the artisans of that era. For example, the fact that the tanning processes (liming, application of mixtures of alum, oil, flour and eggs) described in them are mentioned in other contemporary texts and documents, for instance the municipal byelaws on leather trades which are conserved in diverse European cities, record the use of the same or similar raw materials, tools and technical procedures. In addition, all the processes described (soaking and liming operations on the hide, stripping and depilation, cereal bran treatment, the application of the alum with a mixture of other substances, finishing of the leather by trampling and stretching) are well known and continued to be applied in the same way until the 19th century. They were basically coincident in the samples analyzed, which contributes to re-affirming their value and the unitary character of this type of literature.

The data provided in these texts on the tanning process are highly detailed and, as we have indicated, they give an exact idea of the methods then used by tanners. To proceed with the tanning of a hide, the tanner should first ensure that the hide has been removed "in its season," comes from a healthy animal and has not been exposed to the sun for its drying.[6] The first treatment to which the hides are submitted is their soaking; the aim of this is twofold, on one hand to remove the blood, grease and dirt from the skin (as he proposes) and, on the other, to facilitate its subsequent stripping.[7] One of the recipes in the *Bolonia's Manuscript* recommends that the hides should remain in soak for five days "until the flesh is well macerated," changing the water every day or two days for the bad smell coming out of it; a Florentine text also advises keeping them in water for 5 days.[8] Rosetti suggests leaving them in clear water for four days, while other texts talk of 3 days, changing the water daily;[9] and some other Florentine recipes advocate immersing them in warm water for one day and one night.[10] If the hide is to be tanned keeping the natural fur of the animal, it is better to put it in salt water.[11] Next, stripping is usually done, so that once the hides are drained after soaking, they are stretched out on the stripping bench and stripped with a sharp knife.[12]

6. Bolonia's Manuscript, cap. 341; M. Merrifield, *op. cit.*, p. 563. In some Spanish byelaws of that era, coming from the cities of Ecija and Murcia, this prohibition of exposing recently stripped hides to the sun are also stipulated (M. Martín, *Ordenanzas del concejo de Ecija 1465-1600*, Ecija, 1990, p. 275; J. Torres, "Las ordenaciones al almotacén murciano en la primera mitad del siglo XIV," *Miscelánea Medieval Murciana*, 10, 1983, p. 98).

7. Biblioteca Nazionale, Palatino 934.

8. Bolonia's Manuscript, cap. 341 (M. Merrifield, *op. cit.*, p. 563); Biblioteca Nazionale, Palatino 916.

9. Plictho, p. 1; Biblioteca Nazionale, Palatino 1026; y Biblioteca Riccardiana, ms. 1243.

10. Biblioteca Nazionale, Palatino 867; Biblioteca Riccardiana, mss. 3052 y 3053; Biblioteca Laurenziana, Antinori, ms. 14.

11. Biblioteca Nazionale, Palatino 867; Biblioteca Riccardiana, ms. 3052.

12. Bolonia's Manuscript, cap. 341 (M. Merrifield, *op. cit.*, p. 563); Biblioteca Riccardiana, ms. 3052; Biblioteca Laurenziana, Antinori, ms. 14.

The treatment continues with the immersion of the hides in quick-lime as a previous step to their depilation. For this task, Rosetti recommends the use of well-sifted quick-lime, placed in a wooden tub. Once left in soak for three or four days and washed, the hides are put into water with lime for a further three or four days; they are taken out at least twice a day from the lime and left to drain for one hour before replacing them in the lime water; Rosetti recommends keeping certain hides in lime for fifteen days, lifting up the hide twice a day and letting it drip for two hours, and according to him this operation should be carried out "twice a day, once in the morning and once in the afternoon." In the same way, the Antinori 4 talks about keeping them in lime for fifteen days, a time limit raised by the Palatino 916 to four weeks.[13]

When the hair or fur has become soft enough to be pulled out easily, the hides are removed from the lime bath and left to drain, secured by pegs, for two hours; later they are placed on a trunk or wooden trestle and depilated with a blunt-edged blade. Rosetti indicates, when describing this operation: "depilate them on the trestle with the side of your knife and not with the edge" or "with the side of the iron," the same as the MS Palatino 934 when ordering "remove the hair with a knife on a trestle."[14]

Once depilated, the hides are again immersed in water and lime for 16 or 20 days, airing them every two days; Rosetti recommends 30 days for some types of hides and 12 days for others, although they should always be taken out of the lime and washed every 3 days. And for goat kid hides, he gives a treatment consisting of keeping them for 3 days in the lime solution, taking them out and letting them drain; they are put back into the lime for 6 more days, then they are depilated and returned to the lime for another 3 days, before finishing the depilation and washing them. The Palatino MS 916 also recommends putting the hide in lime for several days until it is well softened. Then it is depilated and put back into the lime for 4 weeks.[15]

The following operation, applied to remove the remains of lime and fur, as well as preparing the hide to receive the tanning better, consisted of treating the hide with certain materials which, on fermenting, produced enzymes acting as tanning agents, i.e. wheat bran and other cereals or the excrements of certain animals. The type of bath most used in the Late Middle Ages in the whole of the Iberian Pensinsula was that based on the fermentation of bran, whose bacterial action, produced by a complex mixture of organic acids and enzymes, eliminated the lime and dissolved the albuminous matter, loosening the skin

13. Plictho, p. 1; Bolonia's Manuscript, cap. 356 (M. Merrifield, *op.cit.*, p. 577); Biblioteca Laurenziana, Antinori, ms. 4; Biblioteca Nazionale, Palatino 916.

14. Bolonia's Manuscript, cap. 356 (M. Merrifield, *op. cit.*, p. 577); Plictho, pp. 1 y 4; Biblioteca Nazionale, Palatino 934.

15. Bolonia's Manuscript, cap. 356 (M. Marrifield, *op.cit.*, p. 577); Plictho, pp. 4, 12 y 14; Biblioteca Nazionale, Palatino 916.

and facilitating the absorption of the tanning.[16] The *Bolonia's Manuscript* advises making a mixture of salt, wheat bread and barley flour, which should not be sifted but "be with the semolina as it comes out of the mill," while Rosetti recommends making a paste with semolina and alum; MS 3052 in the Riccardiana Library also talks about "using semolina" as a prior step to applying the alum. The tanned hides with their hair are laid one on top of the other in this mixture; for 8 days according to the *Bolonia's Manuscript* (which specifies that they should be immersed in this bath for two days, then left to drain for half a day and returned to the bath for another six days), also six according to Rosetti. The treatment was applied to cleanse the lime from the hide ("put it in semolina until the lime is cleaned off," wrote Rosetti) and so that the hair will be secured better, although the depilated hides also received it for fewer days (three, according to the *Bolonia's Manuscript*, instead of the 8 necessary for the hides tanned with hair). The mixture was applied lukewarm on the hide and both Rosetti and the MS Palatino 916, give a recipe consisting of mixing millet flour, wheat flour and salt, trampling the hide and shifting it backwards and forwards in this solution for 4 weeks.[17]

Finally, regarding the tanning materials used to make the hides into chamois leathers, i.e. to apply the so-called mineral tanning on them, it should be noted that almost all the recipes included in these handbooks regard alum as a base. The most common formula was a mixture of alum, salt, flour, eggs and oil. The *Bolonia's Manuscript* contains a recipe made up of two ounces of rock alum, two beaten eggs with their whites and yolks, a handful of flour, salt for a pound of meat, oil to season a stew and hot water, in which first the alum, the flour and the salt should be put, and once these ingredients were mixed, the eggs and oil added; another similar recipe consists of alum, wheat flour, common salt and oil. This mixture was heated and a well beaten egg added to it; and another one, of alum, salt and gum arabic, a mixture to which was added a paste made of flour, oil and eggs. Very similar to these were the ones given by Rosetti, who affirmed that "these things make good chamois leather: hot water, yeast, rock alum, raw salt, olive oil and egg yolk;" one of his recipes consisted of mixing in hot water one ounce of wheat flour and an egg; another one included lard together with a pound and a half of alum and 20 egg yolks.[18]

The mixtures given by the manuscripts conserved in Florentine libraries are very alike. The Palatino 941 recommends mixing alum, egg yolks, ammonia salt, oil and flour in lukewarm water; the MS 1243 of the Riccardiana Library,

16. J. W. Waterer, "Leather," *A History of Technology: The Mediterranean Civilizations and the Middle Ages*, Oxford University Press, 1979, p. 152; in Castilla the operation was called *dar afrechos* and, in Catalonia, *rendir* the leathers with the use of *segó* (bran).

17. Bolonia's Manuscript, caps. 341 y 356 (M. Marrifield, *op.cit.*, pp. 565 y 579); Plictho, p. 1, 4 y 12; Biblioteca Nazionale, Palatino 916; Biblioteca Riccardiana, ms. 3052.

18. Bolonia's Manuscript, caps. 341, 355 y 356 (M. Merrifield, *op. cit.*, pp. 567, 577 y 581); Plictho, pp. 3, 4 y 6.

alum, an egg (one per ounce of alum), oil and flour in lukewarm water, or alum, salt, two eggs and two pinches of yeast. There is a similar recipe in MS 1247 of the same Library: egg yolk, salt, oil and yeast, all mixed together in lukewarm water; while MS 3052 recommends making a composition with alum, salt, flour and eggs, where the hide is soaked for 24 hours.

The alum employed in almost all the recipes is rock alum from the excavations of Rocca de Colonna; the *roca de Colonna* alum was, according to Gual Camarena, the best quality alum in the whole of the Mediterranean, followed by that of Focea. It was extracted in Turkey, near Trebisonda, and it was an alum melted in its own crystallization water and which was left to cool down (*Alum de Roqua ha aytala conaxensa que sien bells trossos e clar e que no y aia terra)*, explains the Hispanic handbook of merchandise edited by this same author, referring to the quality and transparence of this variety.[19] In some cases it was replaced by the so-called *sucarino allum*. This variety of alum (*sucreny)*, also cited by Gual Camarena in his *Vocabulario* (*zucarenyn*, *cuquarini*, *succari*, *suctench*) and which had already appeared in a 1252 document included in *Memorias* by Capmany or in the very Customs of Tortosa at the end of the 13th century, as reproduced in the *Diccionario* by Alcover and Moll, was an artificial mixture of alum with rose water and egg white which acquired the consistency of sugar and was used as an astringent. It could also be substituted by tartar (and appears thus in recipes supplied by the Palatino manuscripts 916 and 1026) but, as in the case of the mordant applied to cloths, it was of a poor quality and it gave more mediocre results.

After finishing the application process of tanning substances, the hides were left to dry in the shade, according to numerous texts, or in the sun, according to others and for how long depended on the season of the year in which the tanning was carried out. As well as drying, the hides were submitted to rubbing, trampling and folding, operations forming part of the currying of the leather and performed by the tanners or curriers to obtain all its plastic properties, i.e., to make it flexible and pliable so that it could be used for making shoes, garments or belts. These crumpling and trampling operations appeared in the texts of Bolonia or Rosetti as "giving the *stroppa*" and "giving the stick," terms discussed and studied by Franco Brunello in his works on the art of leather making in the Veneto.

It is important to point out that the tanning mixtures given by the technical manuscripts cited are also equally documented in many byelaws on trades in the Late Middle Ages. Used without mixing, alum produced stiff, imperfect leather, which had to be softened by beating it so that it was common for tanners, since ancient times, to add salt to it (in a proportion of one third to two

19. *Llibre de conexenses de spicies e de drogues e de aiustaments de pesos, canes e massures de diverses terres*, ff. 37r-v. Cit. M. Gual Camarena, *El primer manual hispánico de mercadería, siglo XV*, Barcelona, Institució Milà i Fontanals, 1981.

of alum) and to spread it with a mixture of egg yolk, flour and oil. In the Crown of Castile, the regulations of glovemakers in Seville and Malaga recommended the addition of flour and eggs to kid and sheep hides tanned with alum "so that the hide becomes loose and soft and appropriate for making gloves."[20] In the same way, and in the Crown of Aragon, the second chapter of the byelaws of 1511 of the city of Segorbe – a town located inland in the north of the kingdom of Valencia – , clearly established that hides should remain in lime between thirty and forty days, and then be immersed in alum for as long as the master artisan thought fit, proceeding subsequently with a treatment of flour and eggs.[21] Finally, in Italy, the first regulations for tanners in Venice, dating from 1271 and studied by Franco Brunello, prescribed, for the mineral tanning of hides, the use of rock alum, excluding others of a scant purity coming from the island of Vulcano (Lipari), and this norm was repeated in many other later statutes.[22] Also, the statute of 1401 for Venetian tanners devotes a whole chapter on depilation with lime, where it is stipulated that hides should be in the lime for one or two days, then removed and subsequently returned to it for one week.[23]

All these instructions are of special value insofar as they permit us to verify the technical validity of the recipes given by the manuscripts, which respond to techniques generally employed in those times and which appear in guild byelaws and statutes, drawn up during the same centuries (14th to 16th), in which the manuscripts were written. In addition, the processes described in Rosetti's handbook which, as we have seen, are identical to those found in 14th and 15th century manuscripts, were conserved until the 18th century and are those which continue to appear in modern technical literature. This demonstrates the technical continuance between the Late Middle Ages and the Renaissance, so that for this profession, as for many others, the 16th century did not signify any revolutionary disruption but only the continuation of methods which had been used since previous centuries. In many cases, since early times, from the ancient world to the Early Middle Ages centuries; this is confirmed by the well-known manuscript *Compositiones ad tingendam musiva* dating from the 8th century, when it alludes to the tanning of hides with alum for making clothes.

For my part, only one doubt remains with respect to these technical recipes from the Late Middle Ages in Italy. Why is only tanning with alum referred to in the manuscript recipes? The Spanish byelaws, the Italian and French statutes

20. *Recopilación de las ordenanzas de la muy noble y muy leal cibdad de Sevilla*, Sevilla, 1527 (reimpr. 1975), f. 233v; *Ordenanzas de la muy noble y muy leal ciudad de Málaga*, Málaga, 1611, f. 50r.

21. J. Aparici Martí, "La manufactura medieval dels cuiros. Les ordenacions tècniques de Sogorb," *Estudis Castellonencs*, 8, 1998-1999, pp. 429-443. Cita en p. 439.

22. F. Brunello, *Storia del cuoio e dell'arte conciaria*, p. 113.

23. F. Brunello, *Concia e tintura delle pelli nel Veneto*, pp. 38-40.

for tanners, thoroughly studied by Franco Brunello, often mention vegetable tanning done with sumac, bark and other plants. In fact, in the guild regulations and statutes, alum tanning is much less spoken about and seems to be a secondary method for working hides. Why, then, is it given such relevance in the manuscripts? Perhaps because they were written by people whose main interest was writing and the preparation of writing and printing material?

APENDIX: ITALIAN RECIPES FOR LEATHER TANNING
RICETTE PER CONCIARE PELLE

Bncf. Pal. 867. Miscellania di XVI secc. Vol. XVI.

[821r] Una pella di capra farla bella camozza. Piglia una pelle caprina, mettila in aqua tepida, lasciala stare un giorno. Poi spremela et metti della aqua con un pugno [821v] di sale al foco et on. 4 di allume di rocha et quando comincia a bollire et fa che rimanga tepida et metteli uno ovo battudo et un poco di olio comune. Et metteli dentro la pelle et menarla con la mano 3 o 4 volte, et dove se ho detto spremerla dice menarla.

Bncf. Pal. 916. Segreti diversi del 1460. XV sec.

[82v] A far concia di pelle. Prima metti in molle in aqua per cinque dí e falle bene distendere col coltello da radere. Factto questo metti le pelli nella chalcina e lasale stare tanto che lasci el pelo e fa bene rustre. E poi rimetti nella detta chalcina per quatro settimane e poi le fa bene radere come ai factto lo choiame e lauale bene si che sieno bene lauate della chalcina e poi la pesta cogli piedi. Factto questo togli le due parti di farina de miglio e la terça parte di farina di grano e sale, e fa uno molticcio e metti dentro e lasale tare per iiii settimane. Poi le togli e mettile a secchare bene e mettile doppie collo rivescio in fuora e quando seranno bene seche bagnale con aqua salsa e mettile al ferro; come se factto gli svanti e falle bene tirare si che sieno concie chome gli svatti, e falle bene tirare. E factto questo togli per ciascheduna pelle due inguastade de lactte e otto huova di gallina e bati bene insieme l'ove e lo lactte e togli un pocho di farina di grano e sale e meschola insieme e mettivi le pelle dentro e lasciavele stare per viii dí e poi lieva lo detto lactte e huova e togli per ogni X pelle libre 1 d'allume de rocha e stempera colla alume della aqua calda e meschola da detta aqua col laccte e huova e che vi stia dentro e lasciale stare dentro tre dí e poi le togli fuora e mettile a sechare come prima. Poi la metti al ferro e falle bene tirare e raschiale al modo che si fa, e sbatti e sara concio (transcr. Pomaro, p. 139).

[83r] A conciare pelli. Habi una pelle di qualanche ragione e mettila in molle e radila e scarnalla bene. Poi togli una scodella di tartaro e una di sal e descola. Poi abi una pentola empila di vino e fa bollire ogni chosa insieme, e quando la pelle é asciutta polla al sole e inbagna con detta drupa une, due o tre volte tanto ti paia factta. E poi stendi in humido, mena con mano, piu le po tignire.

Bncf. Pal. 934. Segreti e Ricette diversi. XV-XVI sec.

[43v] A conzar 1 pele de capreto avanti che'l capreto sia cotto. Tuo'la pele e lavala bene da quel sanguazo e metila a mole; poi lievali el pelo cum 1 cor-

telo con el nervo sopra el cavaletto e lassala sugar, perchè lasa molto meio da piar la conza. Tuo'per cada una pele de capreto on. 1 ½ de lume de rocha e metila a boir in una pignata tanto che la lume se desfazi, e poi temperala aziò la non te schotti. Poi lava bene sta pele ne la ditta acqua de lume e strucholala ben, che ne romagora mancho aqua che poi, e strucholala cum do legneti pizoli. Poi torna a schaldar questa aqua per temperar la conza, la qual se fa così, tuo'mezo pan crudo de bona farina ben tamixada, che sia ben lavado, e impastalo cum 1 ovo e cum una bona cuslier de oio bon, e tempera e meseda ben el levado, l'ovo e l'oio; sia fato modo de sugoli. Poi inpasta questa pelle in questo sugolo molto ben e metila a sugar al sol; e quando la serà suta, non strasuta, abi el tuo raxor e descharnala ben da dreto e da roverso, ed è fatta e bella (transcr. Pomaro, pp. 139-140).

Bncf. Pal. 941. Ricette. XVI sec.

[85v] 39. A far camoza di ogni pelle. R. allume di rocca on. 1 e dissolvi in acqua calda e mettivi entro la pelle e mettivi due torli di ovo, q. 1 di sale armoniaco, on. 1 di butiro, on. 1 di formento, 1 lievito di pane. Poi lava la pelle e lassa asciugar al'ombra, e quando sarà quasi asciutta fregala bene e lassala star nel'acqua che la sia ben bagnata a fine che la si possa ben maneggiare [86r] e quando l'hai al ferro metti prima calcina brodiata dalla banda della carne, e lassa per una ora sotto il letame, e lavali e peli con ferro come un filo o con una costa di bue (transcr. Pomaro, p. 140).

40. Al medesimo. R. pelli che sian ben pelate o rase e ben lavate in acqua tepida dove sia sal comune dissoluto. Poi habbi allume di rocca on. 1 e dissolvi in la detta acqua tepida e poi frega ben le dette pele in la acqua sopradetta e lassavele entro per spatio di 4 hore. Poi fa una mossa in la acqua sopradetta in questo modo, tolli on. 2 di butiro overo olio bono, on. 1 di fermento e 3 torli d'ovi crudi e una octava di sal armoniaco. E tutto mescola molto bene in la sopradetta acqua e lassa stare la pelle quanto ti pare et metti ad asciugar ove non sia vento. E quando sarà quasi asciutta mena bene (transcr. Pomaro, p. 140).

Bncf. Pal. 1026. Ricette e segreti. XV-XVI sec.

[165v] Concia di pelle havuta da Frinzi da Verona. Metti le pelle in molle e tienvele tre dì incircha, e muta loro ogni d`l'aqua, tanto che le tornino in charne bene. Di poi le lava molto bene e col ferro fatto acco le scharna. Di poi per ogni pelle togli on. due d'allume zucherino, on. 1 di gromma di vino biancho, potendo se no di vermiglio e on. ½ di salina e di tutto fa polvere e meschola bene. Poi pigia la pelle a una a una e mettivi su di dicta polvere sutil sutil tanto che la quopra tutta da lato della carne e ripieghale e lascale stare fino in ore 1 e ½ in concha. Di poi piglia el resto di ditta polvere avanzatati e mettivi su del'aqua calda tanto che tu la stemperi e a dischrezion vi metti su

olio e farina, tanto che lo facerà chome ungüento sottile da distenderlo. E di poi apre pelle per pelle e distendivi dicto intriso e di nuovo le ripiega chome sechono e lascalo stare un dì e 1 notte, di poi aprile e distendile chol pelo in terra e laschale tanto stare che le rascughino. E puliscile bene poi chol ferro e son conce (transcr. Pomaro, p. 141).

Conca di pelle havuta da Lorenzo vaiaro. Metti le pelle in molle e lavale molto bene, tanto ne sia uscito el sudiciume e tienle in molle in aqua chiara sotterrate tanto tornino in carne chome fresche; di poi togli on. 1 di salina per pelle e fa salamoia chon aqua tanto vi stieno sotto richoperte e metti dette pelle in dicta aqua insalata e tienvele un dì e una notte. Poi col ferro acco le scharna e tienle sopra la choncha ove erono o sopra un altro vaxo, accochè l'aqua insalata che n'esce mentre le scharni non si perda; e chome l'arai scharnate rimettile in dicta aqua insalata e tienvele un'ora o meno. Di poi a pelle a pelle nele chava e in sul quoio le quopri di farina di miglio o di grano o d'orzo (non dà noia fino in on. 4 di farina per pelle o incircha) e mettile in un'altra choncha distese l'una sopra l'altra, e quando l'a chosì infarinate tutte e messe nella choncha gesta loro adosso l'aqua insalata, se ve n'è restata, e se non ve n'è restata non dà noia perchè la non inchorpo. E tienle chosì quindici dì; ma bisognia che ogni dì le tramuti di una concha in un'altra sanza mancho, se non vuoi rischaldino e infradicisi. E basta le pigli così e gettile sanza gran diligentia di distenderle; di poi chavale, passati detti 15 dì e distendile al sole chol cuoio di sopra e tienvele 5 o 6 ore tanto che sia ascutto el quoio. Poi le rivolgi dal pelo el dì e l'altro tanto sieno ascutte. Di poi chon uno strafinaccolo inméllale sotilmente dal quoio e mettile quoio a paio a paio quoio chon quoio e fanne ruotoli e tienle chosì un dì e una nocte, di poi lo stroppa, cioè le tira e distendi e quel ferro o legno torto che si Chiapa da stroppare, e mettile al sole da lato del pelo. Di poi le pulisci chol ferro da scharnare e son chonce; ma vuolsi puoi schamazarle e liscarle un pocho (transcr. Pomaro, p. 142).

Ricc. 1243. Segreti vari di alchimia. XV sec.

[67v] A volere conciare una pelle o piu per modo si possa lavare. Togli la pelle quando è scorticata e mettila inmolle e sella fassi strancia, ancora la metti inmolle tanto che rinvangha e poi la torei. Poi dalla parte donde sono e peli poni calcina viva stemprata con aqua, poi soprapissa la decta pelle a spatio d'una hora o più e lascia stare. E poi la piglia e pelala con mao se vuoi che da una parte abbia el riccio e dal'altra no e se volessi ch'ella l'avessi da ogni parte quando la pelli abbi una steccha e pelala e faragli lo riccio. Poi piglia la decta pelle e lavala in aqua molto bene. Se la pelle è d'agnello piglia on. iii d'allume e così se fusse Maggiore o due pelli piglia on. vi e se tre pelli on. viiii. Et per ogni oncia d'allume abbi uno ouvo. Et piglia el decto allume e fallo bollire in aqua e lascialo freddare che sia tiepido e poi abbia le uova bene rotte e dibattute e mettili sopra l'allume e uno poco d'olio, et se nello allume volessi mettere uno poco di pasta lievita a bollire sarà buono. E poi abbi uno vaso e poni

ci dentro queste cose e insieme le mesca assai e poi vi metti dentro la pelle e rimenala bene e lascenela stare uno dì e l'altro dì la ne cava e tendila sanza torcere [68r] in luogo dove no sia sole e siani humido. Et se volessi conciare la decte pelle in modo che avessino el pelo, tieni el sopradecto modo quando sono bene scarnate e lavate e torte o vero permute mettile nella decta concia e fanne nel modo sopradecto.

A conciare pelli col pelo da fadere. Togli le pelle col pelo e metti inmolle nell'acqua chiara e lascia stare tre dì e poi radi la pelle dal'lato della carne e scarnala bene e poi la metti nella infrascripta concia, cioè togli la pelle così molle e scarnata e lavata e priemila bene, e poi togli on. ii d'allume di rocho per ogni pelle e una metadella d'aqua per pelle e metti nell'aqua l'allume di rocho e polla a scaldare tanto che si disfaccia l'allume e uno poco di sale. E togli due uova per pelle bene rotte e metti nella decta aqua e togli quanto sono due naci di lievito per pelle e stempra e metti nella decta aqua, e togli uno poco di raditura di sapone bianco quanto è una noce e metti dentro e se vuoi li puoi mettere una metadella di lacte e una meça scodella di farina non stacciata, e metti le pelli in questa concia e lascianele stare dentro [68v] uno dì e una nocte, dipoi le cava e polla a asciugare al sole e maximo a tempo d'inverno e saranno facte belle e buone.

Ricc. 1247. Secreta chimica. XVI sec.

[49r] A conciare una carta de capreto. R. el rosso de uno ovo e uno pizigeto de lume de rocha pesta e tanto de sale comune e tanto di linamento come seria doa noce e 4 ove 6 goze d'olio comune e metti tutti queste cose in uno catino over caldanella e misce simuel bene, avisandoti che'ste vole mettere con queste cose aqua cioè tepida e metti dentro la tua pelle e lassala stare fino a giorni e poi la suga e valla tira che l'avegniera camozata perfetamente e ponala lavare con aqua e fare quello norai.

Ricc. 3052. Miscellania di ricettari antichi. XIV sec.

[25r] Ad conciare pelli da fodarare e fare camoscio e tegniale. Per conciare pelle da foderi, tolle quelle pelli che tu voli, mecteli a mollo nell'aqua e inessa mecte una libra di sale o più se sono più di x pelle agnelline. E la mactina le nectrai e scrullale bene e poi le scarna quanto più puoi e sai. E poi così meçe sciucche e meçe liquide li darai l'alume in questo modo, conponi le pelli l'una sopra l'altra, el carnaccio stia di sopra, e poi per x pelli agi una libra o più d'alume de roccho e meça di sale stenparato in acqua e bagniale a una a una pieçandole e ripolle per una nocte, e se avara preso l'alume strigni la pelle con dui deta se torna bianca avara preso l'alume, se no dalgline più. E poi di questa acqua che resta netride farina quanta sia melglio de melglio e dibasti conessa cinque vuova e falla liquida come di fritelle molto bene menata e questa farina mecte sopra le pelle come facesti prima dell'alume piecandole ad una d'una e

lassale stare per una nocte e uno dì racolte. Poi l'altra nocte si puoi falle stare si pase al sereno e poi a tua posta le concia e scarna al coltello e poi alla litorta.

Per fare chamoscio sença pelo. Prima tolle quelle pelle che voli montonine o de pecora quatro o sei o più, mectele a mollo per uno dì e una nocte, poi le mecte al adova e giecta via ogni sangue e bructura dal carnacio. E puoi un'altra volta le rimecte a mollo per uno dì perche melglio s'amollino. Ma se tu voli che agiano el pelo, mecti nella decta amollatura dui pugnia di sale perche dura molto più lo pelo. Ma per fare camoscio, mecte poi in calcina secondo che bisognia, e se voli salvare la lana meça libra di calcina incrisa abbi come farinata repase le pelli l'una sopra l'altre, poi con'uno pennello di seta porcina pone la calcina e apiccale a una a una e repolle sichè ripelino bene. E fa come se fa per conciare coiame. E poi che seranno scalcinate e messe al adova e lavate bene, lassale stare in aqua chiara una nocte, e poi che sono [25v] bene purgate e scalcinate e tu le mecti nella semula intrisa nell'acqua inuno catino grande per 3 o per 4 dì. E poi da capo un'altra volta le scarna e necta e lava bene, e poi le rimecte nela dicta semula per 4 dì o più secondo el tenpo. Poi da capo le rescarna e premele dall'aqua, poi li da l'alume per 6 pelle grande tolle dui libre d'allume e stenparalo, poi spande le pelle l'una sopra l'altra e dal carnaccio li da l'alume e apiccale e ripolle per una nocte, e non vole essare l'aqua troppo calda, ricordati di mectarci 1 libra di sale. E puoi l'altro di quella alluminata che ti resta intride farina di melglio o di grano, olio come oncie 3, lacte di pecora oncie 1, uno ovo per pelle e levuto come ti pare e omne cosa inpasta bene come di fritelle e mecti dentro tucte le pelli e falle stare 4 dì o meno come ti pare e poi la pone a scioccare. E quando le vorrai stirare mectile in aqua chiara ad una ad una e puoi le pieca e mectale socto a qualche peso che solino per una nocte e poi inanti che sieno sciuche stirale alla torta e poi al ferro e sono facte.

Ricc. 3053. Miscellania di ricettari antichi. XVII sec.

[12r] A cammozare una caprina cruda. R. metti la pelle caprina nell'acqua tiepida e lassa nela stare un giorno, dipoi snervala e metti della acqua a fuoco drendovi un pugno di sal e on. 4 di alume di rocca e fa scaldare tanto che cominci a bollire. [12v] Poi levala dal fuoco e lassala serare si che rimanghi tiepida e mettivi drento un'ovo ben battuto e un poco di olio di oliva e poi vi metti dentro la pelle e rimenala bene tre o quatro volte e sarà fatta.

Laur. Antinori. 14. Ricettario. XIV sec.

[103v] Nota a chonciare xii pelle di chavretto. R. per 12 pelle di chavretto. Prima mettile in molle in acqua per uno dì e una notte, e poi mettile in chalcina per dì xv, e ongni iiii di cavala fuori e mettila a rasciughare per una ora. E poi sille lava bene in acqua chiara chorrente. E poi le scharna e sriva e purgha tanto chella sieno bene nette dalla chalcina. E poi silli metti in acqua di cruscha

fatta chome se ghause qui di dietro e lasciavolo stare per uno dì. Poi sille allumma in allume disfacta in acqua trepida e poi silgli fa un moltizzo fatto in questo modo, cioè per xii pelle si vuole libre una d'allume di roccho e libre otto di farina di scoghale o di milglio o di grano, ma la seghale è migliore, e di queste cose fa uno lievito con l'acqua del'allume a mettivi dentro oncie quatro di sale chomune e oncie tre di butiro con quatro metadelle de lacte per vi rossi d'ovi, cioè tuorla d'ova. E de tutte queste cose fanne un moltizzo nel quale metti dentro le pelle boni nette e purghate dalla chalcina e state nell'acqua della cruscha chom'è detto di sopra. E fa che quando tu le metti nel moltiaio fa che sia chaldo purche non chuocha e li dentro sille fola e per poi molto bene, e ongni tre dí tirale fuori del moltizzo per una ora, poi se chalda le moltizzo e mettivi le pelle dentro pestandole bene sempre cholla piedi, e chosì si fa di tre dì in tre dì fino a dì 15. E poi chavale fuori e non le torciere niente e asciughale, poi le dironpi in uno saccho chon piedi [104r] e selle fusoni troppo dure bangnale chon un pocho d'acqua al modo d'uni spargolo, poi mettile in uno saccho e folale, poi tirale alla stroppa e con le mane.

Chome si fa l'acqua della semola. R. una secchia d'acqua item quarta mezza di semola e mettila nella detta acqua quando ella è un pocho tiepieda beni pungniente, lasciala posare per uno dí e note la fara una grosca, gietta via detta crosca, poi pilglia la chiara di quella e si è perfetta.

Manoscritto di Bologna, XV sec.

[563] 341. Modo da conciare pelle cum lo pelo et senza pelo, cioe pelle de cervo o de lupo o de tasso o de lotrie o de capretti o de capre o d'altri animali, ed é concia probata. Recipe pelle scorticate a stagione e non sieno de bestie che habiano insanita et sieno secche senza sole overo al quanto insalate, et metile in una tina d'aqua et lassale stare li dentro per spatio di 5 dí naturali acio lo carnacio sia ben macero et infra questi 5 dí renova l'aqua doi o 3 volte a le dicte pelle per la puza che fanno.

Poi le cava fora et lassale scolare et scolate che sonno pune l'una sopra al'altra in el banco da scarnare pelle cum carne, intende bene, poi excarna le dicte pelle commo te pare, et pone cusci l'una sopra al'altra acio non te vengano guaste cum lo cortello. Et scarnate che sonno levale dal banco et lassale scolare bene.

Poi tolli uno barile d'aqua et falla bulire et in questa aqua pone libre 4 de sale et commo el sale é disfacto bene et tu la lassa refredare tanto che divente tepida et in [565] questa aqua tepida metice mezo pane de formento et menalo bene per le mano tanto che sia ben disfacto. Poi mecte in la dicta aqua farina de grano, ma é meglio d'orzo, cioé quella farina che te pare, bastevile et che l'aqua dala farina sia un poco spessa. La prima volta sappi che la farina non vole essere staciata ne stamignata, ma vole essere con la remola commo ella viene dal molino.

Facto questo essendo l'aqua tepida, cum la dicta farina mecti dentro le dicti pelle ad'una ad'una et menandole bene cum mano senza extirarle, et fa che lo carnaccio sia desocto bene steso, l'una pelle supra al'altra, et lassale stare in questo modo dentro per spatio de doi dí. Da poi le tira fora et lassale bene scolare per spatio de mezo giorno et la sera le remecte in nela dicta aqua et lassacele stare dentro 3 dí et mistale bene. Et in capo de tre altre dí fate purre a questo modo et remetile purre de dentro et lassale stare per spatio de sei dí in tucto, et questi altre doi dí de sopra, et questo se fa perche el pelo se ferma meglio.

De poi le tra fora de la dicta concia et polle asciugare al'ombra per spatio d'una nocte. Poi le pone ordinatamente l'una sopra al'altra in tabole o asci da scarnare et poi scarna commo te pare, et scarnate che sonno scrullatile bene.

Poi tolli alumi de rocho che sia in peze et non in polvere, che è meglio, et mecti per xii pelle de lupo o de cervo o simili a queste, xii libre d'alume de rocho, che omne pelle ne venga ad havere una libra, e 24 petitti d'aqua, che vengano doi petitti per pelle, et lassa ben disfare lo alumi al foco in questa aqua et fa che l'aqua non bolle cum lo alumi. Poi ce mecte dentro quatro libre de sale et commo é bene disfacto lassa tanto refredare l'aqua che advenga tepida.

Poi mecte in comfetione le dicte pelle et per omne pelle li da uno petitto de la dicta aqua cum lo dicto alumi et sale, et menandole bene per mano in nella dicta aqua tepida per spatio de uno miserere, estirandole et manegiandole ad'una ad'una bene sucte sopra ne la dicta aqua. Poi la goluppa et ponle cum la dicta confetione et ponle da parte et cosí fa a tucti lavanzo de le pelle et lavanzo de l'aqua che te remane, overo confetione, gietala sopra a le dicte pelle et fa che le pelle sieno stesse in la tina l'una sopra al'altra et [567] lassale stare dentro per spatio d'uno dí e d'una nocte. Et sappi che se sonno pelle picollini commo é pelle de capretto, le doi pelle vogliano una libra de alumi. Et de poi tirale fora et lassale scolare pers patio de ½ dí, poi recoglie la dicta scolatura cum questa altra aqua che ce advanzata dele pelle et ripolla da parte.

Et poi per affinare perfectamente le dicte pelle, tolli farina de grano, ma é meglio d'orzo, che sia afiorata, cioé quella farina che te pare che sia bastevile, et stemperala cum l'aqua de lo alumi che reservasti da parte, et sia bene misticata la dicta farina cum la dicta aqua ad modo d'una pasta da fritelle. Et poi in questa pasta ce mecte 16 ova cum le coze et cum tucti chiare et vintelli et rompili bene insiemi et metice uno bichiere d'olio, o manco che piú, et mistica bene insiemi. Poi fa che la dicta aqua sia uno poco calda prima che tu ce mecte le dicte cose et mistica omne cosa bene insiemi.

Poi tolli le dicte pelle ad'una ad'una et indopiale per mezo, cioé che lo pelo venga per de dentro et el carnazo sia de fora, et aciό che el pelo non se habia troppo ad embrutarse per la dicta concia, et metile in la dicta concia, overo pinta, et sia competentemente liquida et mecti dentro le pelle ad'una ad'una et fa che sieno bene impastate da la dicta pinta, et pone l'una sopra al'altra, et se te avanza de la dicta pinta gietala sopra a le dicte pelle et lassale stare per spatio de uno dí et una nocte. Poi tirale fore et polle asciugare al sole o al'ombra, che é melglio, et guarda bene che non le stirasce per veruno modo per infino che non sono sciuche.

Et quando seranno sciuche sfregale bene sopra ad'una stecha de merollo bene tagliente facto a quello mistiero, ació che la farina se ne cagia tucta. Poi le scarna cum lo cortello bene tagliente et scruttale bene cum una vengastra. Poi le remena bene cum mano ació che diventano morbide.

Et sappi che questa concia é meglio d'aprile et de magia che in tutto l'anno et anco de setembre et de octobre. Et sappi che per le pelle picole commo sonno d'agnelli o de volpe se vogliano conciare cum la mestra dele grande cum tucti li modi sopradicti. Et sappi che la concia de pelle senza pelo se vole tenere tucti li modi sopradicti salvo che vogliano essere pellate le dicte pelle in calcina et poi [569] li da la concia ordinatamente commo quelle dal pelo, ma vogliano essere piú remenate assa cum mano perche levano piú bella grana.

342. Concia per una pelle. Havve allume de rocho in polvere once 2, doi ova bene dibatuti, poi tolli uno bono pugno de farina, cioé el fiore, et tanto sale quanto bastasse per insalare una libra de carne et tanto olio quanto condisse una menestra, et una bona fogliecta d'aqua calda. Et mecti in la dicta aqua prima lo alumi bene subtili, poi la farina et poi la sale, et miscola bene. Poi metice li ova et l'olio et mista bene. Et quando l'aqua é calda mecti dentro la pelle depilata et manegiala bene et strocila bene traendola et remitendola in la dicta aqua calda. Poi lassala stare per una nocte o 4 hore almancho. Poi la tra fora senza astirarla et polla asciugare et remenala bene a la stroppa. Poi la pumicia da l'uno lato et l'altro, ed é facto.

344. A fare camoscio cum nervo o senza nervo, cioé scamosciato da omne parte. Tolli uno lingno retondo et grosso quanto la cossa et longo quanto l'homo et al muro appoggialo commo fanno li conciatore de corrame. Et se volesci fare una pelle de capretto subtitamente in uno dí, tolli la pelle che sia fresca et polla insuso questo ligno et cum una costa de cortello per [571] forza de braccio li leva via el pelo et lo nervo. Et se fusse una pelle grande, falla stare in calcina commo fanno li conciatore quando le voglano conciare per corame, et poi l'apoggia al dicto ligno et per forza de la costa li leva lo nervo. Poi la lava bene da la calcina.

Poi tolli 3 fogliecte d'aqua et in la dicta aqua ce pone una oncia e meza d'alume de rocho et mezo pugno de sale comuno, et mecti l'aqua al foco che se disfaccia le dicte cose. Et poi ce pone uno poco d'olio et levalo dal foco, et quando é tepida l'aqua et tu ce pone uno ovo bene dibatuto bene et mistalo bene in la dicta aqua. Poi ce mecti la pelle 4 o 5 volte et da una volta a l'altra lassala uno poco sciugare et l'ultima volta lassala bene sciugare. Poi la mecte a lo lavello o a la stroppa, ed é facto.

345. A fare camoscio senza grasso. Havve lacte, fior de farina et olio lavato cum ranno da capo, ació le pelle non vengano machiate, et mista omne cosa

insiemi cum aqua calda et mecte le pelle in la dicta aqua per 3 dí. Poi le revolge da l'altro lato per 3 altre dí. Poi le pone a sciugare et non le stirare et quando sciuche et tu li da la stecha et stroppa.

346. A fare camoscio bono. Piglia per cescuna pelle once 3 de fiore de farina et uno bichiero de lacte et una oncia de butiro et uno poco de pane de formento, et distempera omne cosa insiemi cum uno poco de ranno da capo molto bene, adó che le dicte cose se incorporano insiemi. Et se fusse poca concia, non ce agiongare se non de lo ranno chiaro. Et lassa stare per 5 dí naturali. Poi lo pone a sciutare et dalli la stroppa.

347. A fare camoscio brevimente. Recipe oncia i de sapone bianco et stemperalo cum lo ranno. Poi mecte le pelle in lo dicto ranno per spatio di quattro dí, et poi le pone a sciutare. Poi le stira a la stecca et seranno bianche et morbide.

348. A fare camoscio che sia bianco et morbido commo una seta. Tolli gracia de porco et strugila in uno pignatto, poi tolli aqua calda et distemperala cum farina, poi ce mecti la dicta gracia et mista bene insiemi. Poi tolli uno altro vaso et stendice le pelle. Poi tolli uno bocale de lacte et mecti sopra a le dicte pelle. Poi mecti la dicta concia et fa che le pelle sieno [573] bene coperte da la concia. Et lassa stare per 5 dí et seranno bianche et morbide.

349. A fare camoscio che arestia morbido sempre mai. Havve lacte, farina d'orzo, olio lavato con lo ranno, adó le pelle pigliano la morbideza, et mista omne cosa insiemi cum aqua tepida. Poi ce pone le pelle piú volte et lassandole apresso che sciucare da una volta a l'altra. Poi le pone a sciutare a l'ombra et dalli la stroppa.

350. A fare camoscio che arestia al'aqua. Tolli 4 ova per pelle et lacte assa, cioé uno bono bichiero per pelle, et uno poco d'olio molto bene menato insiemi. Et poi ce mecti le pelle a molle per spatio de 7 dí et omne dí le remena subtusopra una volta et polle a seccare et dalli la steccha.

351. A scamosciare le pelle. Havve le pelle et mettile a mollo in l'aqua per 5 o 6 dí. Poi le pone a molle in l'aqua tepida per una nocte. Poi le leva via dala dicta aqua et levali via el pelo per forza d'una costa de cavallo. Poi le sciaqua cum aqua chiara multo bene, et poi le pone a scollare uno poco. Poi tolli alumi de rocho et sia bene subtile, et doi ova per pelle, et farina bene stacciata cum uno poco de formento, et mista bene insiemi cum aqua calda ad modo de pasta da fritelli. Et poi ce mecte le dicte pelle per spatio de 3 dí. Poi le tira fora et lassale quasi sciutare. Poi tolli remola et mistala cum aqua calda bene. Poi ce pone le dicte pelle per 3 altri dí. Poi le sciuta bene senza astirarli et dalli la steccha, ed é facta.

[575] 355. A fare camoscio bonissimo. Pilglia la pelle bene scarnata dentro et de fora et poi la creta tucta de farina cum aqua ad modo de pasta da fare cialde, et lassala stare alcuni dí, cioé per dí 3 o piú. Poi la lava bene et mectila in una concha. Poi habbi una pignata nova vitriata et impila d'aqua et polla al

foco, cioé mecti uno mezo d'aqua per pelle et mecti una onzia et meza d'alumi de rocho per pelle. Poi mecti lo dicto alumi a disfare in la dicta pignatta et poi ce pone altratanto sale comuno, et commo sonno bene disfacti et tu leva dal foco la dicta pignata, et mecti [577] l'aqua alumata et salata in una concha. Et commo la dicta aqua é divinuta tepida et tu ce mecti 3 o 4 ova per pelle bene dibatuti et mistali bene cum la dicta aqua. Et poi li mecte uno poco di formento disfacto bene cum la dicta aqua et mective uno poco d'olio, cioé manco che el cuarto d'una foglietta per pelle, et mistica bene omne cosa insiemi. Poi ce pone le pelle et menale bene per la dicta concia et lassa stare per 3 dí le dicte pelle bene coperte dala dicta concia, et pasati i 3 dí cava fora le dicte pelle et spremili bene ad'una ad'una. Poi le remena per mano ad'una ad'una ordinatamente. Poi le pone a sciutare in loco che non habia ne sole ne vento ne fume, et ponele ala stroppa o stecca.

356. A fare concia in camoscio bona et vera et probata. Tolli le pelle stagionate et non sieno de bestie insane et sieno le pelle seche, et metili in uno mastello d'aqua a molle per tre dí. Poi le lava molto bene in lo dicto mastello da omne inmunditia che le pelle havessaro et commo sono benc lavate gietta via quella lavatura.

Poi tolli calcina nova et viva, pollo in lo dicto mastello et distempera la dicta calcina cum aqua molto bene et commo la calcina é ben disfatta et disolta et che ela sia ben brodosa et liquida, et tu ce pone dentro le dicte pelle ad'una ad'una sempre remenando la dicta aqua et calcina, et lassale stare a molle li dentro per 3 dí o 4, o piú o meno secundo le pelle, et per infino a tanto che se pelano bene. Et omne dí, overo omne doi dí al piú, le cava fora una volta da la dicta aqua et calcina et pollc sopra alo dicto mastello per una hora a scolare, poi le retorna in lo mastello commo prima, et commo se pelano bene et tu le pone a scolare in una caviglia molto bene per doi hore.

Poi habbi uno cavallecto da doi pei et mettice suso le dicte pelle ordinatamente l'una supra l'altra. Poi tolli una bastone retratto in forma d'una costa de cavallo et mandato giuso el pelo cum lo dicto bastone molto bene a pelle per pelle.

Poi che sonno ben pellate remectile a molle in lo dicto mastello dove te rimase la dicta aqua et calcina per spatio de 16 o 20 dí et omne capo [579] de doi dí le remena molto bene in la dicta aqua calcinata. De poi 16 o 20 dí et tu le cava fora et portale al'aqua corrente et lavale et spremile molto bene ació la calcina escha fora.

Et commo sonno bene lavate et necti, tolli lo dicto mastello et gietta via quella aqua et calcina et lavalo per modo che sia bene necto et mectice tanta aqua tepida chiara quanto tu crede che le pelle possano ben stare a molle. Poi ce pone dentro tanto de remola grossa che la dicta aqua tepida vengna uno poco spessa. Poi tolli le dicte pelle ben lavate et metile dentro in la dicta aqua remolada ad una ad una et cosí la lassa stare per 3 dí. Poi le cava fora et lavale molto bene al'aqua corrente ació tutta la remola vada via.

Poi porta le pelle bene lavate ad una scala overo ad una caviglia. Poi tolli le dicte pelle ad una ad una et dalli lo torcholo et premile bene che non ce rimagna niente d'aqua et quanto seranno meglio spremute et atorcholate tanto piú bianche viranno. Et se in lo torcolare le pelle facessaro alcune vesiche, apuntale et forale cum uno acho,ació la pelle se possa bene scolare da l'aqua.

Et commo le pelle sonno bene scolate ad una ad una et bene spremute de vantagio stendile cum le mano per tuta la pelle ad una ad una, et pone l'una pelle sopra l'altra ben distesa al collo a le branche et per tucta la pelle. Poi tolli lo dicto mastello bene necto cum tanta aqua tepida quanto tu poi comprendare che le dicte pelle possano bene ricevare, et innanze piú aqua che meno. Poi tolli una oncia d'alumi de rocho bene pisto cum altretanto de sale pisto a misura et non a peso, et meza oncia de gomarabica bene pista. Poi pone le dicte polvere in lo dicto mastello dove é la dicta aqua tepida et remistale bene,ació se disolvano. Poi tolli le dicte pelle ad una ad una bene stese et metile in la dicta aqua tepida dove sono disolute le dicte polvere, spremendole et reimbeverandole et remenandole bene,ació pigliano meglio quella aqua alumata et cusci fa a pelle per pelle.

Et commo le pelle sonno ben remenate et imbeverate et tu la pone a scolare per una hora et ricoglie la scolatura sopra al'altra aqua che te remase de le pelle. Poi tolli farina afiorata tanta quanta te pare bastevi alle pelle et distempera la dicta farina cum la dicta scolatura dele pelle che reservasti et dis- [581] temperala per modo che sia commo pasta da fare fritelli. Poi pone in la dicta pasta una oncia d'olio per pelle, overo uno ovo per pelle, et sappi che quando tu distempere la dicta farina la scolatura vole essere tepida et non calda, et mistica bene insiemi. Poi tolli le dicte pelle ad una ad una et metile in la dicta pasta o compositione et lassale stare per 3 dí naturali al piú.

Poi tolli le pelle commo le venghano senze extirarle de niente et polle in su una corda a secare al'ombra et commo se venghano secando cusci le vieni stirando. Poi le pone ala stecha et remenale bene per mano,ació levano piú bella grana et diventano piú morbide, ed é facto.

Et sappi che omne pelle de capretto o simili a quelle vogliano lo alumi et l'altre cose al peso dicto di sopra. Et se fussaro pelle de castrone o capre o altre simili, vogliano 3 once de alumi per pelle, et cusci 3 once d'olio overo 3 ove per pelle, et una oncia et ½ de gomarabica, et seguita la recepta a lo sopradicto modo.

Plictho de Gioanventura Rosetti, XVI sec.

A incamozzare pelle che staranno all'acqua.

Tolli calcina viua de cogoli e falla tamisare e dapoi mettila nella tina de legno che sia netta e poneli dentro di l'acqua communa e lassala stare cosí a moglie per tre giorni, acció che il foco, cioé il calor della calcina, non brusasse le tue pelle. E piglia quelle pelle che vorrai camociare e mettile a moglie per quattro giorni e forniti li quattro giorni cauale de acqua e scarnale molto bene dalli lochi che li fosse rimasto carne.

E come sonno scarnati ponele nella calcina che hai apparecchiata di sopra e lassale stare nella ditta calcina per insino che tu vederai che le ditte pelle si pellano, e come vedi che le si puo pellare cauale della ditta calcina e pellale sopra il caualetto con la costa d'il tuo cortello, e non con il filo, e come le sera pellate ritornale in questa medesima calcina e lassale stare tre over quattro giorni fino che vedrai che il nervo si possi tirare, e prova se'l nervo si tira leggiero e venendo, tíralo, ma non lo sforzare perche guastaresti le pelle; e se le non lassa il nervo tornele nella calcina e se'l nervo vene facilmente habbi il tuo ferro da scarnare che tagli molto bene e levali via il nervo.

E nota che como la pelle sta nella calcina con il pelo e senza pelo, ti bisogna sempre messedar la pelle nella calcina uno dí sí e l'altro no, cosí messedando la calcina come la pelle, e questo messedare vole esser due volte al dí, l'una la matina l'altra la sera.

E come haverai snervata la pelle ponila nella acqua chiara e lavala molto bene davantaggio acció che la calcina vadi via, perche non la lavando bene la tenta non veneria bene ne bella sopra la pelle. E per due ragione la dei lavare e bene, la prima che ti dissi e l'altra perche la calcina rossega la pelle, si che come intendi come l'haverai lavata con acqua chiara, torzi la pelle e struccola tanto che ne esca l'acqua chiara, e come le ben struccolata e che ne uscisca l'acqua chiara, metti la pelle sopra la stanga per fina che apparecchi quello che diró qui sotto.

Tolli semola de formento e torrai una calderola de acqua e mettila al fuoco e metti dentro in questa acqua lume de rocca quanto seria una nosella, e fa che sia ben pesta e fatta in polvere a disfare con questa acqua, e come vedi che la sará disfatta levala via dal fuoco che la non sia troppo calda e metti questa acqua in uno mastello o tinella, e poi metti dentro la semola e impastala come faresti a fare semola alle galline, e metti questa pelle dentro e fa che questa compositione sia un poco calda, cioé tivida, e lassa stare la pelle in la ditta semola per sei giorni. E ogni giorno messedala e strucolala molto bene, cioé la matina per una volta al dí.

E come vedrai che la semola fara alchune vesighe, leva via la pelle perche la semola havera perso la sustantia che le da dibisogno, e quelle visighe si chiama il fiore. E come haverai cavata fuora la pelle mettila uno poco a sugar tanto che la sia mezza humida, e nota non la sugar al sole, ma ben a l'ombra. E come le uno poco sutta dalli la stroppa e lassala sopra la stanga, e fa che sia stropizzata molto bene e lassala come intendi sopra la stangha fina che farai la conza come intenderai qui sotto e prima.

E ovrai tanta acqua chiara quanta potrai coprir la tua pelle o quante ne concerai, e mettile in una calderola e mettile al foco, e poi torrai onze i de lume de rocca che sia biancha e non rossa, e perche la rossa brusa, torrai sal comun onza mezza e oglio de oliva, onza mezza.

E metti queste robbe in una calderola over paroletto a boglir tanto che la lume e il sale si disfaccia, e dapoi levala dal foco e suoderai questa acqua in uno cadino de legno e lassalo sfredir tanto che possi tenir la mano dentro. Si che la sia tivida piglia fior de farina de formento, cioé falla tamisar tanto che tu cavi il fiore, e mettila dentro in questa acqua e che la sia spessa come uno brodetto de ovi.

E poi torrai la pelle e mettila dentro e lassala stare sei giorni e non piú, ma se pur la lassasti per qualche sforzo, non la lassar piú de otto giorni facendo ogni giorno questo, che la foli con li piedi e calpestrarla una volta al dí. E come sará sei giorni over otto, cavala fuora che la sará camozzata cosí fina quanto si puó fare al mondo ad ogni parangone.

Nota che quando la pelle é asciutta daralli la stroppa e il ferro come sai. E sappi che se la bagni mille siade al dí sempre ritorna al primo suo essere. E intenderai che ogni pelle, si grande come piccola, vole lume de roccha, onze mezza, farina, onze mezza, oglio de oliva e sale, onze mezza.

A incamozzar pelle.

Piglia la pelle e mettila in acqua per una notte, e dapoi scarnela bene dal lato della carne, e dapoi mettila in terra con il pelo in suso e habbi cenere de legne forte e che la sia ben calda, e venerai metendola sopra il pelo della ditta pelle. E metteli tanta cenere che il pelo sia ben coperto si che la sia grossetta, e anderai revolgendola in rodolo si che la assumi come si fa uno scartozzo. E poneli sopra uno contrapeso per una notte, e poi scoprila e valla scarnando con la costa dil ferro come si fa dalla banda della carne.

E dapoi piglia lume de rocca lire una e un quarto e falla disfare in acqua tivida, e poni nella ditta acqua la pelle sopraditta e tirrala bene per ogni verso si che la riceva l'acqua. E dapoi struccola la ditta pelle bene. E dapoi piglia uno poco della ditta acqua calda e mettili dentro vinti rossi de ovi e lire due di grasso, pesta ogni cosa insieme, e metti dentro la pelle la quale serrá incamociata e bellisima.

A camocciar pelle de ogni sorte.

Prima metti la pelle in calcina per giorni trenta, e poi lavala e purgala bene della calcina come si fa. E poi mettila nella semola tanto che la se purghi, e dapoi lavala e struccolala bene e dalli la sua conza.

E dapoi pigliarai farina de formento e uno puoco de levato, e impasta e metti a levare ditta pasta come si fa la communa. E poi piglia lume de rocha, farina, oglio comun, e ongi la pelle una over due volte.

Nota che la pelle di montone vuole esser conciata due volte, cioé dipoi la prima alla seconda scalda anchora la concha et tornala dentro un'altra siada si che sia due per giorni tre, e questo e similmente si fa a fare camoccia.

A camocciar pelle.

Piglia la tua pelle come hai scorticato lo animale piglia la pelle e ponila a molle con acqua commune per una notte e uno giorno. E dapoi piglia la pelle e strucolala bene et revolgi la pelle sopra uno bastone e habbi uno osso che habbia come taglio, come sono alcuni ossi delle spalle che pareno una mezza paletta, e va scarnando la carne che fosse rimasta sopra la pelle dalla banda della carta.

E piglia del'acqua calda si che sia tivida, e levado da fare pan tanto come seria una noce, e tanta lume de rocha che sia per un'altra tanta noce, overo uno rosso de ovo, e tutte queste robe incorpora insieme e metti nella ditta acqua e messeda bene insieme.

E puoi metti la ditta pelle o carta de capretto nell'acqua sopraditta e va tirando e revoltando quella carta e pelle nella ditta acqua, e questo continua per spacio de uno quarto de hora. E dapoi cavala fuora e mettila al sole a sugare e sará finissima camozza.

Queste robbe, cioé acqua calda, levado, lume de rocha, sal crudo, oglio de oliva, rosso de ovo, fa buona camozza.

A incamocciar pelle di capra over di capretto.

Prima piglia la pelle e mettila in acqua corrente per spacio de due hore, e piglia il cortello dalle pelle e con la costa, sopra il legno da scorzare rotondo, scorza la ditta pelle sopra il pelo fregando si che vadi via il pelo sopra il zocho da scarnare.

E poi piglia ovi freschi sei, lume de rocha onze quattro, e sale uno pochetto, e habbi una pignata grande mezza de acqua e fa che la sia tivida, e poi torrai il lume de rocha e il sale e butta nella pignatta. E poi falla boglir e messeda bene con un bastone. Poi levala dal foco e tien messedato con le bacchette tanto che l'acqua non scotti forte.

E poi rompi li ditti ovi e mettili in quella pignatta, e poi piglia il bastone e messeda tanto che se disfaccia, e poi metterai la pelle dentro come l'acqua será fredda.

A incamozzar pelle.

Prima piglia la tua pelle et mettila in moglia. Et poi fa l'acqua calcinata in questo modo: torrai una secchia de acqua et piglia quattro libre de calcina et metti nella ditta acqua et lassala rafredare. Et lava la ditta pelle benissimo. Et poi mettila nell'acqua de calcina con la banda della carne in giuso verso la calcina et fa che ogni giorno levi la pelle della calcina et lassala giozolar per doe hore et lassa cosí giozare per fina che la pelle havera acqua, perche questa acqua torbida la calcina et la fa piú acuta et questo farai per otto giorni, ogni giorno come é ditto leva la pelle et lassa giozzare, et altri otto giorni farai similmente.

Et come haverai passati li altri otto che seranno giorni 15, cavarai fora la pelle et pellala come sai sopra il scagno con la costa d'il ferro. Et fatto questo lavala in acqua chiara benissimo. Et dapoi farai acqua di calcina nova come festi di sopra, et come l'acqua é fredda metti dentro la pelle ben lavata et ogni tre giorni cavarai fuora la pelle et lavala. Come l'haverai lavata tornala nella ditta calcina, et cosí farai per giorni 12. Et dapoi lavala et scarnala.

Et dapoi torrai semola de formento et tanta acqua quanta semola a misura et che l'acqua sia calcinata. Et come sara scarnata la pelle, follela molto bene in acqua fredda overo frescha. Et guarda se l'acqua semollata boglie, cioé se la semola viene di sopra, et come vedi la ditta semola di sopra l'acqua et tu quella volta butterai la pelle dentro come che saperai in la ditta acqua semolata overo calcinata. Et come vedi che la tua pelle beve l'acqua, si che la si domi a fraccar sopra con il dito si come si fa alla cera o cosa semile, all'hora torrai fuora la tua pelle et falla netta dalla semola molto bene da ogni banda con il ferro.

Et poi torna a folar la ditta tua pelle et folala tanto che venghi l'acqua chiara, torzila et dapoi allarghela et distendila. Et nota che se havesti dodese pelle da conzar, piglia mezza secchia de acqua e onze 6 de lume de rocca e lire due de sale, et fa boglir insieme, et come la boglie levala dal foco e lassa sfredir tanto che possi soffrir la mano dentro. Et habbi una concha per ogni pelle et buttali sopra una cazza della ditta acqua et menela molto bene in quella acqua et tirarala poi in longo et in traverso pure nella ditta acqua. Et poi habbi uno altro vaso da gettarla cosí come haverai lavade le ditte pelle, lavandole di una en una.

Item a conoscer se la tua pelle sera conza over non. Nota che quando che li haverai data la lume de rocha, se l'ha receputa la ditta lume, fa in questo modo. Piglia con le mano la pelle che haverai per conza e stringela nel pugno e torzila uno pocho; se la pelle sera rimasta biancha dove la stringesti e rubia dove la torci, será conza, e se non sera conza sappi che la troverai morbida e molesina. E al'hora fa che togli uno puoco de lume de rocha e uno puocho de sal commun, e falla disleguar nella ditta acqua, e da nuovo torna la ditta pelle nella ditta acqua e fa cosí fino che la torna al tuo segno.

Et dapoi lavala si che l'acqua giozzi tutta fuora dilla ditta pelle e come l'é ben scolata distendila al sole tanto che sughi e l'acqua che é scolata della pelle, pigliala, e piglia sei scudelle de farina de formento e dodece ovi freschi e messeda li ovi e farina e impastali insieme, si che sia ogni cosa ben incorporado. E poi torrai l'acqua che cogliesti della pelle che giozzava e fa si come si fa il brodetto de ovi, e farai si presto che ogni cosa sia ben caldo, acció che la tua pelle possi pigliar ben la tua conza. E cosí calda butta dentro la tua pelle e calpestra ben e folla acció che pigli bene la sua conza, e lassala star una notte in la ditta conza e dapoi la mattina cavala fora e lassa giozzolar. E come sono ben scolate valle distendi al sole tanto che siano ben sutte e secche.

E dapoi dalli la tempera e in acqua frescha metendole e lavandole inmediate e folandole con li piedi molto bene. E dapoi pigliale per li orli atorno atorno con il pallo voltando e guazzandola e stropizando bene per longo e traverso. E poi torna anche al pallo e stanzala in longo e in traverso e mettila al sole d'estesa e lassala sugar in locho che non tragga vento, perche il vento magna la conza. E poi tornala a folar sutta e tornala al pallo, e dapoi raspala a tempo quando che la vorrai adoperare.

Item una pelle bechina. Tolli sale e lume de rocha onze tre, sal comun onze una, e questo va cosí fina alle pelle mezane. Item a camozza grossa vecchia, onze sei de lume de rocha e lire due de sale, e secondo che vedi la pelle dalli la lume de rocha, dalli pió o mancho che vedi il bisogno, da tre onze infina a onze sei.

A incamozzar pelle o de capretto o d'ogni altra sorte.

Prima metti a molle la tu apelle per un dí over doi, si che siano mogliate bene e che siano ben lavate e nette. Dapoi mettile in calcina, cioé in una tina di legno. A componer ben la calcina piglia una secchia de acqua et per ogni secchia piglia uno pezzo de calcina viva grosso come seria uno pane grande e metti la dicta acqua nella tinna dove vorrai metter la calcina tanto che sopra avanza alle pelle, et metti dentro la ditta calcina secondo la quantitá si che se disfacci la pietra della calcina, et lassa sfredite l'acqua. Dapoi messeda bene nella tinna quando vorrai mettere dentro le pelle, mettendole sempre dentro a una a una con la banda della carne inverso la calcina, ben distese, et lassale cosí per tre giorni.

E passati li ditti tre giorni, cavale e lassale giozzare, e non perdere quilla acqua che giozza. E dapoi torna a messedar quella acqua colata con quella della tinna e metti dentro le pelle come festi la prima siada, et lassale cosí per cinque over sei giorni overo tanto che le se peli molto bene. E poi cavale e pelale sopra il cavaletto con la costa del cortello.

Et messeda da novo l'acqua calcinada e torna dentro le pelle per tre giorni, e passati cavale fuora e lavale con acqua fresca molto bene, e quando le cavi cavale con la testa in suso e zaffale per le orecchie e cosí le gambuzze e scorlando bene in su e in giu si che le lavi bene, et mettile sopra il cavaletto con il pelo in suso e con una costa de bove, cioé con uno osso perche con il cortello le potresti strazzare, e sempre che stiano a molle acció che si possano pelar meglio.

E fa che habbi un'altro vaso over tinna che possi tenir le ditte pelle dentro coperte et piglia della semola che venghi spesso come mosto solado. Et nota che non si togli altra acqua salvo quella che sono state a moglie. E messeda molto bene la ditta semola con la ditta aqua calcinada, dico quella che é giozzata delle tue pelle prima che le bagnasti in acqua fresca, e lassa stare in questa acqua per hore vintiquattro. E fatto questo ncavale fuora e lavale si che venghi il pastume piú grosso giuso e cosí lavale piú siade con quella acqua che scola delle ditte pelle, calcagnandole e pestegiandole bene circa sei siade, si che l'acqua uscisca fuora delle pelle chiara, ogni siada lavale in acqua chiara. Dapoi struccola bene fuora l'acqua tanto che ti dará il possibile.

E dapoi fa che habbi apparecchiata tanta acqua chiara che possa coprire e pesa la ditta acqua e per ogni lira de acqua vuole due onze de lume de rocha et una onza de sale, e metti la ditta acqua et lume et sale al fuoco tanto che se disfacci il lume, e levala poi dal fuoco e lassala sfredire tanto che la sia tivida. E piglia le pelle a una a una e distendile molto bene e mettile dentro nella tinna e lassa le ditte pelle uno dí e una notte in questa acqua. E dapoi cavale fuora e lassale sgiozzolare, e poi distendile a l'ombra se'l serà de in stade e la invernada al sole, tanto che se impassiscano e quasi mezze sutte. E dapoi stropizzale e tirale molto bene si che siano ben distese.

E dapoi piglia l'acqua colata e mettila al fuoco, per ogni lira della ditta acqua una onza de oglio, e insieme rescalda, e calda lavale via. Item habbi levado per ogni lira quanto sia una nove distemperado con la ditta acqua a poco a poco con farina sfiorada, tanto che sia poco piú spessa che sugoli, e lassalo cosí per spacio de una hora. Dapoi torai il resto d'il residuo di l'acqua che sia uno puoco tivida e mettila sopra questo levado. Item aggiongi per ogni lira de acqua mezza onza de farina e uno ovo, e messeda tutto insieme destendendo dentro per le ditte pelle, e calcagnandole e messedandole acció che la conza vadi per tutto, e lassela dentro doi giorni.

E poi cavala e distendida per il modo sopraschritto e lassala molto ben seccar, e seccate, temperale, bagnale sicando in uno mastello de aqua di subito lavade, distendile in loco humido in terra overo sopra tavole humide per spacio de una hora. E dapoi struccolale e distendile e cosí haverai le pelle ben conze e perfette.

A conzar una pelle de capretto in spacio de doi hore.

Piglia una pelle fresca e gettali sopra uno poco de calcina viva da'l lato della carne, e poi voltala in scartozzo e lassa stare per spatio de mezza hora. E poi habbi uno legno grosso tondo polito e desteñidla suso e con una costa de cavallo va remondando zoso la carne della pelle. E come avería rassato molto bene la pelle, lavala e struccolela bene con doi legni tanto che cavi tutta l'acqua fuora.

E dapoi mettila in la infrascritta conza e lassala star per una pezzo, e piglia del brodo della carne overo del latte overo de acqua tivida una scudella e uno ovo, e sbatti ben insieme, e poi torrai onze 3 de lume de rocca polverizzata che sia cruda e habbi tanto botyro o qualche cosa piú e uno pugnetto de sale e cosí de farina, e poi sbatti ogni cosa ben insieme e venirá a modo de uno brodo e fallo scaldar tanto che sia tivido. E poi conza la pelle e metti dentro in questo brodo e lassa che la pelle s'inbeveri di questa mistura molto bene.

E se la vorrai fare presto lassala poco dentro e cavala fuora e infarinela molto bene e vattene al foco e rimenela bene qua e la, tanto che la sia sutta. E di queste pelle si fa borse e stringhe e quello che ti piacerá, e se la lassi sugar al vento veniranno piú bianche.

Fifteenth-Century Lute Making Techniques and Six Hundred Years of Arabic and Persian Antecedents

Alice Margerum

The only explicit information regarding fifteenth-century lute-making techniques survives in a single folio of incomplete instructions. Written by Henri Arnaut (or Arnault) de Zwolle in the middle of the fifteenth century, Paris Bibliothèque Nationale, MS Latin 7295, folio 132 recto[1] is an annotated diagram that begins, 'Pro composicione lutene.' Whether this work should be considered as a "craft treatise" or not, it contains unique evidence about contemporaneous instrument-making practice. Initially this paper will discuss the manuscript and it's author, and which aspects of lute construction are included and which are omitted in the contents of this folio. Since Arnaut's description is not complete and aspects of his description differ from some construction methods demonstrated by later surviving lutes, Arnaut's text will also be compared with several earlier Arabic and Persian texts that make reference to the way in which the forerunner of the lute, the *'ūd*, was constructed. Despite not being comprehensive, Arnaut's unique text and diagram record sufficient details to indicate that the lute of his time retained features of the medieval *'ūd* as well as displaying aspects more typical of European renaissance lute construction.

In examining a written record, it is important to consider both the author and the audience. Henri Arnaut de Zwolle was a physician, astronomer and astrologer to Philip the Good, Duke of Burgundy, from at least 1432 to 1454. Arnaut's writings are contained within Paris BnF MS Latin 7295, and deal with astronomical, horological and mathematical devices along with a miscellany

1. G. Le Cerf, and E.R. Labande, *Instruments de musique du XVe siècle: les traités d'Henri-Arnaut de Zwolle et de divers anonymes (Ms. B.N. latin 7295)*, Paris, 1932, pp. 32-4 and plate XV, contains a facsimile of the section on musical instruments, Latin transcription, French translation and commentary. I. Harwood, "A Fifteenth-Century Lute Design," *The Lute Society Journal*, II, 1960, p. 3-8, includes reprint of folio 132r, Latin transcription, English translation and commentary.

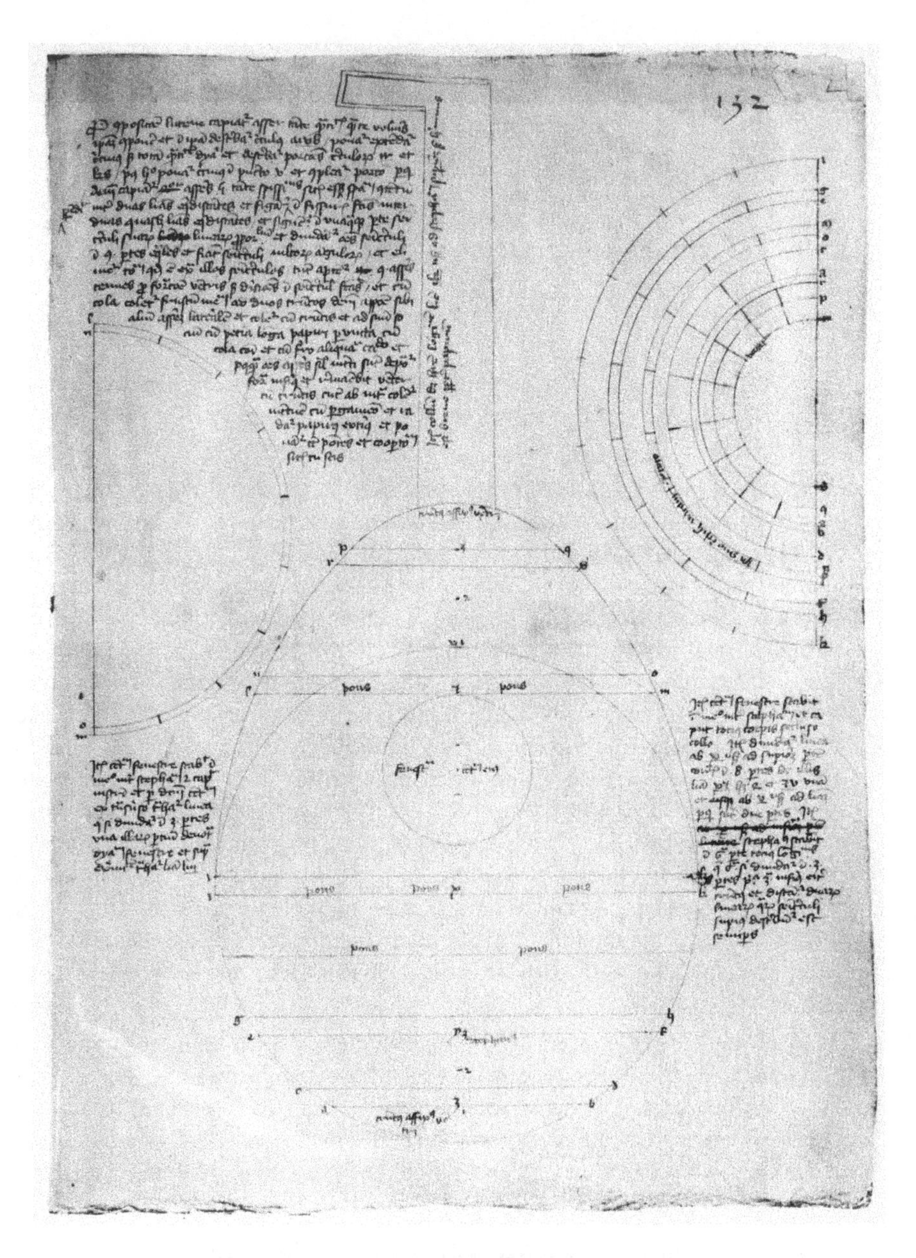

Fig. 1: Henri Arnaut de Zwolle, lute construction, c.1450. MS B.N.F. lat 7295, fol. 132r [from Cerf and Labande, *Instruments de musique du XVe siècle* (Paris, 1932), pl. XV].

of other subjects. Among these, folios 128^{r}-132^{r} are devoted to the construction of specific musical instruments: the clavichord, the organ, another keyboard instrument called the 'dolce melos,' and the lute. It has been suggested that Arnaut might also have made musical instruments, although there is no firm evidence for this.[2] The intended audience for his works is unknown, although they seem most likely to have been for personal use. The current bound volume was compiled from Arnaut's papers, probably during the early sixteenth century.

As the only non-keyboard instrument included, the lute might have caught Arnaut's attention because the stave-built back required unusual construction techniques. Unlike many other parts of Europe, the lute did not become well established in Burgundy until the third decade of the fifteenth century, after the marriage of Philip the Good to Isabel of Portugal. Prior to this, the primary instrument of importance in Burgundy was the harp, which Arnaut mentions only briefly but does not describe. Isabel brought two Spanish lute players to the court at Dijon where they are listed in the accounts from 1433 until 1456, the period during which Arnaut penned his writings on musical instruments.[3]

Arnaut's 'Pro composicione lutene'

The lute folio (Fig. 1) appears to be essentially a diagram with notes added afterwards. This is indicated not only by the way in which the text flows around the diagram but also in that the order-of-working is not sequential. As we shall see, the text begins at the upper left-hand corner, but several necessary steps are omitted there and appear on other parts of the page. The following text is the present author's translation. Although rougher than either that of Le Cerf and Labande or of Harwood, I have tried to stay as close as possible to the vague terminology used in Arnaut's text. Like Harwood, I have chosen to use an imperative tone (as Theophilus does in his *De Diversis Artibus*) rather than passive voice of the original.

The text begins in the upper left-hand corner, telling the reader,

> "For the construction of a lute take hold of a wooden plank sufficient for the size you desire / arrange it and describe the circle *aivb* / Position and extend the pair of dividers to the total diameter and describe the arcs *ir* and *ks* /

2. J. Koster, "Arnaut de Zwolle, Henri," *The New Grove Dictionary of Music and Musicians*, 2nd ed. edited by Stanley Sadie, London, 2001, Vol. 2, p. 34.

3. D. A. Smith, *History of the Lute from Antiquity to the Renaissance*, Lexington VA, 2002, p. 45.

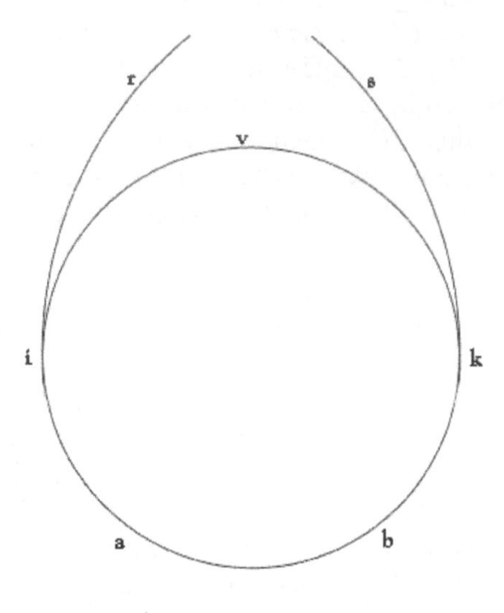

Fig. 2.

> After this, put the dividers into point *v* and complete the arc *pq* then take 5 wooden planks, equal in thickness to the spaces between the sets of two parallel lines ..."

In the opening text, we do not yet have enough information to complete these last steps however. For this, it is necessary to go to the middle of the text at the lower right-hand side of the page:

> "Item: divide a line drawn from *x* to the upper part of the body into 8 parts; *xz* has 4 parts and *zv* one, and from *v* as far as the line *pq* there are two parts."

We now have the points *p* and *q* and can describe arc *pq* (Fig. 3).

This is one of the points for which some modern scholars criticize Arnaut for mathematical inexactitude.[4] The discrepancy is too small to be shown clearly in Figure 3; but the line *pq* occurs at the seventh of these parts but the arc *pq*, shown here by the solid line, does not quite reach the eighth part, which is slightly above it. An arc touching the eighth part, represented here by the dotted line, would lie just outside the points *p* and *q* on the arcs *ir* and *ks* and, if used for the body outline, this would result in bulges at points *p* and *q*. Admittedly, Arnaut never specifies that the arc *pq* crosses through the eighth point at the head of the instrument, although many scholars infer this.

4. G. Söhne, "Lute Design and the Art of Proportion," *Lutes, Viols and Temperaments*, Cambridge, 1984, Appendix 4, p. 109-11.

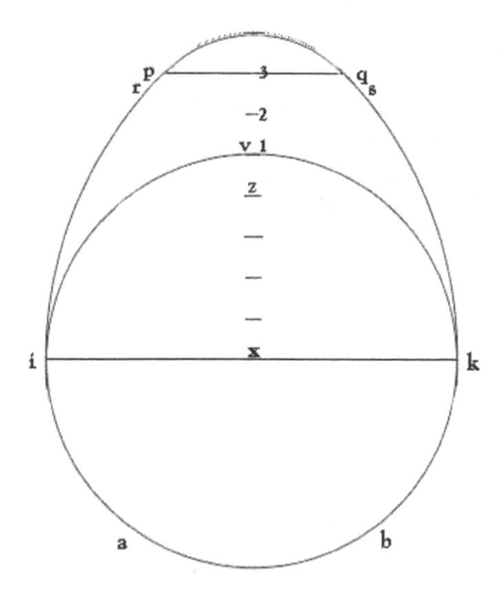

Fig. 3.

The lower right-hand text continues (Fig. 4):

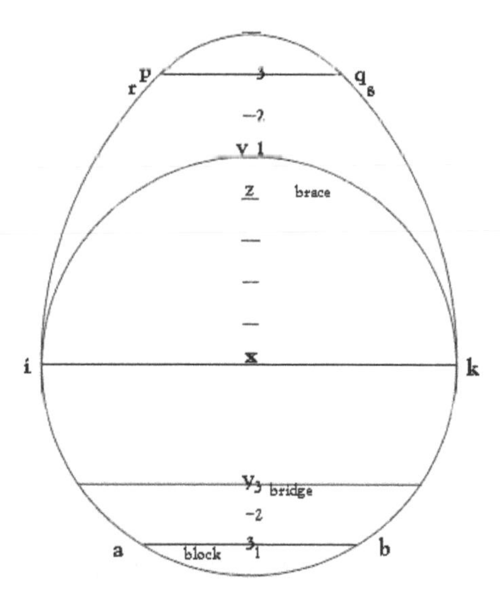

Fig. 4.

"Item: the bridge stands in the sixth part of the whole length, divide that sixth into 3 parts, the first third at the bottom will be the block, and the distance between the pairs of lines [into which the upright planks are placed] is half this third."

The positions of these sets of parallel lines are never specified but, once the previous steps having been completed, it is possible to locate the lines as shown on Arnaut's diagram, using points *y* (the bridge position), *x*, *z* and line *pq* for reference (Fig. 5). This done, it is possible to complete the directions in the upper left-hand text:

> "Then take 5 planks of wood and fix them perpendicularly into the channels made between the sets of lines / and mark on each side (of the planks) semicircles relating to the body width lines to which they are proportional [divide each of the semicircles into 9 equal parts,] and make each of the semicircles many faceted / remove all excess material outside the semicircles."

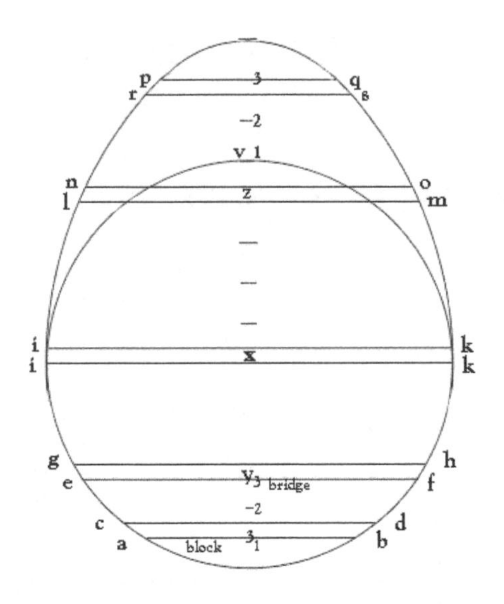

Fig. 5.

This procedure produces a faceted mould with a profile rather like a toast rack (see Figure 9). In the upper right-hand diagram, Arnaut shows the arcs of these uprights but does not show them faceted (Fig. 6).

Three of the arcs shown on Arnaut's page, however, should be ignored, as Arnaut has noted beside them that they are 'vacant' (they have thus been omitted from Fig. 6). Two of these 'empty' arcs correspond to a pair of lines scored across the main outline, above point *v*, which were scribed on the manuscript, but not inked. The corrected arcs *lm* and *no* are depicted at the left-hand side.

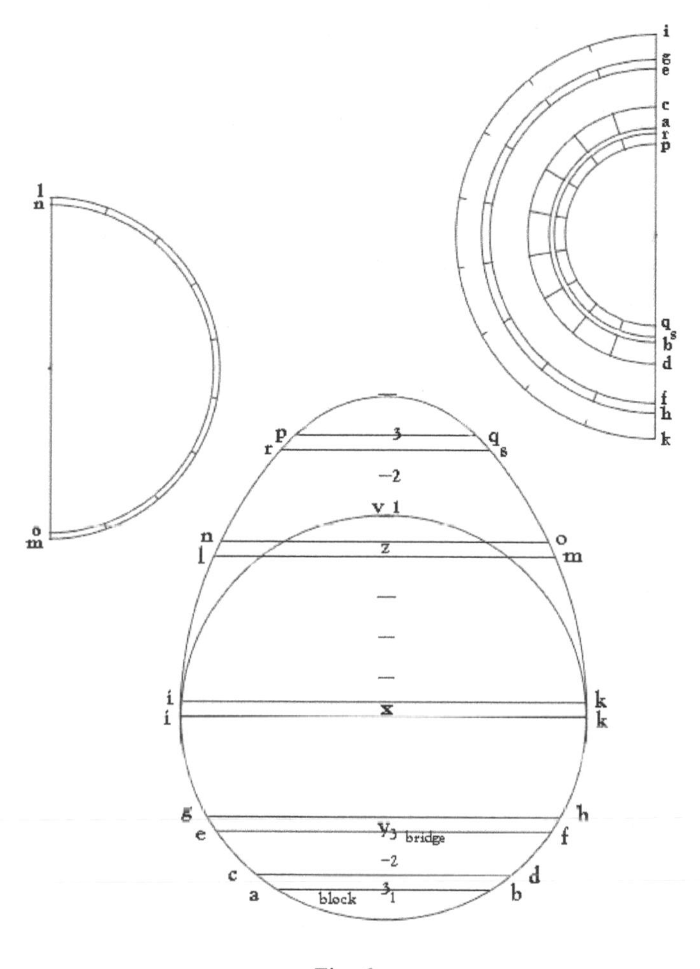

Fig. 6.

The text then continues to describe the actual construction of the lute itself:

> "Then following the divisions on the faceted mould prepare 9 thin planks of wood for the forming of the belly / glue the central stave to the top and bottom blocks / then join another stave to it and to the blocks with a long piece of paper coated with ordinary glue, using a rather warm iron, then when all the strips have been joined together take away the mould from underneath and the body will remain with the blocks. Next, glue the joins inside with parchment, scrape off the outside paper / and put on the braces and the covering, as you know."

The location of this bracing for the soundboard (the face of the instrument) is not described in the text, but does appear on the diagram (see Figs. 1 and 8).

There are two references to the location and size of the soundhole. On the lower, right-hand side of the MS it says:

> "Item: the centre of the soundhole will stand halfway between the bridge and the head of the whole body, excluding the neck."

The lower left-hand text continues:

> "Item: the centre of the soundhole will stand halfway between the bridge and the head of the instrument, and through said centre let a line be drawn cross-wise, divide this into 3 parts, one of those parts denotes the diameter of the soundhole, and at the upper edge draw the line *lm*."

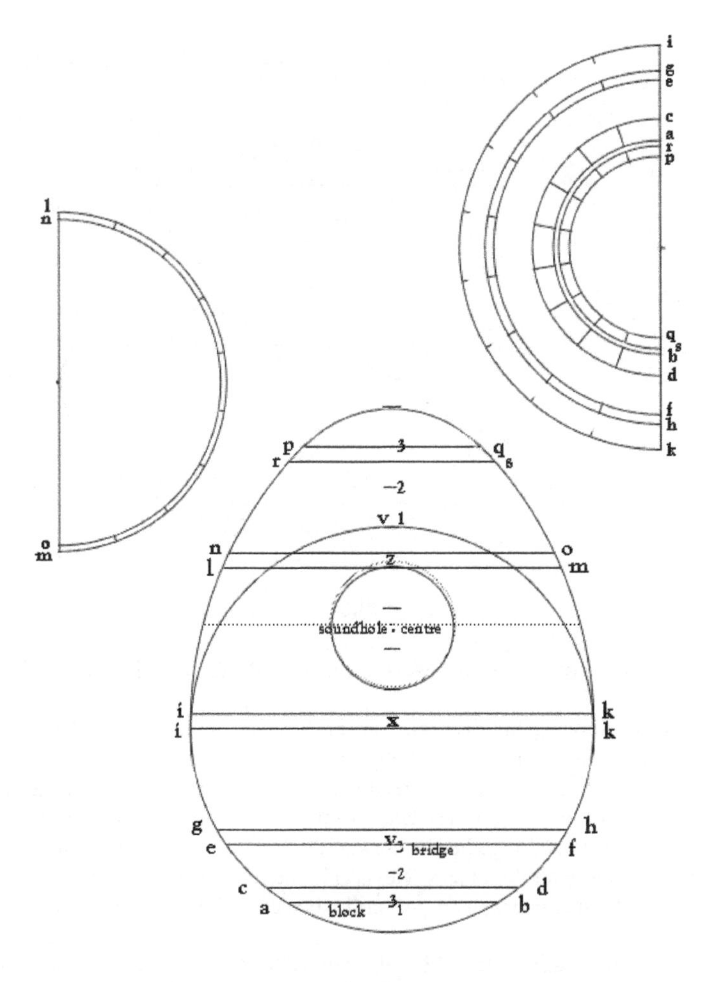

Fig. 7.

Here is another slight inaccuracy for which some modern scholars criticize Arnaut. The width of the soundhole as shown is in whole number proportion to the width, but not in the way that Arnaut describes. The soundhole placement as described, centred halfway between the bridge and the head of the soundbox, should be slightly higher than where it is shown by Arnaut (Fig. 7). The diameter depicted is also smaller than one third of the width at that point. As it appears in Arnaut's diagram, the centre of the soundhole is positioned midway between the second and third of those marks which divide the original circle's radius into five (the divisions marked between points *x* and *v*). The edges of the soundhole circle, however, are shown touching marks one and four, which is wider than described. This makes the diameter of the soundhole in a ratio of 3:10 with the diameter of the original circle *aivb*, not one third of the width at that point. Although line *lm* is not mentioned specifically until this point, it has already been used as reference for the bulkheads of the mould. The diagram shows it passing through point *z*.

There are a few additional bits of information written directly on the diagram itself. Some are just the naming of parts: the blocks (*truncus affixus ventri*), center of the soundhole (*fenestra-centrum eius*), soundboard braces (*pons*), and bridge (*stefanus*). Along the neck of the lute it says, "The neck should be equal in length to line *ik* but here is too short because the paper is not long enough." This is the full extent of Arnaut's text (Fig. 8).

It has been suggested that Arnaut might have made musical instruments or been an enthusiast of the lute. Certainly, aspects of his description are very credible and show knowledge of practical construction methods. However, the confused and incomplete character of the written description undermines the image of Arnaut as a confident or practiced authority on the subject of lute construction. It is possible that, like the clavichord scaling earlier in the manuscript which is credited to 'Baudecetus,' this information was copied from another source. The missing details might have been omitted either because the author had no interest in them or no need to record them. Exterior parts could be easily measured and duplicated from other instruments, whereas the manufacturing method of the back is not so obvious. Perhaps that is also why references to things such as the choice of wood are omitted, because they seemed self-evident.

While Arnaut's diagram is often referred to as the instructions for building a lute it does not provide enough information in itself to complete an instrument.[5] The directions are sketchy and certainly not systematic. Edward Kottick states that, "it could be followed with ease by any modern luthier. To one familiar with lute construction the lack of information about liners, bars, neck,

5. K. Coates, *Geometry, Proportion & the Art of Lutherie*, Oxford, 1985, p. 107-9

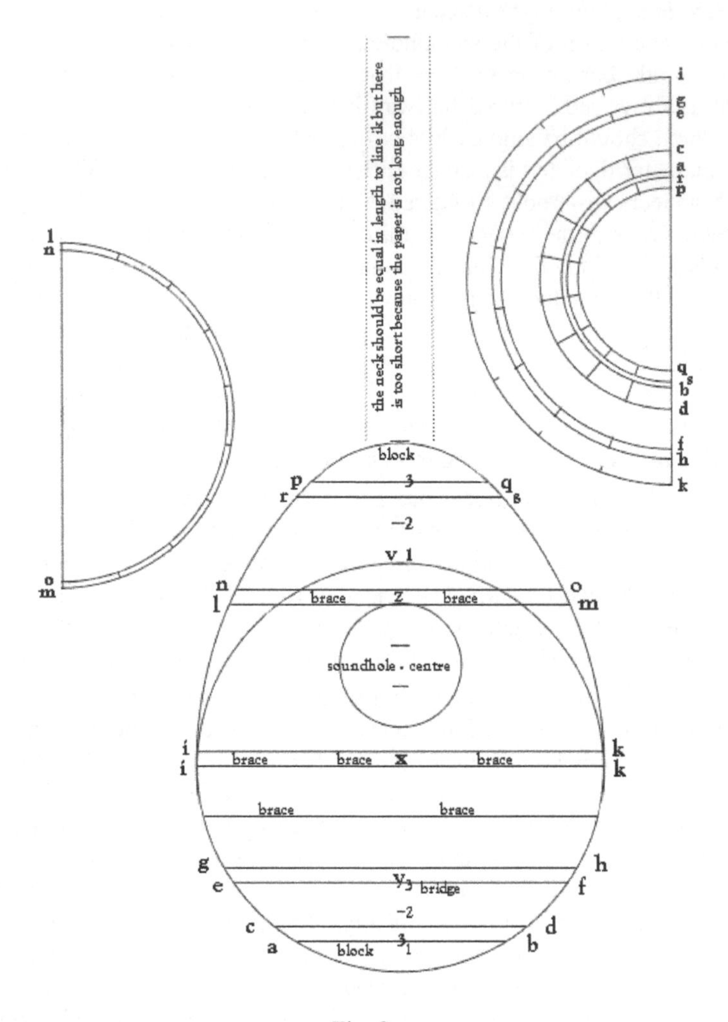

Fig. 8.

fingerboard, peg box, bridge and number of courses ... would seem only a minor obstacle."[6]

The real obstacle to understanding Arnaut's instructions, however, is that there are no surviving lutes from the fifteenth century nor other contemporaneous European texts from which to draw evidence for these missing details. Iconography can indicate shape, size, and the form of an instrument to some extent, but it gives few clues about construction. Information must be drawn

6. E. Kottick, "Building a Fifteenth-Century Lute," *Galpin Society Journal*, Vol. 26, May, 1973, p. 72-83.

from elsewhere. Many of the articles written about reconstructing Arnaut's fifteenth-century lute have attempted to fill in the gaps with information derived from sixteenth-century and later surviving European lutes and written sources. Although later sources and the methods used by modern instrument makers offer useful comparisons, relying upon them too heavily invites anachronisms. Insights into Arnaut's diagram and text might come from some non-European sources that predate his manuscript. Earlier treatises that relate to the direct ancestor of the lute, the *'ūd*, survive in Arabic and Persian.[7] These earlier *'ūd* texts, although they probably did not have any direct influence on Arnaut's work, record aspects of the development of lute-family instruments prior to the fifteenth century, and may offer clues to some of the details omitted by Arnaut.

Apsects of 'ūd Construction in Earlier Arabic and Persian Texts

At least four major music theory treatises dating from before 1400 CE mention aspects of *'ūd* construction.

The earliest surviving work describing the form and construction of the *'ūd* was written in the ninth century CE by Abū Yūsuf Ya'qub ibn Iṣhāq al-Kindī. Hailed as 'The Philosopher of the Arabs,' al-Kindī wrote on a variety of subjects including medicine, optics, and mathematics. Although most of his more than two hundred original works have been lost, he is still considered to have been one of the most important Arabic writers. Several of his surviving works deal both with music theory and musical philosophy. Details of instrument making are contained within *Mukhtaṣar al-mūsīqī fī ta'līf al-nagham wa-ṣinā'at al-ūd*, or the *Short Treatise on Music with Regard to the Composition of Melodies and 'ūd Making.*

The religious and philosophical encyclopaedia *Rasā'il Ikhwān al-Ṣafa*, written in Bosra (now part of Syria) by the Ikhwān al-Safa (The Brothers of Purity) during the second half of the tenth century CE, contains several sections relating to music.[8] This is a work of great influence, often referred to and copied in later sources, including the *Kanz at-tuḥaf*. It contains both music theory and musical philosophy: the chapter on 'The Making of Instruments and Their Tuning,' is followed by another on 'How the Movements of the Spheres Produce Notes Similar to Those of the *'ūd*.'

Abū'l-Ḥusain Muḥammad ibn al Ḥasan, known as Ibn al-Taḥḥan al Mūsīqī, wrote a fascinating treatise, *Ḥāwī al-funūn wa salwat al-maḥzūn*, or *The Col-*

7. My main source regarding these Arabic and Persian texts was the invaluable study by E. Neubauer, "Der Bau der Laute und ihre Besaitungen nach arabishen, persishen und türkishen Quellen des 9. bis 15. Jahrhunderts," *Zeitschrift für Geschichte der arabisch-islamischen Wissenschaften*, Frankfurt am Main, 1993, p. 279-378, additional assistance was provided by Dr. Owen Wright of the School of Oriental and African Studies, University of London.

8. See also A. Shiloah, "The Epistle of the Ikhwan al-Safa," *The Dimension of Music in Islamic and Jewish Culture*, Aldershot, 1993.

lector of Sciences and the Consolation of the Vexed, dated 1057 CE. It is rich in information about the *'ūd* as played in eleventh-century Cairo, but this does not seem to have been a widely circulated text. Because it contains so much practical information, I originally assumed that it was written as a guide for instrument makers. However, Ibn al-Taḥḥan al Mūsīqī was known as a court musician and music teacher. Only recently did I realize that this treatise is probably something more akin to a buyers' guide. Given the circumstances of the author, it is perhaps more likely that it concerns how to choose a well-made *'ūd* rather than guidelines for how to build one.

Copies of an anonymous Persian text, *Kanz at-tuḥaf* or the *Treasure of the Rarities*, written c.1350 CE, have been numerous. By the fifteenth century, this work was also translated into Turkish and further disseminated. It contains fragments of music theory and musical philosophy, some clearly derived from earlier authorities although these are not always specifically credited.

Arnuat's Lute Diagram Compared with the Earlier *'ūd* Texts

These treatises describe some elements of *'ūd* design in earlier centuries, which offer comparisons for the information that Arnaut recorded about lute making in the fifteenth century. Although there is a gap of three to five centuries, and no claim is made here that Arnaut is *directly* influenced by these texts, various features show strong continuity in instrument design throughout the centuries and across cultures. A number of these features stand out including geometrical layout; the use of whole number proportions; the cross-sectional shape of the back; the order of working; and the bridge position.

As in most of the *'ūd* treatises, Arnaut makes clear that his body outline was derived from whole number proportions. By dividing the diameter of his main circle into five, the body width-to-depth ratio is 10:5 (2:1), which is consistent with proportions described by three of the earlier *'ūd* texts; all except Ibn al-Taḥḥan al Mūsīqī's. Arnaut's body length to body width ratio however are a unique 13:10. The convention of describing the *'ūd* in terms of whole number proportions is well established at least as early as al-Kindī. Although it is not until the fourteenth century that this outline clearly appears in depictions. In a fourteenth-century copy of Ṣafi al-Dīn's thirteenth-century music theory treatise *Kitāb al-adwār* (Oxford, Bodleian Library MS Marsh 521), a diagram depicts some of the physical characteristics of the *'ūd*, although these are not mentioned in the text. The illustration of an *'ūd*, shown on folio 157^{v}, shows a clear use of dividers to describe the main circle and arcs for the outline of the instrument, which is quite reminiscent of Arnaut's lute aside from the arc at the top of the body and the length of the neck.[9]

9. Neubauer, "Der Bau der Laute ...," p. 373, illustration 1.

During the fifteenth century, numerous European depictions show a lute outline that appears to be derived from the same method of describing arcs. It is hard to tell when the currency of this outline began to fade since some artists seem to have persisted in portraying older models. Hans Memling, for example, seems to have painted the same lute in many of his paintings from c.1465-85, possibly using a lute he owned as a studio prop. None of the surviving early lutes, however, have this outline, which suggests that in shape at least Arnaut's late was more Akin to an earlier *'ūd.*

The proportions most often quoted with regard to the *'ūd* appear in the tenth-century treatise credited to the Ikhwān al-Ṣafa. This not very specific length:width:depth ratio of 3:2:1, has been interpreted sometimes as the length being equal to the entire length including the neck, and at other times as the length being equal to only the length of the body. Various copies of *Kanz at-tuḥaf* interpret this in either way sometimes altering the wording to make it more specific, although the illustrations rarely match the descriptions.[10]

Given the information provided, it can be assumed that Arnaut's lute is meant to be semi-circular in cross-section. (This assumes that the thickness of the baseboard is taken into consideration when scribing the arcs onto the bulkheads of the mould). A semi-circular cross-section is consistent with the description by al-Kindī, who states that the back should be as if a round body has been split into two halves to make two instruments. This also agrees with the width to depth ratio of the body being 2:1. With regard to the back of Arnaut's lute, Robert Lundberg suggests that 'we may take [Arnaut's] instructions as not literally correct, but as generally correct.' Lundberg subsequently demonstrates that the cross-sections of surviving Renaissance lutes vary greatly.[11]

Arnaut's text is largely concerned with the construction of the mould, which differs from what is know about both *'ūd* and lute construction. His 'toast rack' mould (Fig. 9) is utile and easy to build, but can be harder to use than a solid mould. There are no known historical references to the use of moulds in the construction of the *'ūd,* and many modern types are built without a mould. The top and bottom blocks for the *'ūd* are simply affixed to a board at the correct distance apart and the ribs are built 'in the air' with no support other than these blocks. Later lutes, on the other hand, seem to have been built on a rigid solid form, although no early lute moulds survive. The sixteenth-century verse describing the profession of 'Der Lautenmacher' mentions that the wood is bent over the mould ('Erstlich uber die Form gebogen') but the type of mould is not specified.[12] In the seventeenth century Marin Mersenne implies the use

10. Neubauer, "Der Bau der Laute ...," p. 373-5, illustrations 2-6.

11. R. Lundberg, *Historical Lute Construction*, Tacoma WA, 2002, p. 19 and p. 18-22, see also p. 5.

12. H. Sachs with illustrations by J. Amman, *Eygentliche Beschreibung aller Stände auff Erden,* Frankfurt am Main, fourth edition, 1574. This work is also known as *Das Ständebuch (The Book of Trades)* The third edition of 1568 was reprinted in facsimile by Dover, New York, 1973.

of a full, solid mould but, again, details are not specified.[13] Whether the object shown on the workbench in the accompanying woodcut is meant to represent a solid lute-mould or a completed lute-body, awaiting a neck, is debatable. Arnaut's toast rack mould falls somewhere between the other two styles of construction but, as far as I am aware, is recorded nowhere else until modern times.

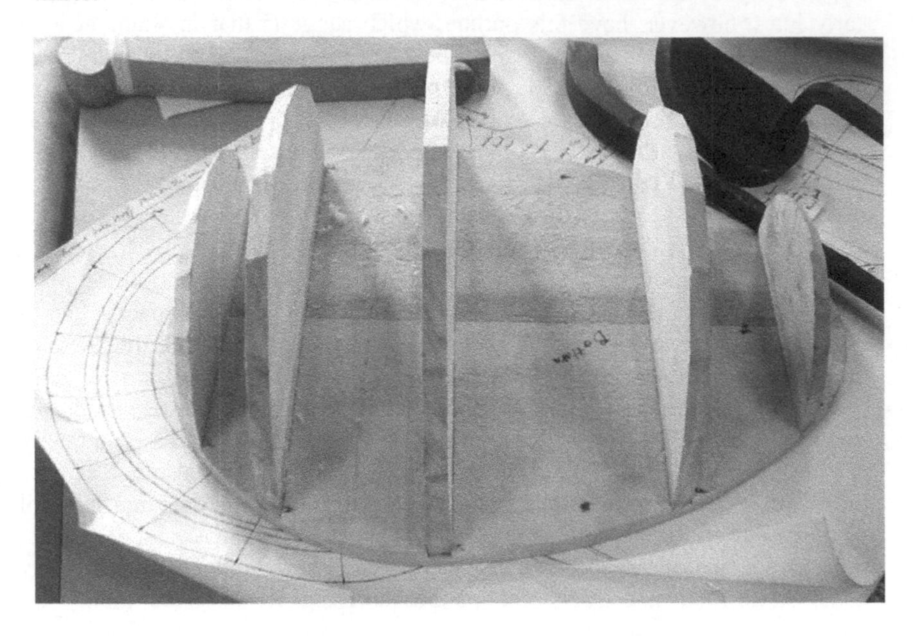

Fig. 9: Lute mould by Andrew Atkinson,
London Metropolitan University, after Henri Arnaut.

The use of an interior lower block is one detail in which Arnaut's lute resembles a modern *'ūd* more than a sixteenth-century European lute. None of the surviving early lutes contain a bottom block. In these survivals, the lower end of the ribs is supported by an exterior 'capping strip' and an interior lining-strip of thin wood. The lower block is still a common feature in *'ūd* construction, however. Unfortunately, none of the *'ūd* texts mentions the interior structure of the instrument.

Both modern *'ūd*-and modern lute-makers begin the construction of the back with the centre stave, as does Arnaut. In a way, this is like building an arch by putting the keystone in place first, and then building downward. It is worth mentioning this because not all carvel-built instruments are constructed

13. M. Mersenne, *Harmonie Universelle, the Books on Instruments*, translated by R. Chapman, The Hague, 1964, p. 77.

in this order. The setar is built upward and inward from the sides to the centre of the back, straight onto the blocks without a mould.

Like Arnaut's, the *'ūd* texts quite consistently describe nine ribs being used to build the back. Ibn al-Taḥḥan al Mūsīqī is unique in mentioning that the finest lutes of his time have eleven ribs and some as many as thirteen. Nine ribs seem to have been the usual number elsewhere and not just part of a conventional description or mystical number association. Al-Kindī specifies nine, and that number continues to appear on lutes through to late fifteenth-century Italian intarsia representations and many later surviving lutes.

Arnaut mentions strips on the inside of the back to reinforce the joins a feature common on both the *'ūd* and lute. Similar strips are mentioned by Ibn al-Taḥḥan al Mūsīqī, but he specifies a high-quality heavy-weight paper should be used rather than parchment. Mersenne in 1636 mentions the use of either vellum or paper for linings. The disadvantage of parchment is that, in addition to being heavier than paper, it tends to be hygroscopic and therefore less stable. Paper is the standard material used for this purpose on surviving Renaissance and Baroque lutes, although parchment appears as well.

One of the most glaring omissions in Arnaut's text is the mention of appropriate timbers for construction. Most of the surviving early Renaissance European lutes have ribs made of a medium to dense hardwood or even of ivory. The soundboards are made of softwood, usually either spruce (*Picea*) or fir (*Abies*). All of the *'ūd* texts refer to the choice of wood in some way, even if it is as simple as the Ikhwān al-Safa's saying that the wood for the *'ūd* should be solid and light and resound when it is struck. Cypress and cedar seem to be the woods most often specified in these texts, within a more vague classification of woods that are medium in density, neither too light nor too heavy. Ibn al-Taḥḥan al Mūsīqī recommends that an *'ūd* should be made entirely out of one wood, if possible, and that it should be old, well-grown cypress. The Persian text of the *Kanz at-tuḥaf* states that the timber should be fully dry when cut and suggests that the best timber is the 'wood of kings', which is brought from over the sea, but that cypress is also good. The fifteenth-century Turkish translator of the *Kanz at-tuḥaf* adds that if neither of these woods is available then one should acquire a timber that is similar to these in hardness, weight and working qualities.

The cypress (*Copressus*) and cedar (*Cedrus*) likely to have been used for medieval *'ūd* construction are denser and harder than some other softwoods, such as the spruce evident in many surviving in the soundboards, but lighter and softer than the woods typically used to construct the backs and necks of lutes. Given that the instrument Arnaut describes displays some characteristics mentioned in the earlier *'ūd* treatises but others found on latex lutes, it is difficult to determine what sorts of woods he would have considered suitable for its construction.

Arnaut says remarkably little about the soundboard, which is a crucial feature in any stringed instrument. Ibn al-Taḥḥan al Mūsīqī suggests that the soundboard should be made from two or three pieces of timber so that it will not split. Most surviving early lutes have thin two-piece fronts with numerous braces. The positions of three braces are specified by Arnaut but details, such as their dimensions, composition or shape, are lacking. None of the *'ūd* sources make clear reference to bracing. Depending upon interpretation, Ibn al-Taḥḥan al Mūsīqī might mention the existence of soundboard braces but says nothing specific about them if he does.

The number, size and placement of soundholes seems to have varied quite a bit in depictions, both for the *'ūd* and the lute. These details are rarely mentioned in textsexcept by Ibn al-Taḥḥan al Mūsīqī who specifies two main soundholes and another small one below the bridge. Arnaut's soundhole placement is not so different from that shown in some northern European mid- to late fifteenth-century lute iconography.

Bridge placement seems to be another feature in which Arnaut's lute resembles an earlier *'ūd*. Arnaut places the bridge at 1/6 of the length of the soundboard from the lower end. This is consistent with *'ūd* bridge placement as mentioned by al-Kindī and the *Kanz at-tuḥaf* as well as that depicted in the manuscript Oxford, Bodleian MS Marsh 521. Ibn al-Taḥḥan al Mūsīqī, who gives measurements rather than proportions, describes a bridge position much closer to the lower end, equivalent to less than 1/12. The bridge position of early surviving lutes tends to be 1/8 of the soundboard length or less, although a placement approximating to 1/6 can be seen in some mid-fifteenth-century representations.

None of the texts offer any guidance for the practical details of the bridge construction such as height, width or string spacing. Ibn al-Taḥḥan al Mūsīqī merely warns against decoration of the bridge with heavy materials such as gold, ivory, ebony or precious stones, which is a sensible precaution, since excess weight could act like a mute dampening the vibration of the soundboard.

Finally, with regard to the neck, Arnaut only mentions its length, and it is far longer than in any *'ūd* description or later surviving lute. The neck length of an *'ūd* is usually stated as either equal to half of the body width or one-third the string length. The length of Arnaut's lute neck is equal to the lute's width and therefore is almost half of his string length. Although this is also greater than many contemporaneous European lute depictions, similarly long necks appear on lutes in earlier fifteenth-century paintings by Giovanni Boccati, Carlo Crivelli, Stefano di Giovanni dit Sassetta, Pietro di Domenico da Montepulciano and the Master of Cologne. None of the texts indicate how wide the neck should be, nor how it is attached to the body. Arnaut's drawing of the neck does indicate some taper to the neck width but this is not described. Only Ibn al-Taḥḥan al Mūsīqī specifies that the back of the neck should be rounded in cross-section and finely made so that it is comfortable under the hand when one is playing. None of the texts makes any reference to a fingerboard.

Arnaut makes no mention fretting at all, whereas, entire chapters were written about the proper tuning and fretting of the *'ūd* in these and other music theory texts.[14] The basis for these frettings schemes is a Pythagorean diatonic system of intervals, with three of the frets common to almost all the texts: those yielding a Pythagorean major second ('index-finger fret'), major third ('ring-finger fret') and perfect fourth ('little-finger fret') above the open string.[15] The position of the 'middle-finger fret' varies from author to author, sometimes representing a Pythagorean minor third or a smaller or larger interval. In some cases, a text might describe more than one 'middle-finger fret,' each for use in a different type of music. Oddly, Ibn al-Taḥḥan al Mūsīqī, who specifies the importance of using a pair of dividers to locate the fret positions accurately, is one of the few sources who doesn't offer instruction as to where these positions should be (and also claims that he does not need to use frets).

The width of the neck of Arnaut's lute would be somewhat determined by the number of strings, and whether the strings were single or grouped into pairs (that is, single or double courses, in modern parlance). Given the available evidence, Arnaut's lute is most likely to have had four or five courses of strings but the number of strings is not mentioned. The use of five courses on the *'ūd* is documented as early as the tenth century by al-Fārābi. Ibn al-Taḥḥan al Mūsīqī acknowledges that five courses were used elsewhere, but not in Egypt, where the usual number was four. Ibn al-Taḥḥan al Mūsīqī is the only author to specify pegbox length, or to even mention the pegbox or pegs. In mentioning the number of pegs he clearly indicates the number of strings per course, in this case eight or ten pegs for four and five courses, offering the first unambiguous reference of the use of double-strung courses. Arabic and Persian treatises agree that each course of strings should tuned a fourth higher than the previous.[16] Stringing the *'ūd* in double courses seem to have been standard but is rarely documented in texts. Italian depictions of the early fifteenth century show both five-course and four-course lutes. If Arnaut's lute were strung like a later fifteenth-century lute it might have had five courses (tuned in fourths and a third) with the highest course single-strung. If it were strong like an earlier *'ūd* it would have had four or five double-courses tuned entirely in fourth. Either way, Arnaut's lute is unlikely to have had six courses, since evidence for this does not appear for the lute or *'ūd* until around 1475.

14. L. Manik, *Das arabische Tonsystem im Mittelalter*, Leiden, 1969.

15. The notable exception being in another treatise by al-Kindī, *Risāla fi 'l-luḥun wa 'l-nagham* (Treatise on the Melodies and the Notes), which gives an unique fretting arrangement in chapter 1, but implies an entirely conventional one in chapter 2.

16. In *Risāla fi 'l-luḥan wa 'l-nagham* al-Kindī lists a primary tuning in fourths but mentions some other tunings in common use which involve retuning the lowest string only.

The stringing material used on Arnaut's lute is likely to have been gut, although silk is also possible. Al-Kindi specifies that the two treble courses are strung with silk and two bass courses gut. The *Kanz at-tuḥaf* describes five courses of either silk or gut. The Ikhwān al-Ṣafa specify that all four courses are silk. While unusual, silk strings are mentioned in European literature as early as the fourteenth century.[17]

In terms of the surface finishing of the lute or the *'ūd*, none of the texts refer directly to the application of a protective finish to stet. It is possible that the instruments were not varnished, oiled or finished in any way, or they might indicate that this step was not considered part of the *design*. Cypress and cedar are naturally resistant to insect and fungal attack, so may not have needed a finish. Ibn al-Taḥḥan al Mūsīqī does caution that an *'ūd* must be protected against extreme conditions and suggests the use of rosewater or other essence for protection. By the time in which Arnaut wrote this work waxes, resins, oils, paints and varnishes were used on woodwork but it is unclear which, if any, of these would have been used on lutes.

Although none of these treatises dealing with *'ūd* or lute construction offer complete instructions, by comparison to each other they offer indications of when and where the structure and details of this family of instruments changed and when they remained constant. Curiously, Arnaut leaves out some of the same details omitted by the earlier authorities on the *'ūd*; it may be that these were details which all the authors considered unimportant or self-evident.

The Purposes, and Importance, of Arnaut's Lute Diagram

Although Arnaut's lute diagram records invaluable details about the craft of lute making in the fifteenth century, it should not itself be considered as a 'craft treatise.' It seems more likely that the method of construction was a curiosity to Arnaut, rather than the lute being the author's vocation or passion. He seems primarily interested in two aspects of the lute: the geometrical layout, and the method of constructing the back of the lute. According to Katherine Eagleton, the astronomical instruments described by Arnaut in the same manuscript, are not devices that would have been in common use but were objects displaying novelty, some even being unworkable.[18] Arnaut seems to have been concerned with recording intriguing designs of interesting instruments whether they were horological, astronomical or musical. Arnaut's motivation for preserving this information is less important than that it was preserved.

17. C. Page, *Voices and Instruments of The Middle Ages*, Berkeley CA, 1986, p. 216.

18. Personal communication during the 'International Symposium on Craft Treatises and Handbooks,' Cordoba, October 2005.

The importance of Arnaut's single folio on the construction of the lute is that, although incomplete, it is the only source that records details of the instrument at a time when it was beginning to display some of the characteristics of the typical renaissance European lute but still retained some features of the medieval *'ūd*. To better understand the lute of that time it is worth considering the earlier Arabic and Persian texts that relate to the structure of these similar instruments, although I am not suggesting that Arnaut necessarily had any direct knowledge of these earlier *'ūd* treatises. The physical object that Arnaut described, displays some characteristics recorded in earlier Arabic and Persian descriptions of the *'ūd*, but which are not found in later European lutes, there by demonstrating some of the gradual changes in this cross-cultural tradition of musical instrument making.

"... BUT THE FROG WILL DIE." NOTIONS ON LANGUAGE, SERIOUSNESS AND LITERARY TRADITIONS IN THE *LIBER ILLUMINISTARUM* (TEGERNSEE, 15th CENTURY, CGM 821)

Manfred Lautenschlager

In the first decade of the 15th century Heinrich Wittenwîler, a nobleman and advocate at the bishop's court of Constance in Southern Germany, wrote a satirical, didactic verse epic called *The Ring.*[1] In the early part of it we learn of a young and vulgar peasant called Bertschi, who falls in love with Mätzli, a grotesquely ugly country wench from the neighbouring village. As an unwelcome wooer, Bertschi soon finds his beloved locked away by her family in an attic. Out of reach of his pleas to Mätzli he has to resort to the written word to communicate his wishes and intentions to her, which is all the more difficult for him since he can neither read nor write. But with the help of the barber-surgeon in his village, Bertschi equips himself with a love letter full of courtly ideas and ideals, contrasting sharply with his verbal suggestions to the barber. The barber ties the letter to a brick and at night throws it through the open window into the barn in which he knows Mätzli to be locked away. The brick promptly strikes Mätzli badly on the head, but recovering from the blow the next morning, she finds herself with two things that utterly baffle her: a severely wounded head and a letter she is unable to read. Hiding the letter, she allows herself to be taken to the local doctor, to whom she presents her head and then her letter. Now the doctor treats her head then reads and interprets the letter to Mätzli, writes an answer for her in courtly terms and then proceeds to teach her thoroughly what making love beyond the written word is all about.

1. *Heinrich Wittenwilers Ring*, ed. Edmund Wießner, Leipzig 1931 (repr. Darmstadt 1973), commentary vol. Leipzig 1936; on this edition is based the new edition by Horst Brunner which includes a modern German translation, Stuttgart 1991 (RUB 8749); there is also an English translation of the full text of Wittenwiler's *Ring* by George Fenwick Jones, *Wittenwiler's Ring and the Anonymous Scots Poem Colkelbie Sow. Two Comic-Didactic Works from the Fifteenth Century.* Chapel Hill 1956 (University of North Carolina Studies in Germanic Languages and Literatures. 18); repr. New York 1969.

Having indulged in this practical lesson a number of times, the doctor soon finds out he has to find a way of "restoring" Mätzli's virginity in case she does end up marrying Bertschi.

So here we see the author Wittenwîler dealing with utterly serious matters while presenting these to his audience in a most entertaining and burlesque way. To make this double intention even more explicit to the reader, he decided to mark the passages he means to be funny with a green line to the left of his verses, and passages to be read as serious with a red line.[2] Hence most of Bertschi's wooing is in green, the barber's letter (a paradigm of love lyric) in red, conveying the letter with the brick to Mätzli and her awakening from her swoon the next morning in green again, but red again when she finds the letter she cannot read, etc.

Now the doctor, on seeing Mätzli, first treats her head, properly washing her wound with vinegar and ashes, onions and sea-salt, all of which seem as sweet as honey to Mätzli in love, hence: a green line accompanies this passage, which changes to red for the words: "Thus love turns gall into honey, whereas such honey-love will turn to most bitter gall in the end" (vv. 2075-78). So this truism seemed to be more serious to Wittenwîler than the quite reasonable and practical treatment of the wound on Mätzli's head; but a red line marks the doctor's advice later on suggesting to Mätzli how best to feign virginity in case she should marry Bertschi. This red and serious passage reads (vv. 2215-33): "Now go to your nephew and provide yourself with petals of the lily, buds of the cypress and take galls with it and boil it well; this concoction you then apply frequently to your lower parts and speak at the same time aloud: 'Bring back my good luck!' Then you will feel your vagina narrowing and contracting strongly.[3] But do understand me well: the galls have to be taken from the tree, not the goat, and all three ingredients have to be of the same weight. – And for the second trick, take a fish-bladder and fill it with pigeon's blood: this will be of good use for you in the night when they will put you to bed with him."

Now, for the modern reader, it may seem strange that these 15th century doctor's recipes tinged with magic are marked to be read as deadly serious, whereas a quite reasonable disinfection of a wound before dressing it, serves

2. Cf. Der *Ring*, vv. 36-41.

3. This verse has led George Fenwick Jones, the editor of the English version of the *Ring*, to believe that the doctor was trying to initiate an abortion; but in fact the recipe here is in chime with a group of similar prescriptions from the *Trotula* treatises, where various herbs with constringent properties have the same effect of narrowing the vagina and thus making it easier for the woman to feign virginity (*The Trotula. A Medieval Compendium of Women's Medicine*. Ed. and transl. by Monica H. Green, University of Pennsylvania, Philadelphia 2001; here esp. recipes [190] -[195] "Glanzlichter der Buchkunst," pp. 144-146), also in case of an intended abortion, containing herbs like sage, rue, juniper etc. that were in use for the purpose, are conspicuously absent from Mätzli's recipe (cf. John Riddle, *Eve's Herbs. A History of Contraception and Abortion in the West*. Cambridge (Mass.) 1997). Also the author's sarcastic view on peasant life is more consistent with the narrative when he has the doctor sell off Mätzli, having made her pregnant, as a virgin to her future husband, providing her with a recipe for that purpose.

as a humorous counterpoint to a serious aphorism on love. What we can learn from Wittenwîler here is that an early 15th century author can well make use of medical treatises as a genre of literature and put them into a humorous context – but without necessarily making fun of the recipes themselves.

We should be alert to this when dealing with craft treatises that contain recipes presenting the full range of the artes mechanicae, such as medicine, hunting, cultivation of crops and trees, along with the many varieties of preparing colours, glues, ink etc. One of these manuscripts is the *Liber Illuministarum* from the monastery at Tegernsee in southern Germany, compiled by the librarian Konrad Sartori of that monastery in the years around 1500.[4] This book is one such wide-ranging collection, predominantly of workshop practices for illuminators and painters, welders and forgers and the like, but it also contains long passages of purely medical recipes with a strong bias on magic.

One such group of recipes might have been of use to Mätzli's doctor, as it deals with various ways of preventing a woman from conceiving a child. These are recipes that mainly work on the grounds of either sympathetic or contagious magic (as Sir James George Frazer put it[5]) and which read as follows:

mag[764]: "When you take a boy's milk-tooth that has fallen out and wrap it in a leaf of silver and hang this up above the woman's bed, she cannot become pregnant."

The same happens if you hang a parcel containing either hare's droppings *mag*[778c], or the ring-finger of a miscarried child *mag*[777] over the bed, or if you tie two testicles of a weasel onto the woman's thigh *mag*[791]. The latter will only be effective, though, if the knot holds firm, the recipe assures us. This remark seems to hint that the collector of such recipes does indeed expect to address a mixed audience whom he intends to entertain with witty asides about information that credulous readers may take for granted.

Similarly the billy-goat during the mating period is widely considered to be a paradigm of acrimony and inner heat most manifest in his blood, urine or fat; hence a dose of a very strong male substance of this or a similar kind, directly consumed by a woman, can act as an antidote to her female virtues or performance and also serve as a contraceptive when dutifully consumed over a prescribed period of time. Thus, a woman cannot get pregnant who either drinks the urine of a billy-goat *mag*[788a] or the blood of a hare *mag*[788b], or who frequently rubs the urine of a wolf on herself.

Now, even when told or transcribed with tongue in cheek, these recipes formally preserve all the seriousness of and faithfulness to their magical princi-

4. *Der "Liber illuministarum" aus Kloster Tegernsee. Edition, Übersetzung und Kommentar der kunsttechnologischen Rezepte*. Ed. Anna Bartl, Christoph Krekel, Manfred Lautenschlager, Doris Oltrogge, Stuttgart 2005.

5. Sir James George Frazer, *The Golden Bough, part I: The Magic Art*, vol. I, ch. III., pp. 52-219; 3rd ed. London 1911.

ples, and it is often only when a full sequence of them fills the pages of a manuscript, as in some parts of the *Liber Illuministarum*, that the shear variety of these concise and yet fantastical recipes makes them highly amusing to peruse. The basic principles of magic must be taken as granted and provide the serious foundation to the reading, yet the ingenuity and, in some cases, novelty of their application to daily life is most entertaining and almost always calls for an imagined green line.

How can you make someone, preferably a woman, speak out and answer your questions truthfully while she is asleep? "Take the fat from a hare and wrap it up and put this beneath the pillow at night, so the person will speak up and tell you plainly the truth" *mag*[581]. Or: "Take the heart of a dove and the head of a frog, dry them and crush them to powder; if you put this onto the woman's breast while she sleeps, she will tell you everything you want" *mag*[837]. And this recipe wisely adds: "But make sure you remove this powder again before she wakes up, otherwise she will go mad." Now, this aside seems to cover the full scale of these recipes in action, between irony and tragedy, and the recipe itself looks like the nucleus of a 15th century novel or tale.

There are a small number of recipes in the manuscript, though, that seem to have been recorded for the mere fun of it, or to play practical jokes on other people, for instance, the one advising: "If you take some hairs that grow around a male donkey's genitals, grind them well, put this in a glass of wine and then give it to another person, he will immediately be forced to fart" *mag*[802a]. – Or: "If you take the eggs of ants and grind them and put them into water or wine, this will have the same effect" *mag*[802b]. Now, who on earth would find pleasure in such a prank, one is inclined to ask? For me, the provenance of this recipe smells strongly of a cathedral or monastery school, where pupils would delight in playing such tricks on their superiors, and it is not difficult to imagine the chuckles while potion charged in this way is standing readily at hand for its intended user at the table or during a ceremony.

There is a further category that seems to be intended for quite a different atmosphere. These are funny, perhaps even strange recipes that try to create miraculous apparitions or imaginary sceneries for people congregating in one room and witness various experiments that are carried out for their entertainment. One that is straightforward to us today with our knowledge about the range of spectral colours in natural light, is the one contained in the chapter about the various qualities of gems and stones, which says: "The stone called Iris is translucent and has sharp edges like rock-crystal and, when put in sunlight, projects the colours of the rainbow onto the wall" [891].

A similar, yet more artificial effect seems to have been intended, when we learn that "In a dark room, take a glass-bowl filled with water and fresh white of egg, or filled with rainwater and, when lit up (from behind), this produces objects like castles, stags, etc." So here we obviously have a group of people in a dark room together, observing projections from a water bowl against a wall and interpreting the various shapes and apparitions known to them from their experience, not unlike watching different cloud formations in the sky. In this context, incidentally, rainwater means a more or less muddy type of water, taken from brooks and rivulets after heavy rainfall, as we learn from a marvellously illustrated late 14th century manuscript of the *Tacuinum sanitatis* (Cod. Vindob. Ser. nova 2694, fol. 89v: Aqua pluvialis; Electio: recepta a bona terra).[6]

A similar congregation of people in a room may have witnessed the following experiment, about which it is far from clear whether it was for anything more than pure entertainment, or what its purpose or effect may have been *mag*[845d]. In the first of a sequence of recipes a powder is produced from the seeds of a type of rose called "eglerisa," mustard-seeds and "pes mustele," i.e. "weasel's foot," which I take to be another, still unidentified plant rather than the actual paw of a weasel. Now this powder can be used for various purposes: a) it can prevent a tree from producing fruit, b) it can serve as bait for fish, c) it can bring a certain fish back to life again, and d) "if you put that very same powder into a lamp and light it, all people gathered around will look very black" (*nigerrimi* in Latin), and finally e) "if you take that powder, mix it with olive oil and sulphur and paste a wall with it, at sunshine a house thus prepared will look as if on fire."

Now this sequence of recipes clearly shows us a group of people meeting to indulge in all sorts of experiments which they seem to carry out for their entertainment, without intentionally harming anyone or being in any way mischievous; hence the interest in collecting these recipes seems to be both lighthearted and aimed at achieving special effects that were seriously meant to work and impress other people.

This interest in special effects, especially with colours, is common to these experimental recipes and to most of the other recipes in the craft treatises recorded with no other intention than the serious business of producing practicable materials for painters and illuminators. Because even here we find people experimenting with colours and inks that shine at night, for instance, trying to make use of fluorescent materials such as those obtained from rotting wood or even from the ephemeral glow-worm. Producing colours with a metallic lustre or glitter is another intention we clearly find [60]: for instance, a thin varnish made of gold and minium to coat other colours to make them look like iron, or that makes other metals look like gold. In Latin this is called a "color martirisatus," a colour that looks like Mars, a tag for iron. And, of this colour, it is said that it "corizat omnes alios colores," i.e. causes other colours to shine brightly. In these new and experimental recipes, we find new vocabulary and expressions coined to characterize what hitherto has never been described

6. The facsimile edition of 1966 has now been included and reedited in the series "Glanzlichter der Buchkunst:" *Tacuinum sanitatis in medicina*. Kommentar von Franz Unterkircher. Graz 2004.

either in Latin or in the vernacular. Thus in another recipe [75], which tries to convey the effect of peacock-like colouring, a new German word "engelseri" is used, and may have been specially coined for the purpose. It is possibly intended to compare this colour to what one would expect the feathers of an angel to look like.

Now with this type of recipe writing, as with the mainstream of notes and treatises originating from a workshop background, we have to be aware that however unusual or experimental they may seem at a first glance, they usually have to be taken seriously and "at their word:" especially when we encounter neologisms, hapax legomena, or even a shift in the meaning of words, used in different, hitherto unfamiliar ways. Unless they are put in a context of parties aiming at merry-making and practical jokes, a man like Wittenwîler would undoubtedly have marked them with a serious red line. Thus their intentions have to be taken at face value if we truly endeavour to experience the spirit in which they were written, as in the case of a man who comes home from the woods with a poisonous mushroom, prepares a perfect meal and eats it all. Is there any chance of saving him? As the *Liber illuministarum* would advise *med*[565]: "A tree-frog, or several of them, tied fast upon the navel of the patient, will extract poisonous materials that have been eaten – but the frog will die." – Which is a most serious matter for either party concerned.

LIST OF CONTRIBUTORS

Luís U. Afonso, Instituto de História da Arte, Faculdade de Letras, Universidade de Lisboa, Portugal.
luis.afonso@fl.ul.pt

Spike Bucklow, Hamilton Kerr Institute, Fitzwilliam Museum, University of Cambridge, United Kingdom.
sb10029@cam.ac.uk

Silvia A. Centeno, The Jewish Theological Seminary, 3080 New York, NY 10027, USA.

Mark Clarke, Universidade Nova de Lisboa, Portugal.
mark@clericus.org

Ricardo Córdoba de la Llave, Departamento de Ciencias de la Antigüedad y de la Edad Media, Universidad de Córdoba, Plaza del Cardenal Salazar, 3, 14003 Córdoba, Spain.
rcllave@uco.es

Teresa Criado Vega, Departamento de Ciencias de la Antigüedad y de la Edad Media, Universidad de Córdoba, Plaza del Cardenal Salazar, 3, 14003 Córdoba, Spain.
teresa_criado61@yahoo.es

António J. Cruz, Escola Superior de Tecnologia de Tomar, Instituto Politécnico de Tomar, Estrada da Serra, 2300-313 Tomar, Portugal / Centro de Investigação em Ciência e Tecnologia das Artes (CITAR), Universidade Católica Portuguesa, R. Diogo Botelho, 1327, 4169-005 Porto, Portugal.
ajcruz@ipt.pt

Consuelo Dalmau Moliner, Facultad de Bellas Artes, Universidad Complutense, Madrid, Spain.
consuelodalmau@art.ucm.es

David Edge, The Wallace Collection, Hertford House, Manchester Square, London W1U 3BN, United Kingdom.

Gertrude M. Helms, Independent Scholar.
tru.helms@gmail.com

Stefanos Kroustallis, Associate Professor in History of Science, University of Castilla-La Mancha, Spain.
stefanos.kroustallis@gmail.com

Manfred Lautenschlager, Department of German Mediaeval Studies, University of Erlangen, Germany. Manfred.
Lautenschlager@t-online.de

Eva López Zamora, Ph. D. in Fine Arts and painting conservator-restorer.
evlopezz@hotmail.com

Débora Matos, Department of Digital Humanities, King's College London, Room 210, 2nd floor, 26-29 Drury Lane, London, WC2B 5RL, United Kingdom.
debora.matos@kcl.ac.uk

Alice Margerum, Independent Scholar and Harpmaker (www.carvedstrings.com). 310 Mason. Ave., Hancock, Michigan, 49930, USA.
carvedstrings@gmail.com

Doris Oltrogge, Cologne Institute for Conservation Science, Fachhochschule Köln, Germany.
doris.oltrogge@fh-koeln.de

Cheryl Porter, Independent Researcher.
chezzaporter@yahoo.co

Margarita San Andrés Moya, Facultad de Bellas Artes. C/ Greco nº 2. Universidad Complutense de Madrid, (28040) Madrid, Spain.
msam@ucm.es

Natalia Sancho Cubino, Facultad de Bellas Artes. C/ Greco nº 2. Universidad Complutense de Madrid, (28040) Madrid, Spain.
nataliasanchocubino @gmail.com

Nellie Stavisky, The Jewish Theological Seminary, 3080 New York, NY 10027, USA.
nestavisky@jtsa.edu

Ad Stijnman, Rijksmuseum Amsterdam, Netherlands.
a.stijnman@rijksmuseum.nl.

Nicolas Thomas, Institut National de Recherches Archéologiques Préventives (INRAP), Laboratoire de médiévistique occidentale de Paris (LAMOP-UMR 8589), France.
nicolas.thomas@inrap.fr

Arie Wallert, Rijksmuseum Amsterdam, Netherlands.
a.wallert@rijksmuseum.nl

Steven A. Walton, Department of Social Sciences, Michigan Technological University, 209 Academic Office Building, Houghton, Michigan 49931, USA.
sawalton@mtu.edu

Alan Williams, The Wallace Collection, Hertford House, Manchester Square, London W1U 3BN, United Kingdom.
allan.williams@wallacecollection.org

Lightning Source UK Ltd.
Milton Keynes UK
UKHW021837030320
359713UK00003B/162